Nanochemistry

Nanochemistry
2nd Edition

G.B. Sergeev
Laboratory of Low Temperature Chemistry
Chemistry Department
Moscow State University
Moscow 119899
Russian Federation

K.J. Klabunde
Department of Chemistry
Kansas State University
Manhattan, Kansas 66506
U.S.A.

AMSTERDAM • BOSTON • HEIDELBERG • LONDON
NEW YORK • OXFORD • PARIS • SAN DIEGO
SAN FRANCISCO • SYDNEY • TOKYO

Elsevier
Radarweg 29, PO Box 211, 1000 AE Amsterdam, The Netherlands
The Boulevard, Langford Lane, Kidlington, Oxford OX5 1GB, UK

Copyright © 2013 Elsevier B.V. All rights reserved.

No part of this publication may be reproduced, stored in a retrieval system or transmitted in any form or by any means electronic, mechanical, photocopying, recording or otherwise without the prior written permission of the publisher

Permissions may be sought directly from Elsevier's Science & Technology Rights Department in Oxford, UK: phone (+44) (0) 1865 843830; fax (+44) (0) 1865 853333; email: permissions@elsevier.com. Alternatively you can submit your request online by visiting the Elsevier web site at http://elsevier.com/locate/permissions, and selecting *Obtaining permission to use Elsevier material*

Notice
No responsibility is assumed by the publisher for any injury and/or damage to persons or property as a matter of products liability, negligence or otherwise, or from any use or operation of any methods, products, instructions or ideas contained in the material herein

British Library Cataloguing in Publication Data
A catalogue record for this book is available from the British Library

Library of Congress Cataloging-in-Publication Data
A catalog record for this book is available from the Library of Congress

ISBN: 978-0-444-59397-9

For information on all Elsevier publications
visit our web site at store.elsevier.com

This book has been manufactured using Print On Demand technology. Each copy is produced to order and is limited to black ink. The online version of this book will show color figures where appropriate.

Working together to grow
libraries in developing countries

www.elsevier.com | www.bookaid.org | www.sabre.org

ELSEVIER   BOOK AID International   Sabre Foundation

Cover credit:
Adapted from Stoeva, Zaikovski, Prasad, Stoimenov, Sorensen, Klabunde. Langmuir 21, 10280–10283 (2005). Scheme 1, Figure 3, copyright, American Chemical Society.

To: N.S. Merkulova and Linda M. Klabunde

Contents

Preface ... xi

1. **Survey of the Problem and Certain Definitions** ... 1

2. **Synthesis and Stabilization of Nanoparticles** ... 11
   2.1 Chemical Reduction ... 13
   2.2 Reactions in Micelles, Emulsions, and Dendrimers ... 18
   2.3 Photochemical and Radiation-Chemical Reductions ... 22
   2.4 Cryochemical Synthesis ... 27
   2.5 Physical Methods ... 38
   2.6 Particles of Various Shapes and Films ... 43

3. **Solvated Metal Atom Dispersion (SMAD) for Making Metal Nanoparticles** ... 55
   3.1 Experimental Techniques ... 55
   3.2 Aggregation of Metal Atoms or Reactive Molecules in Low-Temperature Matrices/Solvents ... 56
      3.2.1 Control of the Gold–Tin (Au–Sn) Bimetallic System ... 57
         3.2.1.1 Experimental Results on Au Atom–Sn Atom Clusters in Cold Solvents ... 58
      3.2.2 Reactivity of Aggregates (Nanoparticles or Nanocrystals) ... 61
      3.2.3 Trapping and Stabilization ... 61
   3.3 Examples of Useful Synthesis ... 61
      3.3.1 Gold Nanoparticles ... 61
      3.3.2 Silver and Copper ... 63
      3.3.3 Other Metals ... 63
      3.3.4 Binuclear Compounds ... 63
   3.4 Digestive Ripening or "Nanomachining" ... 64
   3.5 Rods, Wires, and Stars ... 69

4. **Experimental Techniques** ... 75
   4.1 Electron Microscopy ... 76
      4.1.1 Transmission Electron Microscopy ... 77
      4.1.2 Scanning Electron Microscopy ... 77
   4.2 Probe Microscopy ... 78
   4.3 Diffraction Techniques ... 81
      4.3.1 X-ray Diffraction ... 81
      4.3.2 Neutron Diffraction ... 82

	4.4	Miscellaneous Techniques	82
		4.4.1 EXAFS	82
		4.4.2 X-ray Fluorescence Spectroscopy	82
		4.4.3 Mass Spectrometry	83
		4.4.4 Photoelectron Spectroscopy	83
		4.4.5 Nuclear Magnetic Resonance (NMR) Spectroscopy	83
		4.4.6 Ultra Violet–Visible Spectrometry (200–800 nm)	84
		4.4.7 Dynamic Light Scattering	84
	4.5	Comparison of Spectral Techniques Used for Elemental Analysis	85
5.	**Cryochemistry of Metal Atoms and Nanoparticles**		**89**
	5.1	Reactions of Magnesium Particles	90
		5.1.1 Grignard Reactions	90
		5.1.2 Activation of Small Molecules	93
		5.1.3 Explosive Reactions	96
	5.2	Silver and Other Metals	100
		5.2.1 Stabilization by Polymers	101
		5.2.2 Stabilization by Mesogenes	110
	5.3	Reactions of Rare-earth Elements	115
	5.4	Activity, Selectivity, and Size Effects	122
		5.4.1 Reactions at Superlow Temperatures	122
		5.4.2 Reactions of Silver Particles of Various Sizes and Shapes	132
	5.5	Theoretical Methods	137
		5.5.1 General Remarks	137
		5.5.2 Simulation of the Structure of Mixed Metallic Particles	138
		5.5.3 Simulation of Properties of Intercalation Compounds	143
		5.5.4 Simulation of Structural Elements of Organometallic Co-condensates	145
6.	**Chemical Nanoreactors**		**155**
	6.1	General Remarks	155
	6.2	Alkali and Alkaline-Earth Elements	160
	6.3	Transition Metals of Groups III–VII in the Periodic Table	169
	6.4	Elements of the Group VIII of the Periodic System	179
	6.5	Subgroups of Copper and Zinc	191
	6.6	Subgroup of Boron and Arsenic	198
7.	**Assemblies Involving Nanoparticles**		**209**
	7.1	Assemblies Involving Nanoparticles	209
	7.2	Forces between Nanoparticles	215
		7.2.1 Attraction Forces	215
		7.2.2 Theory of NP Interaction Potentials	215
		7.2.3 Nanocrystal Superlattices	216

8.	**Group of Carbon**	**221**
	8.1 Fine Particles of Carbon and Silicon	221
	8.2 Fullerenes	223
	8.3 Carbon Nanotubes	225
	8.3.1 Filling of Tubes	226
	8.3.2 Grafting of Functional Groups. Tubes as Matrices	227
	8.3.3 Intercalation of Atoms and Molecules into Multiwalled Tubes	229
	8.4 Graphene	230
	8.5 Carbon Aerosol Gels/Turbstratic Graphite/Graphene	231
9.	**Organic Nanoparticles**	**235**
	9.1 Introduction	235
	9.2 Methods for the Preparation of Nanoparticles	237
	9.2.1 Physical Methods	237
	9.2.1.1 Mechanical Grinding of the Original Substance	237
	9.2.1.2 Laser Ablation	239
	9.2.2 Chemical Methods	242
	9.2.2.1 Solvent Replacement	242
	9.2.2.2 Antisolvents for Precipitation	244
	9.2.2.3 Chemical Reduction in Solution	245
	9.2.2.4 Ion Association	246
	9.2.2.5 Synthesis of Nanoparticles in Water–Oil Emulsion	247
	9.2.2.6 Photochemical Method	248
	9.2.2.7 The use of Supercritical Fluids	248
	9.2.2.8 Cryochemical Synthesis and Modification of Nanoparticles	251
	9.3 Properties and Application of Organic Nanoparticles	257
	9.3.1 Spectral Properties	257
	9.3.2 Quasi-one-dimensional Systems	260
	9.3.3 Drugs and Nanoparticles	263
	9.4 Conclusion	269
10.	**Size Effects in Nanochemistry**	**275**
	10.1 Models of Reactions of Metal Atoms in Matrices	276
	10.2 Melting Point	278
	10.3 Optical Spectra	281
	10.4 Kinetic Peculiarities of Chemical Processes on the Surface of Nanoparticles	287
	10.5 Thermodynamic Features of Nanoparticles	289
	10.6 Magnetic Properties	293
	10.7 Electrical/conducting Properties	294

11.	Nanoparticles in Science and Technology	299
	11.1 Catalysis on Nanoparticles	299
	11.2 Oxide Reactions	311
	11.3 Semiconductors, Sensors, and Electronic Devices	314
	11.4 Photochemistry and Nanophotonics	323
	11.5 Applications of CNTs	326
	11.6 Nanochemistry in Biology and Medicine	329
	11.6.1 DNA-modified Nanoparticles	336

Conclusion	347
Index	355

Preface

Nanoscience and nanotechnology represent one of the main directions of natural science of the twenty-first century and are being actively and rapidly developed. Nanoscience deals with the search and description of fundamental phenomena, relationships, and properties typical of small-scale particles of the nanometer size. Nanotechnology implements the achievements of nanoscience in new processes, materials, and devices. In nanoscience and nanotechnology, the fundamental and applied problems are intertwined, and the latest achievements of theoretical and experimental physics, chemistry, biology, material science, and technology are used.

Nanoscience is a multibranch direction of natural science that combines the features typical of living organisms and the inorganic world.

Nanochemistry forms an important part of nanotechnology, because a lot of processes and syntheses of new materials start from atoms, molecules, clusters, nanoparticles. Thus, on the one hand, chemistry and nanochemistry deal with the initial stage on preparation of various materials and, on the other side, for nanoparticles of different elements, unusual chemical reactions have been observed, and products with unusual chemical properties have been synthesized. The phenomena that depend on the number of particles involved in the reaction are studied by nanochemistry.

Such phenomena associated with the dependence of the chemical activity on the size of involved particles are referred to as the size effect. The experimental and theoretical development of the latter determines the progress in many directions and applications of nanochemistry.

The development of nanoscience and nanotechnology and, hence, of nanochemistry, proceeds very rapidly. This is evidenced by the appearance of the second and new editions of many monographs.

The monograph by professor G.B. Sergeeev was first published in Russian in 2003; its extended English version was published by Elsevier in 2006. For the second edition, professors Sergeev and K. Klabunde decided to join their efforts in order to more completely reflect the state in the art in nanochemistry. This favored considerable widening and renewal of the covered material.

Three new chapters, which deal with preparation of solvated dispersions of metal atoms used in the synthesis of nanoparticles, the self-assembling of nanoparticles, and the control over their size, and also a chapter on the synthesis and properties of organic nanoparticles were added. Besides these new chapters, new paragraphs and sections were supplemented to virtually all chapters of the first edition. The new material constitutes about 40% of the total content.

Many chapters of the first edition have not yet lost their significance as regards both modern science and education and are fully included in the second edition.

The book "Nanochemistry" is of interest for those who deal with problems of both nanoscience and nanotechnology or is interested in the latter one.

In preparation of the second edition, inestimable assistance was rendered by N.S. Merkulova, the wife of G.B. Sergeeev, to whom the latter is extremely grateful. Likewise, K. Klabunde is indebted to his wife, Linda Klabunde, for her valuable contributions, assistance, and patience, which allowed this Second Edition to be completed.

Chapter 1

Survey of the Problem and Certain Definitions

Nowadays, we are witnessing the development and advancement of a new interdisciplinary scientific field—nanoscience. Despite its name, it cannot be associated solely with miniaturization of the studied objects. In fact, nanoscience comprises closely interrelated concepts of chemistry, physics, and biology, which are aimed at the development of a new fundamental knowledge. As was shown by numerous examples in physics, chemistry, and biology, a transition from macrosizes to those of 1–10 nm gives rise to qualitative changes in physicochemical properties of individual compounds and systems.

The historical aspect of the formation and development of independent fundamental directions of nanoscience and the prospects of their application in different branches of nanotechnology were discussed in detail in numerous reviews.[1-4] Numerous books and articles by Russian scientists who had a great influence on the progress in studying small-scale particles and materials can be found in Ref. 3. Their contribution was acknowledged to a certain extent by the 2000 Nobel Prize, which was awarded to Zh. I. Alferov for his achievements in the field of semiconducting heterostructures.

In the past 10–15 years, the progress in nanoscience was largely associated with the elaboration of new methods for synthesizing, studying, and modifying nanoparticles and nanostructures. The extensive and fundamental development of these problems was determined by nanochemistry. Nanochemistry, in turn, has two important aspects. One of these is associated with gaining insight into peculiarities of chemical properties and the reactivity of particles comprising a small number of atoms, which lays new foundations of this science. Another aspect, connected to nanotechnology, consists of the application of nanochemistry to the synthesis, modification, and stabilization of individual nanoparticles and also for their directed self-assembling to give more complex nanostructures. Moreover, the possibility of changing the properties of synthesized structures by regulating the sizes and shapes of original nanoparticles deserves attention.

The advances in recent studies along the directions mentioned are reflected in several reviews and books.[5-13] A special issue of the journal *Vestnik Moskovskogo Universiteta* was devoted to the problems of nanochemistry.[14]

The dependence of physicochemical properties on the particle size was discussed based on optical spectra,[15] magnetic properties,[16,17] thermodynamics,[18] electrochemistry,[19] conductivity, and electron transport.[20,21] Different equations describing physical properties as a function of the particle size were derived within the framework of the droplet model.[22] A special issue of *Journal of Nanoparticle Research* is devoted to the works of Russian investigators in the field of nanoscience.[23] Many aspects of synthesis, physicochemical properties, and self-assembly have been reviewed.[24]

In nanochemistry, which is in a stage of rapid development, questions associated with definitions and terms still arise. The exact difference between terms such as "cluster," "nanoparticle," and "quantum dot" has not yet been formulated in the literature. The term "cluster" is largely used for particles that include small numbers of atoms, while the term "nanoparticle" is applied for larger atomic aggregates, usually when describing the properties of metals and carbon. As a rule, the term "quantum dot" concerns semiconductor particles and islets, the properties of which depend on quantum limitations on charge carriers or excitons. In this book, no special significance will be attached to definitions, and the terms "cluster" and "nanoparticle" will be considered as interchangeable.

Table 1.1 shows some classifications of nanoparticles, which were proposed by different authors based on the diameter of a particle expressed in nanometers and the number of atoms in a particle. These classifications also take into account the ratio of surface atoms to those in the bulk. A definition given by Kreibig[25] is similar to that proposed by Gubin.[26] It should be mentioned that a field of chemistry distinguished by Klabunde[12] pertains, in fact, to particles measuring less than 1 nm.

Nanoparticles and metal clusters represent an important state of condensed matter. Such systems display many peculiarities and physical and chemical properties that were never observed earlier. Nanoparticles may be considered as intermediate formations, which are limited by individual atoms on the one hand and the solid phase on the other. Such particles exhibit the size dependence and a wide spectrum of properties. Thus, nanoparticles can be defined as entities measuring from 1 to 10 nm and built of atoms of one or several elements. Presumably, they represent closely packed particles of random shapes with a sort of structural organization. One of the directions of nanoscience deals with various properties of individual nanoparticles. Another direction is devoted to studying the arrangement of atoms within a structure formed by nanoparticles. Moreover, the relative stability of individual parts in this nanostructure can be determined by variations in kinetic and thermodynamic factors. Thus, nanosystems are characterized by the presence of various fluctuations.

Natural and technological nanoobjects represent, as a rule, multicomponent systems. Here again, one is up against a large number of different terms such as "nanocrystal," "nanophase," "nanosystem," "nanostructure," and "nanocomposites," which designate formations built of individual, separate nanoparticles.

Chapter 1 Survey of the Problem and Certain Definitions

TABLE 1.1 Classification of Particles by their Sizes

(a)

U. Kreibig[25]

Domain I Molecular clusters	Domain II Solid-state clusters	Domain III Microcrystals	Domain IV Bulk particles
$N \leq 10$ Indistinguishable surface and volume	$10^2 \leq N \leq 10^3$ Surface–volume ratio ≈ 1	$10^3 \leq N \leq 10^4$ Surface–volume ratio < 1	$N > 10^5$ Surface–volume ratio < 1

(b)

K. Klabunde[12]

Chemistry			Nanoparticles			Solid-state physics	
Atom	$N = 10$	$N = 10^2$	$N = 10^3$	$N = 10^4$	$N = 10^6$		Bulk matter
Diameter (nm)		1	2	3	5	7	10 >100

(c)

N. Takeo (*Disperse Systems,* Wiley-VCH, 1999, p. 315.)

Superfine clusters	Fine clusters	Coarse clusters
$2 < N \leq 20$ $2R \leq 1.1$ nm Indistinguishable surface and internal volumes	$20 < N \leq 500$ $1.1 \text{ nm} \leq 2R \leq 3.3 \text{ nm}$ $0.9 \geq N_s/N_v \geq 0.5$	$500 < N \leq 10^7$ $3.3 \text{ nm} \leq 2R \leq 100 \text{ nm}$ $0.5 \geq N_s/N_v$

(d)

G.B. Sergeev, V.E. Bochenkov (Physical Chemistry of Ultradispersed Systems: Conference Proceedings, Moscow, 2003, pp. 24–29.)

Chemistry of atoms		Nanochemistry Number of atoms in particle					Chemistry of solid-state
Single atoms	10	10^2	10^3	10^4		10^6	Bulk
Diameter (nm) 1		2	3	5	7	10	>100

For instance, nanostructure can be defined as an aggregate of nanoparticles of definite sizes, which is characterized by the presence of functional bonds. In the reactions with other chemical substances, such limited-volume systems can be considered as a sort of nanoreactors. Nanocomposites represent systems where nanoparticles are packed together to form a macroscopic sample in which interactions between particles become strong, masking the properties of individual particles. For every type of interaction, it is important to know how the properties of a sample change with its size. Moreover, it should be mentioned that with a decrease in the particle size, the concept of phase becomes less clear: it is difficult to find boundaries between homogeneous and heterogeneous phases, and between amorphous and crystalline states. At present, the common concepts of chemistry, which define the relationships such as composition–properties and structure–function, are supplemented by the concepts of size and self-organization, giving rise to new effects and mechanisms. Nonetheless, despite all achievements of nanochemistry, we still cannot give a general answer to the question how the size of particles of, e.g. a metal, is related to their properties.

Metallic nanoparticles measuring less than 10 nm represent systems with excessive energy and a high chemical activity. Particles of about 1 nm need virtually no activation energy to enter into either aggregation processes, which result in the formation of metal nanoparticles, or reactions with other chemical compounds to give substances with new properties. The stored energy of such particles is determined first of all by uncompensated bonds of surface and near-surface atoms. This can give rise to unusual surface phenomena and reactions.

The formation of nanoparticles from atoms involves two processes, namely, the formation of metal nuclei of different sizes and the interactions between the formed particles, which generate the formation of assemblies that possess a nanostructure.

Virtually all methods of nanosynthesis produce nanoparticles in nonequilibrium metastable states. On the one hand, this factor complicates their investigation and application in nanotechnologies aimed at the development of stable devices. On the other, nonequilibrium systems allow carrying out new unusual chemical reactions, which are difficult to predict.

Elucidation of the relationship between the size and chemical reactivity of a particle is among the most important problems of nanochemistry. For nanoparticles, two types of size effects are distinguished.[27] One of these is their intrinsic or internal effect, which is associated with specific changes in superficial, bulk, and chemical properties of a particle. The other, external effect, represents a size-dependent response to external factors unrelated to the internal effect.

Specific size effects manifest themselves to a great extent for smaller particles and are most likely in nanochemistry, where irregular size–properties dependencies prevail. The dependence of activity on the size of the particles taking part in a reaction can be associated with the changes in the particle properties in the course of its interaction with an adsorbed reagent,[28] correlations between geometrical and electron shell structures,[29] and symmetry of boundary orbitals of a metal particle with respect to adsorbed-molecule orbitals.[30]

As mentioned above, nanochemistry studies the synthesis and chemical properties of particles and formations with sizes below 10 nm along one direction at least. Moreover, most interesting transformations are associated with the region of ca. 1 nm. Elucidation of mechanisms that govern the activity of particles with sizes of 1 nm and smaller is among the major problems of modern nanochemistry, despite the fact that the number of particles is a more fundamental quantity as compared with their size.

The dependence of chemical activity on the size of reacting particles is explained by the fact that properties of individual atoms of elements as well as of clusters and nanoparticles formed from atoms differ from the properties of corresponding macroparticles. To understand and roughly analyze the size-dependent chemical properties, we can compare the reactivities of compact substances, nanoparticles, and clusters of species.[31] The demarcation lines between sizes of such formations vary from element to element and should be specified for each case.

In nanochemistry, the interaction of every particle with the environment has its own specifics. When studying individual properties of such a particle, attention should be focused on qualitative changes in particle properties as a function of its size. Moreover, the properties of isolated nanoparticles are characterized by a wide statistical scatter, which varies in time and requires special studies.

The internal size effect in chemistry can be caused by the changes in the particle structure and the surface-induced increase in the electron localization. Surface properties affect the stabilization of particles and their reactivity. For small numbers of reagent atoms adsorbed on the surface, a chemical reaction cannot be considered as in infinite volume, due to the commensurable surfaces of a nanoparticle and a reactant.

Reaction kinetics in small-scale systems with limited geometry differs from classical kinetics, because the latter ignores fluctuations in concentrations of reacting particles. Formations containing small numbers of interacting molecules are characterized by relatively wide fluctuations in the number of reactants. This factor gives rise to a time lag between the changes in reactant concentration on the surfaces of different-size nanoparticles and, as a consequence, to their different reactivities. Kinetics of such systems is described based on a stochastic approach,[32] which takes into account statistical fluctuations in the number of reacting particles. The Monte-Carlo technique was also used for describing the kinetics of processes that occur on the surface of nanoparticles.[33]

In nanoparticles, a considerable number of atoms pertain to the surface, and their ratio increases with a decrease in the particle size. Correspondingly, the contribution of surface atoms to the system's energy increases. This has certain thermodynamic consequences, for example, a size dependence of the melting point, T_m, of nanoparticles. The size, which determines the reactivity of particles, also gives rise to effects such as variations in the temperature of polymorphous transitions, a solubility increase, and a shift of chemical equilibrium.

Experiments and theoretical studies on thermodynamics of small particles testify that the particle size is an active variable, which, together with other thermodynamic variables, determines the state of the system and its reactivity. The particle size can be considered as an equivalent of the temperature. This means that nanoscale particles can enter into reactions untypical of bulk substances.

Moreover, it was found that variations in the size of metal nanocrystals control the metal–nonmetal transition.[34] This phenomenon is observed for particles with diameters not exceeding 1–2 nm and can also affect the reactivity of the system. The activity of particles also depends on interatomic distances. Theoretical estimates by the example of gold particles have shown that average interatomic distances increase with a decrease in the particle size.[35]

As a rule, nanoparticles, free of interactions with the environment, can exist as separate particles only in vacuum, due to their high activity. However, using the example of silver particles with different sizes, it was shown that optical properties are identical in vacuum and upon condensation in an argon medium at low temperatures.[36] Silver particles were obtained by mild deposition in solid argon. Spectra of clusters comprising 10–20 silver atoms resembled those of particles isolated in the gas phase by means of mass spectrometry. Based on these results, it was concluded that deposition processes have no effect on the shape and geometry of clusters. Thus, the optical properties and reactivity of metal nanoparticles in the gas phase and inert matrices are quite comparable.

A different situation is observed for nanoparticles obtained in the liquid phase or on solid surfaces. In the liquid phase, the formation of a metal nucleus of a particle from atoms is accompanied by interaction of particles with the environment. The interplay of these two processes depends on many factors, most important of which are the temperature and the reagent ratio in addition to physicochemical properties of metal atoms and the reactivity and stabilizing properties of ligands of the medium. The interaction of atoms and metal clusters with a solid surface is an intricate phenomenon. The process depends on the surface properties (smooth facet of single crystals and rough and developed surfaces of various adsorbents) and the energy of particles to be deposited.

As mentioned above, the main problem of nanochemistry is to elucidate the relationship between the size and chemical activity of particles. Based on the experimental data available, we can formulate the following definition: size effects in chemistry are the phenomena that manifest themselves in *qualitative* changes in chemical properties and reactivity and depend on the number of atoms or molecules in a particle.[37]

It is difficult to regulate the sizes of metal nanoparticles that are often poorly reproducible, being determined by their preparation method. The mentioned factors limit the number of publications containing an analysis of the effect of particle size on its reactivity. In recent publications, such reactions were most actively studied in the gas phase, and experimental studies were supplemented by a theoretical analysis of the results obtained.

Chemical and physical properties of metal nanoparticles formed of atoms were observed to change periodically depending on the number of atoms in a particle, its shape, and the type of its organization. In this connection, attempts were undertaken to tabulate the electronic and geometrical properties of clusters and metal nanoparticles by analogy with the Mendeleev Periodic Table. As was shown by the example of sodium atoms, Na_3, Na_9, and Na_{19} particles are univalent, while halogen-like clusters Na_7 and Na_{17} exhibit enhanced activity. The lowest activity is typical of particles with closed electron shells, namely, Na_2, Na_8, Na_{18}, and Na_{20}.[38] This analogy, which was demonstrated for small clusters with properties determined by their electronic structure, makes it possible to expect the appearance of new chemical phenomena in reactions with such substances.

For sodium clusters containing several thousand atoms, periodic changes in the stability of particles were also revealed. For Na particles containing more than 1500 atoms, the closed-shell geometry prevails, which resembles that of inert gases.

It was noted[38] that the size of particles containing tens of thousand atoms can affect their activity in a different manner. Sometimes, the key role is played by electronic structures of each cluster; otherwise, the geometrical structure of the electronic shell of the whole particle has a stronger effect on the reactivity. In real particles, their electronic and geometrical structures are interrelated and it is not always possible to separate their effects.

The problem of elucidating the dependence of chemical properties on the size of particles involved in a reaction is closely linked with the problem of revealing the mechanisms of formation of nanoscale solid phases during electrocrystallization. Interactions of atoms in the gas and liquid phases or upon their collision with a surface first of all give rise to small clusters, which can later grow to nanocrystals. In the liquid phase, such nucleation is accompanied by crystallization and solid-phase formation. Peculiarities of the formation of nanoscale phases during fast crystallization were considered on qualitative and quantitative levels.[39,40] Nanochemistry of metal particles formed by small numbers of atoms demonstrates no pronounced boundaries between phases and the questions of how many atoms of one or other elements are necessary for spontaneous formation of a crystal nucleus that can initiate the formation of a nanostructure have not yet found the answer.

In nanochemistry, when studying the effect of the particle size on its properties, the most important factors are the surface on which the particle is located and the nature of the stabilizing ligand. One of the approaches to solving this problem is to find the symmetry energy of the highest occupied molecular orbital and/or the lowest unoccupied molecular orbital as a function of the particle size. Yet another approach is based on finding such a shape of nanoparticles that would allow the optimal conditions for the reactions to be reached.

To date, nanochemistry of some elements of the Periodic Table was studied in sufficient detail, while other elements were studied incompletely.

From our viewpoint, in the next 10–15 years, the role of nanochemistry in the development of nanotechnology will increase; this is why in the following chapters we will discuss in detail the synthesis, chemical properties, and reactivity of atoms, clusters, and nanoparticles of different elements in the Periodic Table.

REFERENCES

1. Roco, M. C.; Williams, S.; Alivisatos, P., Eds. *Nanotechnology Research Directions: IWGN Workshop Report-vision for Nanotechnology in the Next Decade*; Kluwer: New York, 1999; pp 1–360.
2. Alferov, Zh. I *Semiconductors* **1998**, *32*, 1–14.
3. Petrunin, V. F. Ekaterinburg, UrO RAN. Physical Chemistry of Ultradispersed Systems. *In Proceedings of Conference (Russ.)*, 2001; 5–11.
4. Andrievsky, R. A. *Russ. Chem. J.* **2002**, *46*, 50–56.
5. Gleiter, H. *Acta Mater.* **2000**, *48*, 1–29.
6. Pomogailo, A. D.; Rozenberg, V. I.; Uflyand, I. E. *Metal Nanoparticles in Polymers*; Khimiya: Moscow, 2000; pp 1–672.
7. Roldugin, V. I. *Russ. Chem. Rev.* **2000**, *69*, 821–844.
8. Bukhtiyarov, V. I.; Slin'ko, M. G. *Russ. Chem. Rev.* **2001**, *70*, 147–160.
9. Sergeev, G. B. *Russ. Chem. Rev.* **2001**, *70*, 809–826.
10. Summ, B. D.; Ivanova, N. I. *Russ. Chem. Rev.* **2000**, *69*, 911–924.
11. Klabunde, K. J., Ed. *Nanoscale Materials in Chemistry*; Wiley: New York, 2001; pp 1–292.
12. Klabunde, K. J. *Free Atoms, Clusters and Nanosized Particles*; Academic Press: San Diego, New York, Boston, London, Sydney, Tokyo, 1994; p 311.
13. Sergeev, G. B. *Chemical Physics in Front of XXI Century (Russ.)*; Nauka: Moscow, 1996; 149–166.
14. *Nanochemistry. Special issue*, Vestn. Mosk. Univ. Ser. 2, Khim. **2001**, *42*(5), 300–368.
15. Kreibig, U.; Vollmer, M. *Optical Properties of Metal Clusters*; Springer: Berlin, 1995; pp 1–532.
16. Suzdalev, I. P.; Buravtsev, Yu. V.; Maksimov, V. K.; Imshennik, V. K.; Novichikhin, S. V.; Matveev, V. V., et al. *Ros. Khim. Zhurn.* **2001**, *XLV*, 66–73.
17. Binns, C. *Surf. Sci. Rep.* **2001**, *44*, 1–49.
18. Wang, Z. L.; Petroski, J. M.; Green, T. C.; El-Sayed, M. A. *J. Phys. Chem. B* **1998**, *102*, 6145–6151.
19. Gorer, S.; Ganske, J. A.; Hemminger, J. C.; Penner, R. M. *J. Am. Chem. Soc.* **1998**, *120*, 9584–9593.
20. Alivisatos, A. P. *J. Phys. Chem.* **1996**, *100*, 13226–13239.
21. Doty, R. C.; Yu, H.; Shih, C. K.; Korgel, B. A. *J. Phys. Chem. B* **2001**, *105*, 8291–8296.
22. Lakhno, V. D. *Clusters in Physics, Chemistry, Biology (Russ.)*; RHD: Izhevsk, 2001; 1–256.
23. Sergeev, G.B. *J. Nanopart. Res.* **2003**, *5*(5-6), 529–537.
24. Schmid, G., Ed. *Nanoparticles: From Theory to Application*; Wiley-VCH: Weinheim, 2005; pp 1–444.
25. Kreibig, U. *Z. Phys. D Atom. Mol. Clust.* **1986**, *3*, 239–249.
26. Gubin, S. P. *Chemistry of Clusters (Russ.)*; Nauka: Moscow, 1987, p 263.
27. Opila, R. L., Jr.; Eng, J. *Prog. Surf. Sci.* **2002**, *69*, 125–163.
28. Winter, B. J.; Parks, E. K.; Riley, S. J. *J. Chem. Phys.* **1991**, *94*, 8618–8621.
29. Groenbeck, H.; Rosen, A. *Chem. Phys. Lett.* **1994**, *227*, 149.
30. Groenbeck, H.; Rosen, A. *Phys. Rev.* **1996**, *B54*, 1549–1558.
31. Haynes, C. L.; Van Duyne, R. P. *J. Phys. Chem. B* **2001**, *105*, 5599–5611.
32. Khairutdinov, R. F.; Serpone, N. *Prog. React. Kinet.* **1996**, *21*, 1–30.
33. Zhdanov, V. P.; Kasemo, B. *Surf. Sci. Rep.* **2000**, *39*, 25–104.
34. Vinod, C. P.; Kulkarni, G. U.; Rao, C. N.R. *Chem. Phys. Lett.* **1998**, *289*, 329–332.
35. Haberlen, O. D.; Chung, S. C.; Stenek, M.; Rosch, N. *J. Chem. Phys.* **1997**, *106*, 5189–5201.

36. Harbich, W. *Phylos. Mag.* **1999,** *B79,* 1307–1311.
37. Sergeev, G. B. *Ros. Khim. Zhurn.* **2002,** *46,* 22–29.
38. Heiz, U.; Schneider, W. -D. In *Metal Clusters at Surface—Structure, Quantum Properties, Physical Chemistry*; Meiwes-Broer, K. -H., Ed.; Springer: Berlin, 2002; pp 237–273.
39. Melikhov, I. V. *Russ. Chem. Bull. (Russ.)* **1994,** 1710–1718.
40. Melikhov, I. V. *Inorg. Mater.* **2000,** *36,* 278–286.

Chapter 2

Synthesis and Stabilization of Nanoparticles

Chapter Outline

2.1 Chemical Reduction 13
2.2 Reactions in Micelles, Emulsions, and Dendrimers 18
2.3 Photochemical and Radiation-Chemical Reductions 22
2.4 Cryochemical Synthesis 27
2.5 Physical Methods 38
2.6 Particles of Various Shapes and Films 43

Metal atoms exhibit a high chemical activity, which is retained in their dimers, trimers, clusters, and nanoparticles containing large numbers of atoms. The study of such active particles is possible as long as various stabilizers are used; hence, the problems concerning the synthesis of such nanoparticles and their stabilization should be considered in tandem. At present, there are many methods for synthesizing particles of various sizes. Insofar as metals constitute the majority of elements in the Periodic Table, we will consider them as examples, based on studies published for the most part in the past decade.

In principle, all the methods for synthesizing nanoparticles can be divided into two large groups. The first group combines methods that allow preparation and studies of nanoparticles but do not help much in the development of new materials. They include condensation at superlow temperatures, certain versions of chemical, photochemical, and radiation reduction, and laser-induced evaporation (Figure 2.1).

The second group includes methods that allow preparation of nanomaterials and nanocomposites, based on nanoparticles. These are, first of all, different versions of mechanochemical dispersion, condensation from the gas phase, plasmochemical synthesis, and certain other methods.

The above division reflects another peculiarity of methods under consideration, which is expressed as follows: the particles can either be built from

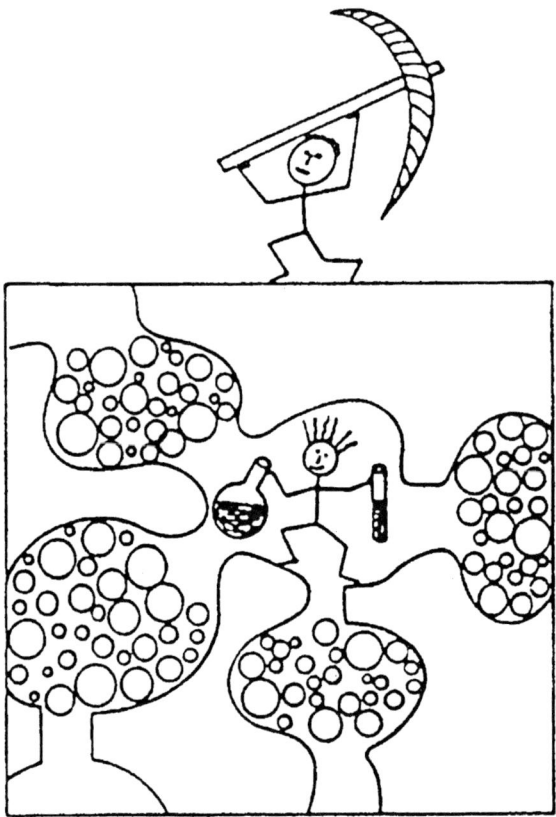

FIGURE 2.1 Two approaches to the synthesis of nanoparticles. A comparison of nanochemistry and nanophysics.[1]

separate atoms (an approach from the "bottom") or by various dispersion and aggregation procedures (an approach from the "top"). The approach from the "bottom" largely pertains to chemical methods of preparation of nanosize particles, whereas the approach from the "top" is typical of mechanical/physical methods. For example, grinding of a solid material can be done on large scale. However, grinding and pulverizing have limits because, as particle size decreases, chemical reactivity increases. This eventually leads to the back reaction of particles, necking, and coalescence. Thus, consider the extreme of water droplets. These droplets never spontaneously split apart, but do spontaneously coalesce to form larger droplets. The same is true of metal droplets as well as solid particles. Because of these increased surface energies and reactivities, grinding and pulverization are not good methods for reaching below about 50 nm, and certainly not good for attaining monodispersity (all particles the same size). The most satisfactory results are for solids with very high lattice energies, such as magnesium oxide and other ceramics. The least satisfactory results are for low melting, low lattice energy solids, such as zinc metal or magnesium metal.

A modification that helps stabilize small particles as they form, is adding an active surface ligand, called a chemo-modified grinding. But even this approach has not proven to be very successful, at least for exacting studies. Nonetheless, if large amounts of nanomaterials are needed, and there are no requirements for monodispersity or ligand stabilization, grinding/pulverization of bulk solids is a viable synthetic method.

Of course, "bottom-up" methods of synthesis yield more control in nanoscale synthesis, but usually are more expensive than "top-down." Bottom-up means building nanoparticles from their constituent fundamental building blocks, such as atoms or reactive small molecules. So, in bottom-up synthesis, there is a need for a molecular precursor that can be suddenly changed to a fundamental building block. For example, a soluble stable metal salt could be reduced to form metal atoms, which rapidly aggregate to nanoparticles. Another example is to hydrolyze a soluble metal alkoxide to an insoluble metal hydroxide, which rapidly aggregates to nanoparticles. A third example is to use thermal energy to liberate atoms from bulk metal, and allow the atoms to aggregate in a controlled fashion.[1] Naturally, the proposed division is rough and schematic. Preparation of nanoparticles from atoms allows individual atoms to be considered as the lower limit of nanochemistry. Its upper boundary corresponds to atomic clusters, whose properties no longer undergo qualitative changes with an increase in the number of constituent atoms, thus resembling the properties of compact metals. The number of atoms that define the upper boundary is unique for every element in the Periodic Table. It is also of paramount importance that the structures of equal-size nanoparticles can differ if they were obtained by using different approaches. As a rule, dispersion of compact materials into nanosize particles retains the original structure in resulting nanoparticles. In particles formed by aggregation of atoms, the positions of atoms can be different, which affects their electronic structure. For example, a particle measuring 2–4 nm can demonstrate a decrease in the lattice parameter. The above factor poses a problem of the necessity of analyzing the law of conservation of chemical composition at the nanolevel.

2.1 CHEMICAL REDUCTION

Nowadays, the attention of many scientists is focused on the development of new methods for synthesis and stabilization of metal nanoparticles. Moreover, special attention is paid to monodispersed particles. Chemical reduction is used most extensively in the liquid phase, including aqueous and nonaqueous media. As a rule, metal compounds are represented by their salts, while aluminohydrides, borohydrides, hypophosphites, formaldehyde, and salts of oxalic and tartaric acids serve as the reducers. The wide application of this method stems from its simplicity and availability.

As an example, we consider the synthesis of gold particles. Three solutions are prepared: (a) chloroauric acid in water, (b) sodium carbonate in water, and (c) hypophosphite in diethyl ether. Then, their mixture is heated for an hour up to 70 °C. As a result, gold particles of 2–5 nm diameters are obtained. The major drawback of this method is the large amount of admixtures contained in a colloid system formed of gold nanoparticles, which can be lowered by using hydrogen as the reducer.

In the general case, the behavior of a metal particle in solution is determined by the potential difference $\Delta E = E - E_{redox}$, where E is the equilibrium redox potential of the particle and E_{redox} is the corresponding solution potential. Particles grow when $\Delta E > 0$ and dissolve when $\Delta E < 0$. For $\Delta E = 0$, an unstable equilibrium is established. The situation is complicated by the fact that the redox potential of a metal particle depends on the number of atoms. In this respect, the chemical reduction occurs in systems thermodynamically and is kinetically unstable. Chemical reduction is a multifactor process. It depends on the choice of a redox pair and concentrations of its components as well as on the temperature, pH of the medium, and diffusion and sorption characteristics.

Recently, the processes in which a reducer simultaneously performs the function of a stabilizer became widely used. Among such compounds are numerous N–S-containing surfactants, thiols, salts of nitrates, and polymers containing functional groups.

Reagents most frequently used as the reducers of metal ions are tetrahydroborates of alkali metals (MBH_4), which operate in acidic, neutral, and alkaline aqueous media. Alkali-metal tetraborates can reduce most cations of transition and heavy metals, which is explained by the high redox potential of MBH_4 (1.24 V in alkaline medium) as compared with the standard potentials of many metal ions, which lie in the interval $-0.5 \leq -E \leq = -1.0$ V.[42] Reduction of metal ions was shown to involve the formation of complexes with bridge bonds M···H···B, which favors the subsequent hydrogen-atom transfer with the break of the bridge bond, followed by a redox process with the breakage of a B–H bond to give BH_3. The obtained borane undergoes hydrolysis and catalytic decomposition on the surface of metal particles.

Syntheses of metal nanoparticles in liquid media involved using hydrazine hypophosphite and its derivatives and also various organic substances as the reducers.[6] Certain problems concerning the kinetics and mechanism of formation of metal nanoparticles in liquid-phase redox reactions were analyzed.[2] The analysis was based on the analogy with crystallization processes and topochemical reactions of thermal decomposition of solids and also with reactions of the gas–solid type. However, as correctly reasoned, the analogies of such a kind and the results obtained based on a formal description of the kinetics of chemical reduction should be viewed with caution. The peculiarities of the kinetics and mechanisms of complex and multifactor processes such as the redox synthesis, the growth, and stabilization of metal nanoparticles require further research. Chemical interactions in the reduced-metal-ion–reducer system can be associated with the transfer of an electron from the reducer to the metal ion via the formation of an intermediate complex, which lowers the electron-transfer energy. A so-called electrochemical mechanism, which also involves electron transfer but occurs with direct participation of the surface layer of growing metal particles, was discussed.[3]

Spherical silver nanoparticles measuring 3.3–4.8 nm were synthesized by the reduction of silver nitrate by sodium borohydride in the presence of tetraammonium disulfide.[44] Dibromidebis[(trimethyl ammoniumdecanoylamino)ethyl] disulfide was used as the stabilizer. Particles obtained were characterized by intense light absorption in the wavelength region of 400 nm, which corresponds to the silver plasmon peak and points to the metallic nature of particles. By studying the effect of the medium on the stability of particles, it was found that the latter are aggregated in the presence of sulfuric and hydrochloric acids. The stability of silver particles also depended on the pH of the medium: in aqueous media with pH 5–9, the particles remained stable for a week. An increase or a decrease in the pH resulted in fast aggregation and deposition of silver particles. The effect of the latter factor on the stability of gold particles was less pronounced.

It is shown that small positively charged silver clusters, stabilized in the form of polyacrylate complexes ("blue silver"), can be prepared by the partial oxidation of the products of borohydride reduction of the Ag^+ cations in aqueous polyacrylate solutions.[4,5]

Particles of controlled sizes (1–2 nm) were obtained by using an amphiphilic polymer poly(octadecylsiloxane) as the matrix.[6]

Hybrid materials based on polyelectrolyte gels with oppositely charged surface-active substances (surfactants) were used as nanostructured media for the reduction of various platinum salts with sodium borohydride and hydrazine. It was shown that the reduction with sodium borohydride mainly yields small platinum particles with radii of ca. 2–3 nm, while the reduction with hydrazine produces particles measuring ca. 40 nm.[7]

For cobalt nanoparticles, the mechanism of formation, electron spectra, and reactions in aqueous media were studied.[8] The radiation-chemical reduction of cobalt ions in aqueous solutions of $Co(ClO_4)_2$ and HCOONa produced spherical cobalt particles with diameters of 2–4 nm. As a stabilizer, sodium polyacrylate with a molecular mass of 2100 u was used. The radiolysis produced solvated electrons e^-_{aq}, hydroxyl radicals, hydrogen atoms, and $CO_2^{-\cdot}$ radicalions:

$$H_2O \rightarrow e^-_{aq}, H, OH,$$

$$OH(H) + HCOO^- \rightarrow H_2O(H_2) + CO_2^{-\cdot}$$

Hydrated electrons and radical ions $CO_2^{-\cdot}$ reduced Co^{2+} ions to give cobalt nanoparticles with an absorption peak in the wavelength region of 200 nm. By using pulsed radiolysis, it was shown that these processes follow the autocatalytic mechanism.

Radiation chemical reduction of Ni^{2+} ions in aqueous $Ni(ClO_4)_2$ solutions containing isopropanol, which was carried out in the presence of polyethylene, polyacrylate, and polyvinyl sulfate, produced metal sols formed by spherical particles with diameters of 2–4 nm. Nickel nanoparticles, which were easily oxidizable by O_2 and H_2O_2, formed sufficiently stable nickel–silver nanosystems[9] with silver ions.

Spherical copper particles measuring 20–100 nm were obtained by γ-radiolysis of aqueous $KCu(CN)_2$ solutions in the presence of either methanol or 2-propanol as the scavengers of hydroxyl radicals.[10]

The formation of silver particles during γ-radiolysis of silver nitrate solutions in water, ethanol, and 0.01 M $C_{12}H_{25}OSO_3Na$ was studied. The fractal dimensions of particle aggregates in these solutions were 1.81, 1.73, and 1.70, respectively.[11] A synthesis of stable nanoparticles (average size of 1–2 nm) of platinum, rhodium, and ruthenium in organic media as a result of heating colloidal solutions of corresponding metal hydroxides in ethylene glycol was described.[12]

Monodispersed particles of amorphous selenium were prepared in ethylene glycol via the reduction of selenious acid with hydrazine. The preparation of particles with controllable surfaces was based on varying the temperature in a range from −10 to +60 °C. This interval includes the temperature of glass transition of selenium, which provided control over the rate of incorporation of iron oxide into selenium particles.[13]

Silver particles with sizes ranging from 2 to 7 nm were synthesized by electrochemical dissolution of a metal anode (silver plate) in an aprotic solution of tetrabutylammonium bromide in acetonitrile.[14] This process was shown to depend on characteristics such as the current density and the cathode nature. Thus, at high current densities under nonequilibrium conditions, particles of irregular shapes can be formed. With an increase in the current density from −1.35 to −6.90 mA/cm², the particle diameter decreased from 6 ± 0.7 to 1.7 ± 0.4 nm. The reduction of silver ions stabilized by tetrabutylammonium bromide resulted in the formation of metal nanoparticles and their deposition on cathodes made of either platinum or aluminum. Figure 2.2 illustrates this process. On a platinum cathode, spherical silver nanoparticles were deposited. The deposition on an aluminum cathode produced films. An analysis of optical spectra of nanoparticles in the course of their synthesis made it possible to conclude that this process involves an autocatalytic stage. Moreover, it was shown that the half-width of a peak corresponding to the surface plasmon of a particle depends linearly on $1/R$ (R is the particle radius) and the plasmon band shifts to lower frequencies with a decrease in the particle size. Modern problems of nanoelectrochemistry were surveyed in Ref. 15.

Organic solvents are preferred for the preparation of nanoparticles. They perform stabilizing functions. Such solvents or surfactants play a key role in the synthesis of nanoparticles. They are bound to the surface of growing nanocrystals via polar groups, form complexes with species in solutions, and control

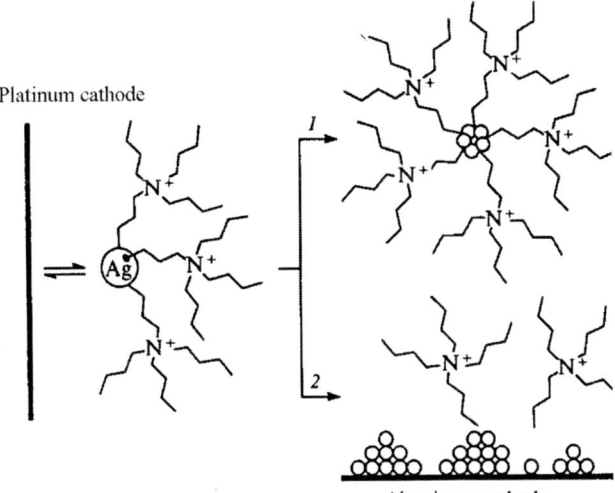

FIGURE 2.2 Illustration of competition of two processes:[55] (1) formation of silver particles, (2) deposition of particles and film formation.

their chemical reactivity and diffusion to the surface of a growing particle. All the mentioned processes depend on the temperature, the surface energy of a nanocrystal, concentration of free particles in solution and their sizes, and the surface-to-volume ratio of a particle.

It is possible to exercise control over the formation of different-size particles by regulating the dependencies mentioned above. Recently, these principles were applied for synthesizing various nanoparticles with relatively narrow size distributions, e.g. oxides such as Fe_2O_3[16] and $MnFe_2O_4$,[17] metal alloys such as $CoPt_3$,[18,19] and semiconductors such as CdS and CdTe,[20] InAs and InP,[21] and Ge.[22]

At present, attention is drawn to anisotropic properties and chemical reactivity of different facets of nanocrystals, which have different atomic densities, polarity, and number of dangling bonds. The efforts of scientists have also focused on synthesizing particles of different shapes. Shape-controlled growth has been demonstrated for titanium dioxide,[23] cobalt,[24] nickel,[25] CdTe,[26] and ZnTe.[27]

Recently, progress in synthesizing various core–shell structures was observed. Such structures were described for the systems CdSe/CdS,[28] CdTe/CdSe and CDSe/ZnTe,[29] FePt/Fe_3O_4,[30] Pt/Co,[31] and Ag/Co.[32]

In many cases, the core–shell structures are formed as symmetrical systems; however, it is also possible to grow one material in an asymmetric fashion onto another one. Such heterostructures were demonstrated for CdSe–CdS systems containing spherical CdSe cores and rod-like CdS shells.[33] Of great promise are also heterodimers that combine several properties such as fluorescence and magnetic behavior in one material. CdS–FePt particles measuring 7 nm pertain to such systems.[34] Furthermore, a gold deposit could be grown on the ends of nanosize CdSe tetrapods.[35] Preparation of metal and metal-oxide nanoparticles is considered in detail in reviews.[36,37]

Stabilization processes accompany the preparation of different nanoparticles. To prevent aggregation, active particles should not be allowed to contact one another. This is achieved by either the presence of like charges or by steric repulsion of hydrophobic chains of the stabilizer.

2.2 REACTIONS IN MICELLES, EMULSIONS, AND DENDRIMERS

Synthesis of metal nanoparticles and their compounds involves using micelles, emulsions, and dendrimers as a sort of nanoreactors that allow synthesizing particles of definite sizes. Nanoparticles of crystalline bismuth measuring less than 10 nm were obtained by the reduction of aqueous solutions of bismuth salts. This process took place inside inverse micelles based on sodium diisooctylsulfosuccinate (conventional designation AOT).[38] Mixing of an isooctane solution of AOT with a definite amount of aqueous solution of $BiOClO_4$ led to the formation of inverse micelles. A micellar solution of $NaBH_4$ was prepared in the same fashion with the same ratio $w=[H_2O]/[AOT]$. Both solutions were mixed under an argon atmosphere. Bismuth particles precipitated upon mixing and aging of the resulting mixture for several hours at room temperature. The liquid phase was separated in vacuum, and the dry deposit was dispersed in toluene. A dark solution thus obtained contained, according to powder X-ray diffraction (XRD) and scanning tunneling microscopy (STM) tests, bismuth particles measuring 3.2 ± 0.35 nm for $w=2$ and 6.9 ± 2.2 nm for $w=3$. Antioxidation protection of crystalline particles by polymers increased the particle size to 20 nm.[39]

Reduction of rhodium salts in water in the presence of an amphiphilic block-copolymer styrene–ethylene oxide and an anionic surfactant, e.g., sodium dodecylsulfate, produced rhodium particles with diameters in the range of 2–3 nm stabilized by the block-copolymer.[40] Luminescent nanomaterials based on yttrium oxide doped with europium were synthesized by employing nonionic- inversed microemulsions based on polyethylene oxide and other ethers.[41]

At present, the active search for macromolecules that can serve as matrices in synthesizing nanoparticles is in progress. In such methods, the stabilizer molecules interact with surfaces of metal particles thus affecting their growth. For example, reduction of bivalent copper in the presence of poly-N-vinylpyrrolidone produced particles with diameters of 7 ± 1.5 nm at a temperature of 11 °C and 102 nm at 30 °C.[42] Interesting results were obtained when studying the temperature dependence of the stability of already formed nanoparticles. Heating of copper particles formed at 11 °C up to 30 °C resulted in the loss of protective properties of their polymeric coatings, which enhanced aggregation and accelerated oxidation of metal particles. An opposite result was achieved by cooling a system formed at 30 °C up to 11 °C. In the latter case, the particles did not aggregate, their sizes remained unchanged, and their resistance to oxidation increased. Competition of different temperature-dependent processes is reflected not only in the stability of particles but also in their size distribution.

An original method of utilizing high-pressure polyethylene for stabilization of metal nanoparticles has been proposed.[43] Polyethylene has voids capable of stabilizing nanoparticles. The accessibility of voids is enhanced by the dispersion of the polymer in heated hydrocarbon oil. Oil molecules penetrate into polymeric globules making them more accessible for metal-containing compounds. Thermal destruction of the latter yields metal particles. This method allows powders of metal-containing polymers to be obtained. The metal concentration and nanoparticle composition can be changed in a wide range. The usage of polymers as nanoreactors has been analyzed in Ref. 44.

In recent years, much attention was drawn to monodispersed colloidal particles of polymeric materials due to their potential applications in biosensors,[45,46] nanophotonics,[47] colloidal lithography,[48] and as porous membranes[49] and seed particles for core–shell and hollow structures.[50] Polystyrene-based colloidal particles arrayed in porous structures were synthesized.[51,52]

Molecularly imprinted polymers have attracted attention as receptors for recognition of biomolecules.[53,54] A method of preparation of monodispersed polypyrrole nanowires with 100 nm diameter, which was based on using nanoporous alumina membranes and silica nanotubes as the templates, was developed. Alumina membranes were removed by dissolution in diluted phosphoric acid, while silica nanotubes were dissolved in HF solution.[55]

Photochemical reduction of Ag^+ ions in the presence of dendrimers with terminal amino or carboxy groups produced silver particles with an average diameter of ~7 nm.[56] A possible mechanism of particle formation is as follows:

$$Ag^+ + [dendrimer]\text{-}COO^- \xrightarrow{h\nu} Ag^0 + [dendrimer]\text{-}COO \rightarrow [dendrimer] + CO_2$$
$$Ag^+ + [dendrimer]\text{-}NH_2 \xrightarrow{h\nu} Ag^0 + [dendrimer]\text{-}NH_2^+$$

It is possible to control the particle size by altering the nature of dendrimers. Currently, for stabilization of metal nanoparticles, dendrimers based on polyamidoamines and their different modifications are used. A dendrimer represents a highly branched macromolecule, which that includes a central nucleus, intermediate repetitive units, and terminal functional groups.[57] Dendrimers present a new type of macromolecules which that combine the high molecular mass and low viscosity of their solutions, a three-dimensional molecular shape, and the presence of a spatial structure. The size of dendrimers varies from 2 to 15 nm, and they represent natural nanoreactors. Dendrimers with small numbers of intermediate units exist in the "open" form, whereas those involving many units form spherical three-dimensional structures. Terminal groups of dendrimers can be modified with hydroxy, carboxy, and hydrocarboxy groups.

An example of using dendrimers as microreactors for synthesizing metal nanoparticles is shown in Ref. 58. Monodispersed spherical polyamidoamine dendrimers are permeable for low-molecular-weight reactants. Thus, an addition of $HAuCl_4$ to an aqueous solution of a dendrimer with primary and ternary amino groups results in the appearance of a protonated dendrimer with as the counter-ion. Reduction of anions by sodium borohydride produced 1–5 nm gold particles. By varying either the concentration ratio (D) of counter-ions to terminal amino groups or the diameter (generation) of the dendrimer, one can control the particle size. Reduction of gold ions in a dendrimer of the ninth generation (G.9) yielded spherical gold particles measuring 2.5, 3.3, and 4 nm for $D = 1:4$, 1:2, and 1:1, respectively. For $D = 1:1$, gold particles measuring 2, 2.5, and 4 nm were formed in G.6, G.7, and G.9 dendrimers, respectively. Reduction of gold and silver salts in the presence of modified dendrimers produced particles with an average diameter of 2–6 nm. A spectroscopic study has revealed the autocatalytic mechanism of this reaction.[59]

The interaction of a fourth-generation dendrimer N_3P_3-$(((OC_6H_4CHNN(CH_3)P(S)(OC_6H_4CHNN(CH_3)COCH_2CH_2CH_2SH)_2)_2)_2)_6$ with $Au_{55}(PPh_3)_{12}Cl_6$ in dichloromethane yielded ligand-free clusters Au_{55} with a diameter of 1.4 nm, which were organized in well-shaped microcrystals $(Au_{55})_x$.[60] The stability of an unligated Au_{55} cluster can be attributed to the perfect packing and geometry of full-shell clusters, as was demonstrated by treating ligand-free Au_{55} clusters in an oxygen plasma without any visible oxidation.

Dendrimers of different generations with various terminal functional groups proved to be suitable templates for synthesizing mono- and bimetallic particles of small sizes. Different poly(amidoamines) were most popular as dendrimers. With these dendrimers, 1–3 nm gold nanoparticles were synthesized.[61,62] A large volume of results were reported on the preparation of palladium nanoparticles and their use in various catalytic processes. Size-selective hydrogenation of α-olefins and polar olefins on palladium nanoparticles was considered in detail.[63,64] Synthesis, properties, and surface immobilization of palladium and platinum nanoparticles with average sizes of 1.5 nm and 1.4 nm incorporated into a poly(amidoamine) dendrimer (G_4-NH_2) with terminal amino groups were analyzed.[65]

In recent years, dendrimers were actively used for preparation of different bimetallic nanoparticles measuring several nanometers. Moreover, if metal salts were simultaneously added to a dendrimer before the reduction of the latter with sodium borohydride, their alloy was synthesized. Sequential addition of metal salts gave rise to various core–shell nanoparticles. Using the example of bimetallic Pd–Pt particles, it was shown that the size of dendrimer-encapsulated particles depends on the metal–dendrimer and metal–metal ratios.[66] Bimetallic Pd–Rh particles incorporated into a fourth-generation dendrimer were used for partial hydrogenation of 1,3-cyclooctadiene.[67]

Bimetallic particles based on gold and silver were studied most comprehensively. These materials exhibited different optical properties, which depended on the percentage of elements in the composition and their geometrical organization.

The fact that either alloys of random compositions or core–shell structures could be formed deserves attention. The formation and properties of Au–Ag alloy particles with sizes less than 10 nm[68] and their structure as a function of their size[69] were discussed.

DNA-modified core–shell Ag–Au nanoparticles were synthesized by using tyrosine as a pH-dependent reducing agent.[70,71] It was shown that core–shell bimetallic Pd–Au particles can be obtained by selective reduction of the shell metal immediately on the core of the first metal by mild reducing agents such as ascorbic acid.[72] In a similar way, Pd–Pt particles measuring 1.9±0.4 nm were synthesized.[73] Both types of bimetallic particles exhibited high catalytic activity as compared with mixtures of corresponding monometallic particles toward hydrogenation of allyl alcohol.

To determine the chemical properties of mono- and bimetallic particles measuring ~2 nm more accurately, a method of extraction of particles from the dendrimer into the organic phase by n-alkanethiols was developed.[74–76]

Dendrimers were actively used for preparation of various core–shell bimetallic particles measuring 1–3 nm. Based on dendrimeters as the templates, core–shell particles of PdPt with a size of 1.9±0.4 nm[114] and PdAu measuring 1–3 nm[113] were synthesized. PdRh particles were also prepared and used in catalytic hydrogenation of 1,3-cyclooctadiene.[108] PdPt and PdAu particles displayed higher activity in hydrogenation of allyl alcohol as compared with individual metals. Different types of AuAg particles with sizes from 1 to 3 nm were synthesized.[77]

Dendrimers were used for synthesizing supported nanoparticles of platinum,[78] palladium, and gold[79] measuring 1–2 nm, and also CuO[80] and Fe_2O_3.[81] In some cases, enhanced catalytic activity was observed.

It was shown that dendrimers with encapsulated nanoparticles can be used as precursors in preparation of heterogeneous bimetallic catalysts Pt–Au measuring less than 3 nm.[82] The preparation of a fourth-generation dendrimer with terminal amino groups and encapsulated palladium and gold nanoparticles G4-NH_2($Pd_{27.5}Au_{27.5}$) measuring 1.8±0.4 nm was described. The synthesis was carried out in a methanol solution and involved the reduction of K_2PtCl_4 and $HAuCl_4$ with $NaBH_4$.[83] Then, the dendrimer with encapsulated bimetallic nanoparticles was deposited on TiO_2 powder. To remove dendrimer, the powder was annealed at 500 °C first in oxygen flow and then in hydrogen flow. Particles deposited on titanium dioxide had an average size from 1.8 to 3.2 nm and contained 48±3% Pd and 52±3% Au, which corresponded to the equimolar reagent ratio in the initial mixture. Titanium dioxide with deposited bimetallic particles was used in catalytic oxidation of carbon monoxide. This catalytic system exhibited a synergetic catalytic effect and had higher activity as compared with individual palladium and gold.

A reaction of terminal amino groups of poly(amidoamine), a fourth-generation dendrimer, with tert-thiophene dendrons was accomplished.[84] The synthesized compound was used for stabilization of palladium and gold nanoparticles.

Syntheses in porous structures closely supplement the methods that employ micelles and dendrimers for synthesizing nanoparticles. The prospects of using mesopores for preparation of different nanosize materials were discussed.[41,85] Nanoparticles of silver and silver sulfide were obtained in nanosize voids of perfluorinated ionomeric membranes.[86] Reduction of metal ions in the presence of aminodextran and styrene resulted in the formation of spherical polystyrene particles with diameters of 2.0 μm covered with gold and silver islets measuring 5 to 200 nm.[87]

At present, porous inorganic materials like zeolites are actively used for the formation of metal nanoparticles. Solid zeolites with pores and channels of strictly definite sizes represent convenient matrices for stabilization of nanoparticles with desired properties. Two basic methods are used for fabricating nanoparticles in zeolite pores. The first method is associated with direct adsorption of metal vapors in thoroughly dehydrated zeolite pores.[41,88] The other, more popular method involves chemical transformations of precursors incorporated into pores and representing metal salts, metal complexes, and organometallic compounds. A similar method allowed nanowires with 3 nm diameters and with a length several times greater to be fabricated in the channels of molecular sieves.

The high thermal and chemical stability of zeolites with incorporated nanoparticles makes it possible to consider them as the most promising catalysts.

Nanoporous metal oxides also can find wide application as heterogeneous catalysts. Recently, a synthesis of mesoporous magnesium oxide with a periodically ordered pore system and a narrow-pore-size distribution was proposed. The system retained its structure during repeated heating up to 800 °C.[89] The synthesis of highly porous uranyl selenate nanotubes was also described.[90]

Mesoporous silica was proposed for use in the preparation of functional polymer–silica composite hybrid materials.[91] Various monomers of the vinyl series (styrene, metacrylic acid, etc.), binders such as divinylbenzene, and initiators of radical polymerization were adsorbed on the walls of silica pores with a diameter of 7.7 nm and were polymerized. Polymerization occurred on the walls and decreased the pore diameter to 6.9 nm. A synthesized material synthesized based on styrene was sulfonated by concentrated sulfuric acid and used as the acid catalyst in esterification of benzyl alcohol with hexanoic acid, which confirmed its high selectivity.

In addition to inorganic porous materials, organic and, particularly, polymeric porous materials were synthesized. A method of preparation of thermostable polymers with pores measuring from 1 to 50 nm was developed based on incorporating inert solvents such as tetrahydrofuran and using a step-wise cross-linked polymerization in the absence of interphase separation.[92,93]

2.3 PHOTOCHEMICAL AND RADIATION-CHEMICAL REDUCTIONS

Synthesis of metal nanoparticles influencing the chemical system by high energies is associated with the generation of highly active strong reducers like electrons, radicals, and excited species.

Photochemical (photolysis) and radiation-chemical (radiolysis) reductions differ in energy. Photosynthesis is characterized by energies below ~60 eV, whereas radiolysis uses energies of 103–104 eV. The main peculiarities of processes occurring under the action of high-energy radiation concern the nonequilibrium energy distributions of particles, the overlap of characteristic times of physical and chemical processes, which is of prime importance for chemical reactions of active species, and the presence of multichannel and nonsteady-state processes in the reacting systems.[94]

Photochemical and radiation-chemical reduction methods have advantages over the chemical reduction method. Owing to the absence of impurities formed when chemical reducers are used, the former methods produce nanoparticles of high purity. Moreover, photochemical and radiation-chemical reductions make it possible to produce nanoparticles under solid-state conditions and at low temperatures.

Photochemical reduction in solution is employed most frequently for synthesizing particles of noble metals. Such particles were obtained from solutions of corresponding salts in water, alcohols, and organic solvents. In these media, under the action of light, the following active species are formed:

$$H_2O \rightarrow e^-_{aq} + H + OH$$

By reacting with alcohols, a hydrogen atom and a hydroxyl radical produce an alcohol radical:

$$H(OH) + (CH_3)_2CHOH \rightarrow H_2O(H_2) + (CH_3)_2COH$$

A solvated electron interacts with, e.g., a silver atom, reducing the latter to metal:

$$Ag^+ + e^-_{aq} \rightarrow Ag^0$$

In the course of photoreduction, in the initial period of photolysis, the UV absorption spectrum reveals bands at 277 and 430 nm, which correspond to clusters and silver nanoparticles measuring 2–3 nm, respectively.[95] With the increase in photolysis time, the absorption-band maximum can shift to both short and long wavelengths. The short-wave shift points to the decrease in the average size of silver particles, whereas the long-wave shift corresponds to the presence of aggregation processes.

Photoreduction gives rise to light-induced formation of not only nanoparticles of definite sizes but also of greater aggregates. The effect of light was studied by the example of gold particles in acetone, ethanol, and isopropanol.[96–98] Illumination with a mercury lamp light led to broadening and disappearance of a band of gold surface plasmon at 523 nm. As a result, the band at 270 nm became more intense, and a new band appeared at 840 nm. The authors associated the shift of the plasmon band with the dipole–dipole interaction of particles in aggregates. Upon 20-h photolysis, complete precipitation of gold particles was observed. The aggregation rate was shown to depend on the nature of solvent and the light wavelength. The UV radiation effect is stronger as compared with that of visible light. The effect of light wavelength was associated with the strengthening of van der Waals forces and light-generated changes in the Coulomb interaction of surface charges.

Light-stimulated aggregation of silver particles was studied.[99,100] The aggregation mechanism was attributed to the appearance of light-generated particles with opposite charges. The appearance of such particles was caused by the exchange of electric charges formed as a result of photoemission. Such an exchange, being associated with the dependence of Fermi energy on the particle size, operates through a dispersed medium and results in leveling of potentials of different-size particles. The exchange gives rise to long-range electric forces, which that favor the approach of particles to one another up to the distances at which van der Waals forces responsible for aggregation appear.

Photoreduction of silver nitrate in the presence of polycarboxylic acids allowed the methods of controlling the shape and size of particles to be developed. Spherical and rod-like silver particles were obtained.[101]

Synthesis of silver nanoparticles in nanoemulsions by radiation reduction was described.[102,103] The authors managed to narrow the size distribution of particles by using the exchange of substance between microemulsion droplets.

Due to its availability and reproducibility, the radiation-chemical reduction for synthesizing nanoparticles progressively gains wide acceptance. In the liquid phase, the stages associated with spatial distribution of primary intermediate products play a key role in the production of metal nanoparticles. In contrast to photolysis, the distribution of radiolysis-generated intermediate particles is more uniform, which allows particles with narrower size distributions to be obtained.

The method of pulsed radiolysis made it possible to synthesize active particles of metals in unusual oxidation degrees.[104] The reaction with a hydrated electron ,whichthat has a high reduction potential, proceeds as follows:

$$M^{n+} + e_{aq}^{-} \rightarrow M^{(n-1)+}$$

The presence of a single electron in the outer orbital of an atom or a metal ion determines their high chemical reactivity. Lifetimes of such species in water run to several micro- or milliseconds. The optical properties of such metal particles are defined by their reduction potentials. For instance, in isoelectric series of metals with similar electronic configurations, the ionization potential increases with an increase in the metal-ion charge.[104] Moreover, the wavelength corresponding to the maximum of light absorption shifts to the short- wavelength region. Whereas silver and gold atoms that are in the beginning of the period exhibit reductive properties, their isoelectron analogs in the end of the period, viz., trivalent tin and lead, are strong oxidants. Within the same subgroup, the potential of isocharged ions increase with the increase in the period number.

In the course of radiation-chemical reduction, first, atoms and small metal clusters are formed, which is followed by their transformation into nanoparticles. For their stabilization, additives similar to those used in chemical reduction were introduced. The joint use of pulsed γ-radiolysis and spectrophotometry allowed studying the initial stages of the formation of the metal particles that represented the simplest type of charged clusters such as and . Further interaction of clusters, the mechanism of which is still unclear, produced metal nanoparticles. Steady-state and pulsed radiolysis techniques allowed quite a number of nanoparticles of various metals to be obtained.[105,106]

The methods for synthesizing bimetallic and trimetallic metal nanoparticles with the core–shell structure, which employ radiolysis, were developed. Nanoparticles that consist of two or more different metals are of special interest in view of developing materials with new properties, because on the nanolevel, one can obtain such intermetallic compounds and alloys that can never be formed as compact metals.

Radiation reduction of salt solutions produced nanoparticles, which that comprised two[107] or three[108] metals. Au–Hg particles were synthesized in two steps. First, gold particles measuring 46 nm were synthesized by radiation-chemical reduction. Then, $Hg(ClO_4)_2$ and isopropanol were added to the gold sol, which resulted in deposition of mercury ions on gold particles, after which mercury ions were reduced by free radicals formed during radiolysis.

Palladium particles with 4 nm diameters and a narrow size distribution were obtained by the reduction of Na_2PdCl_4 salt in the presence of sodium citrate as a stabilizer. The addition of $K_2Au(CN)_2$ to a sol of palladium particles in methanol and their γ-irradiation resulted in the reduction of gold ions. In the process, no separate Au gold particles were formed, and all gold was deposited on palladium particles to form an external layer. Moreover, a silver layer could also be deposited onto Pd–Au particles. Synthesized particles consisted of palladium cores and two shells, namely gold and silver. Such multiplayer clusters are of interest for studying femtosecond electronic processes.[109,110]

Silicates modified by organic compounds and used as matrices and stabilizers allowed single-step syntheses of sols, gels, and bimetallic nanoparticles to be performed.[111] STM studies of Pt–Pd particles have shown that they consist of palladium cores coated with platinum shells. Thin silicate films containing bimetallic nanoparticles were used in electrocatalytic oxidation of ascorbic acid. The latter example shows that the presence of two metals can strongly and often unpredictably change the properties of nanoparticles. A second metal deposited on a standard metallic catalyst allows unique surfaces with new properties to be obtained.

To obtain bimetallic gold–palladium nanoparticles, a sonochemical method was applied.[112] The particles were synthesized from aqueous solutions of $NaAuCl_4 \cdot 2H_2O$ and $PdCl_2 \cdot 2NaCl \cdot 3H_2O$ in the presence of sodium dodecylsulfate, which served as both stabilizer and reducer. The synthesized bimetallic particles synthesized represented cores of gold atoms surrounded by shells of palladium atoms. The dimensions of cores and shells for different gold-to-palladium ratios were measured by means of high-resolution electron microscopy. X-ray spectroscopy allowed determination of gold and palladium contents in the particles. Given the density, mass, and initial ratios, it is possible to estimate the sizes of cores and shells. Table 2.1 compares experimental and calculated data. As seen from the table,

TABLE 2.1 Core (gold) Diameter and Shell (Palladium) Thickness in Bimetallic Au–Pd Particles, nm[112]

	Experiment		Calculations	
Au/Pd ratio	Core diameter	Shell thickness	Core diameter	Shell thickness
1:1	6.0	1.0	6.4	0.8
1:4	5.0	1.5	4.8	1.6

the sonochemical synthesis of bimetallic particles allows one to exercise control over sizes of cores and shells by changing the concentrations of Au^{3+} and Pd^{2+}. Bimetallic Au–Pd particles exhibited strong catalytic activity toward hydrogenation of pent-4-enic acid.

A comparative study of Au, Ni, and bimetallic Au–Ni nanoparticles supported by amorphous carbon, which were fabricated by laser evaporation of the corresponding pure metals and the alloy, was carried out.[113] Studies performed by different techniques have shown that the particles had a diameter of 2.5 nm and narrow size distribution; the composition of bimetallic particles corresponded to that of the original alloy.

Fe–TiH particles were obtained by mechanosynthesis from a mixture of iron with titanium hydride.[114] Nanosize compositions of iron–tungsten with a tungsten content of 2–85 at.% were synthesized by the joint reduction of a mechanical mixture of iron hydroxide ($FeOOH \cdot nH_2O$) and tungstic acid (H_2WO_4) in the course of its 1-h exposure to a hydrogen flow at 740 °C. The particles obtained were studied using XRD and Mössbauer spectroscopy techniques.[115]

Different versions of the sol–gel synthesis were classified with chemical methods of fabrication of oxide and sulfide nanoparticles.[116] The scheme of synthesis of nanooxides of metals is as follows:

$$M(OR)_n + xH_2O \to M(OH)_x(OR)_{n-x} + xROH \text{ (hydrolysis)}$$

$$M(OH)_x(OR)_{n-x} \to MO(n/2) + (2x-n)/2 \cdot H_2O + (n-x)ROH \text{ (condensation)}$$

where M is a metal and R an alkyl group. The process was catalyzed by the changes in the pH initial solution. In acidic media, linear chains were formed, whereas in alkaline solutions, branched chains appeared. By substituting the corresponding sulfides $M(SR)_n$ for alkoxides of metals and treating them with hydrosulfide, nanoparticles of metal sulfides could be obtained.

At present, new methods for synthesizing nanocrystalline oxide materials, which apply different compounds in subcritical and supercritical states, are actively developed. Most widely used compounds are carbon dioxide and, recently, water.[117–119] Nanoparticles of metal oxides were also obtained by changing their radius from the micron level to the nanolevel by dissolving them in electrolytes.[120]

A method of rapid expansion of supercritical solutions into a liquid solvent was extensively used for preparation of nanosize particles of metals, semiconductors, and their conjugates with biomolecules.[121] This method was used for synthesizing particles of average diameter less than ~50 nm from CO_2-soluble polymers.[122] Rapid expansion of supercritical solutions in liquids was used for production of the finely divided nonsteroidal antiphlogistic drugs Ibuprofen and Naproxen.[123] The solution was rapidly expanded in water, poly(N-vinyl-2-pyrrolidone) with a molecular weight ~40,000 u, or sodium dodecylsulfate. In water, homogeneous particles of Ibuprofen (α-methyl-4-(2-methylpropyl)benzoacetic acid) formed a suspension in 15 min. In polyvinylpyrrolidone, Ibuprofen particles of an average diameter of 40 nm were stable for several days; in sodium dodecylsulfate, the particle size was 25 nm.

2.4 CRYOCHEMICAL SYNTHESIS

The high activity of metal atoms and small clusters in the absence of stabilizers results in their aggregation to larger particles. The aggregation process proceeds without activation energy. Stabilization of active atoms of virtually all elements in the periodic table was realized at low (77 K) and superlow (4–10 K) temperatures by the method of matrix isolation.[124] The essence of this method consists in using inert gases at superlow temperatures. Gases most generally employed as the matrices are argon and xenon. Vapors of metal atoms are condensed together with a large (usually thousandfold) excess of inert gas on a surface cooled to 4–12 K. The high dilution by an inert gas and the low temperatures virtually rule out diffusion of metal atoms; thus, they are stabilized in the condensate. Physicochemical properties of such atoms are studied by various spectral and radiospectral techniques.[125]

For a chemical reaction to occur at low temperatures, the active particles stabilized in the condensate should be mobile. In principle, the matrix isolation and chemical reactions are conflicting processes. The stabilization of active particles eliminates their reactions and, vice versa, the presence of a chemical reaction implies the absence of stabilization. When studying samples obtained by the matrix isolation method, their heating made it possible to realize quite a number of new and unusual chemical reactions between atoms and certain chemical substances, especially when introduced into low-temperature condensates.

In the general form, such transformations can be illustrated by the following scheme:

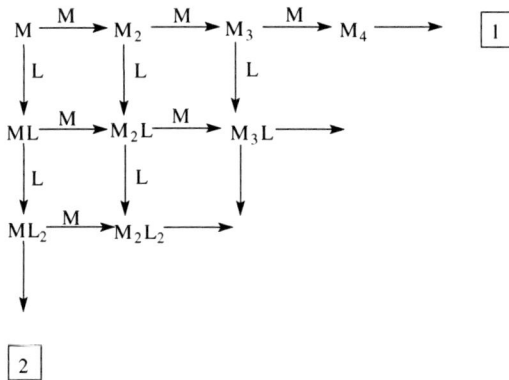

Here M is a metal and L a chemical compound (ligand). This scheme shows consecutive and parallel competitive reactions. Direction 1 reflects processes of aggregation of metal atoms to give dimers, trimers, and nanoparticles; direction 2 corresponds to interactions of atoms with ligands followed by the formation of complexes or organometallic compounds.

The processes in low-temperature condensates described by the above scheme are nonequilibrium and dependent on many factors, including the metal–ligand ratio, the cooled-surface temperature, the condensation rate, the pressure of reagent vapors in the cryostat, and the rate of sample heating. The following factors affect most strongly the formation of nanoparticles during cryocondensation: the rate at which the atoms reach the cooled surface, the rate at which the atoms lose their excessive energy via the interaction with the condensate, and the rate of the removal of clusters from the zone of active condensation of atoms. Preparation of metal nanoparticles by co-condensation on cold surfaces allows introduction of various additives that can change the physicochemical properties of the system.

Certain peculiarities of synthesizing samples for cryochemical reactions involving atoms, clusters, and metal nanoparticles should be noted.

Metal atoms can be synthesized by various heating techniques. Alkali, alkali-earth, and certain other metals are easily evaporated. Their vapors can be obtained by direct heating. As a rule, heating is furnished by using a low-voltage (5 V) transformer, which provides high (up to 300 A) current densities. The inlet tubes are cooled. A metal sample to be vaporized may be shaped as a wire, spiral, or ribbon. Highly conductive metals (Cu, Ag, and Au) are usually vaporized in Knudsen cells165 by direct or indirect heating. Given the temperature T and the pressure P in the cell and the outlet hole dimensions, we can estimate the evaporation rate by using

$$N5 \ P/(2pMRT)1/2$$

where N is the number of moles of substance evaporated in a second per cm^2 of the outlet hole area, M the molecular mass, and R the universal gas constant.

The numerical solution of equations that describe the temperature variations makes it possible to calculate the temperature profile in the Knudsen-cell chamber. Indirect heating provides a better temperature uniformity in the cell. The evaporation temperature of a metal is determined by means of optical pyrometers. The condensate temperature is measured by thermocouples (copper–gold), solid-state thermometers (GaAs diodes), capacitive low-temperature glass–ceramic transducers, and hydrogen and pentane thermometers, which operate based on the pressure–temperature dependence.

The control over the flows of gases or vapors is exerted by means of needle valves calibrated according to variations in the pressure drop in a vessel of known volume. The material for valves is chosen with allowance made for the properties of substances present and their possible corrosion. Different types of rheometers and Bernoulli pressure gauges are employed for measuring the pressure drop in a capillary through which the gas flows. The pressure drop is directly related to the flow velocity and is determined by means of pressure gauges. The flow velocities used in the method of matrix isolation fit approximately within an interval of 0.1–0.01 mmol/h. For such deposition rates, the temperature of a surface on which a sample is condensed increases by fractions of a degree in several hours.

When studying the chemical reactions in low-temperature condensates, it is essential that no chemical reactions occur during the sample synthesis. Elimination of gas-phase reactions is achieved by using a molecular beam mode. As compared with the cryoreactor dimensions, the free path λ should be greater to avoid collisions and gas-phase reactions. The quantity λ is approximately related to the pressure by a relationship $\lambda = K/p$, where K is a constant. Below, the dependence of pressure on the free path is shown.

Pressure P (mmHg)	10^{-2}	10^{-3}	10^{-4}	10^{-6}
Free path λ (cm)	0.5	5	50	Several tens of meters

The use of a Knudsen cell requires taking into account the Knudsen number $K_n = 2\lambda/R$, where R is the characteristic size. Pressures of the order of magnitude of 10^{-3} mmHg are sufficient for $\lambda \gg R$ and $K_n \gg 1$. For such a pressure and $T = 300$K, the transition of the flow from a continuous mode to a molecular one occurs when $\lambda \ll R$ and $K_n \ll 1$.

The evaporation rate of particles from a Knudsen cell can be found from tabulated vapor pressure versus temperature dependences. The simplest way to determine the amount of evaporated metal is weighing the metal sample before and after the experiment. Optical techniques can be used if all evaporated substances fall in the vicinity of transmission and absorption bands of a sample. The quartz crystal microbalance technique is used for measuring the amount of deposited metal. The operation of such microbalance is based on a linear dependence of the frequency of quartz-crystal oscillations on the deposit mass. Standard quartz crystals have a frequency of 5000 kHz, a diameter of 8 mm, and a thickness of 0.3 mm. Deposition of an additional mass shifts the resonance frequency. The crystal sensitivity is 5×10^{-10} ng/Hz. The dependence of the resonance frequency f on the mass m is defined by the following equation:

$$\Delta f = fK \, \Delta m / S$$

where Δf is the frequency shift, S is the surface area, and K is a constant dependent on the thickness and density (2.65 g/cm^3) of a quartz crystal. The crystal is fixed on a support together with two electrodes of a generator. Resonance-frequency variations are recorded by a frequency meter, and the balance is preliminarily calibrated.

More serious problems are seen during determining the amount of the ligand. Chemical substances have, depending on their nature, different accommodation coefficients which characterize the fraction of particles irreversibly adsorbed on the surface. For example, only 15% CO_2 molecules are condensed upon the first collision.

Condensation of substances from the gas phase is accompanied by liberation of latent heat of fusion, L_f, which is absorbed by a thermostat. This heat is ejected through the already deposited matrix layer. Thus, the latent heat of fusion and the thermal conductivity λ of the matrix material are important characteristics. They determine the rate of matrix formation and the time the stabilized particles take to aggregate.

The temperature difference between the surface and the base of the matrix layer can be assessed if one assumes the establishment of a steady state.[124] The thickness l of a layer, which was deposited on a surface with an area S at a condensation rate n (mol/sec) in time t, is described by the expression

$$l = nt/\rho S$$

where ρ is the molar density of the substance. The rate of heat liberation is $Q_1 = nL_f$ cal/sec. Heat removed through the matrix layer is determined by the expression

$$Q_2 = S\lambda(T - T_0)/l$$

where λ is the thermal conductivity, and T and T_0 are temperatures of the layer and support surfaces. At a steady state, $Q_1 = Q_2$ and

$$nL_f = S\lambda(T - T_0)/l$$

Substituting the expression $l = nt/\rho S$, we obtain

$$T = T_0 + L_f n^2 t / \lambda \rho S^2$$

Thus, the surface temperature increases linearly with time and by a quadratic law with the increase in the substance deposition rate. The mobility of metal atoms in the condensate may also be influenced by the emission from a source at a temperature of 1000 °C and higher. Such an effect should be checked experimentally.

Several special cryoreactors were developed for cryochemical synthesis of atoms, clusters, and nanoparticles. Figure 2.3 illustrates a cryoreactor used for both matrix isolation and spectroscopic studies of active metals in the temperature interval 12–70 K. Figure 2.4 presents a scheme of a cryoreactor used for condensation at the liquid-nitrogen boiling point (77 K) and higher.

The main part of this unit represents a polished copper cube pre-cooled to 77 K. Upon the deposition of substances under study, a sample is turned by 180°, and IR reflectance spectra are measured. The cryoreactor makes it possible to obtain spectra at different temperatures and exercise precise control over the sample temperature. Salt and quartz windows fixed between hollow copper holders make it possible to obtain UV and optical transmission spectra. Special cryoreactors were designed for recording electron paramagnetic resonance (EPR) spectra. The scheme of one of our cryoreactors for the preparation of thin film materials is presented in Figure 2.5.

As an example, we show a setup developed at Oxford University (Figure 2.6). One advantage of this setup is the possibility of synthesizing grams of substances in several hours of its operation. Such a setup was actively used for synthesizing rare-earth elements (REE) compounds. Japanese scientists used a simpler approach.[126] A ligand is evaporated and then condensed on the walls of a 1-l vessel cooled by liquid nitrogen. After being conditioned for 60 min, the condensate is slowly heated and analyzed. As a rule, this procedure allows new organic compounds to be obtained.[127,128]

A reactor shown in Figure 2.7 makes it possible to synthesize several milligrams of substance. The condensation of vapors of metals, ligands, and, if necessary, stabilizers proceeds on the glass-vessel walls cooled by liquid nitrogen. At the end of the condensation, the sample is heated and accumulated at the vessel bottom, from where it can be gathered for further studies without breaking vacuum. Thus, various types of sols or organodispersions of metals were obtained. This cryostat was modified with the aim of obtaining systems containing nanoparticles of two different metals. Similar but partially modernized reactors allowed metal vapors to be condensed into cold liquids at the bottom of a cryostat.

One advantage of such reactors is the relative simplicity of experiments. In similar reactors, by either simultaneous or successive condensation of vapors of metals and various ligands, quite a number of new organometallic compounds were synthesized.[126] However, for cryoreactors described above, determining the reagent ratio is a difficult task, which is their drawback. A similar drawback is typical of the Green cryoreactor, which is widely used for cryosyntheses (Figure 2.8c and d)[129] and, in principle, represents a sort of a rotor evaporator with a rotating flask immersed in a cold bath.

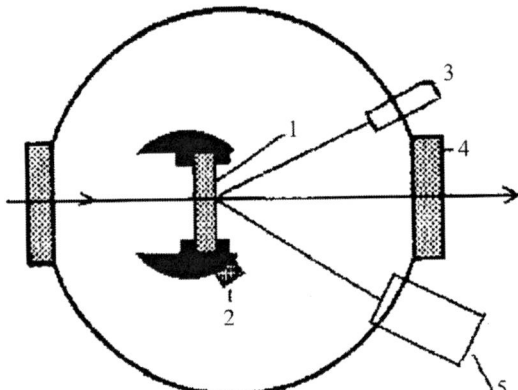

FIGURE 2.3 Window-level cross section of a cryostat designed for a temperature range of 12–70 K: (1) window with a formed sample, (2) quartz microbalance, (3) gas inlet, (4) external windows, and (5) evaporator of metal.

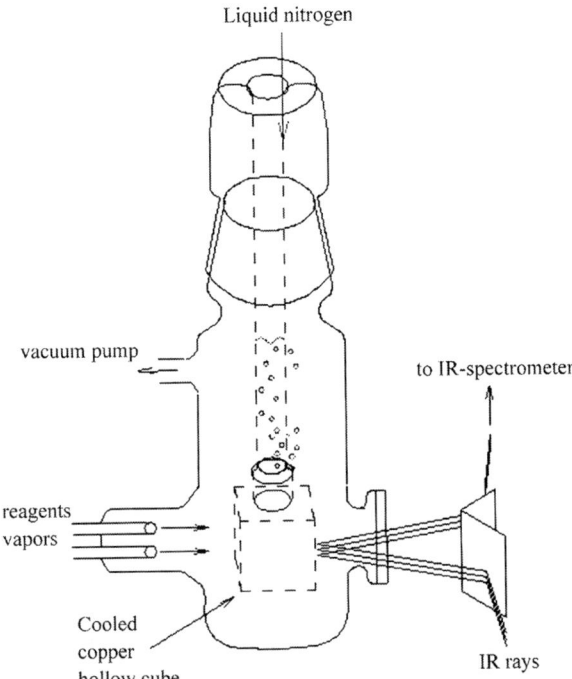

FIGURE 2.4 Cryostat for low-temperature IR spectroscopy at 77 K.

An interesting cryochemical synthesis of composition materials was proposed in the early 1990s by Nanophase Technologies Corp. (USA).[130] Figure 2.9 illustrates the reactor scheme. Two metals are evaporated in vacuum and condensed on a pick-up finger cooled with liquid nitrogen. After a certain time, the condensate is scraped off with a special tool and accumulated at the

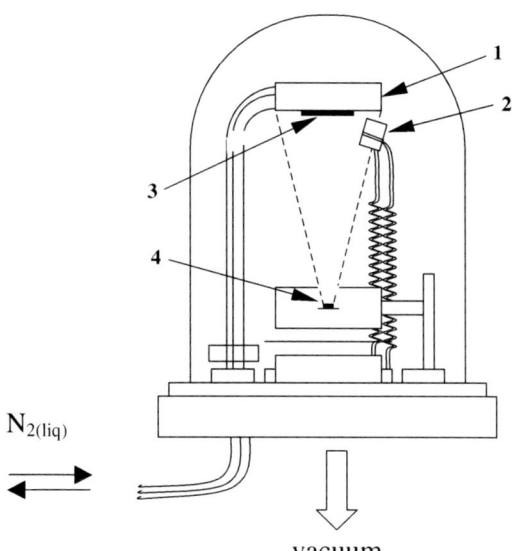

FIGURE 2.5 Experimental setup scheme for low-temperature physical vapor deposition. (1) Liquid nitrogen-cooled substrate holder; (2) quartz microbalance; (3) substrate: glass, alumina; and (4) evaporating metal.

reactor bottom. The condensate is pressed at low and high pressures to obtain a bimetallic vacuum-consolidated nanocomposite. The cryoreactor efficiency was only 50 g/h, but valuable materials obtained made this process economically attractive in the early 1990s.

Vacuum Metallurgical Co. Ltd. (Japan) has developed a semicommercial unit for the low-temperature synthesis of nanoferromagnetic materials, isolated metal nanoparticles, and ceramic and film materials (Figure 2.10). This unit employed modern methods of nanoparticle stabilization such as surfactants and supersonics. It has two chambers, namely, for synthesizing and collecting nanoparticles and combines aerosol techniques with cryochemistry.

The Japanese scientists proposed an interesting method that combines low temperatures with jet aerosol techniques.[131] Figure 2.11 illustrates the scheme of an apparatus that can be used at different pressures. A helium flow carries metal nanoparticles away from heating chambers. In a setup with pressures above 0.5 kPa, an organic solvent (hexane) is added to the helium flow, and the mixture is condensed in a trap cooled with liquid nitrogen.

In another version, at pressures below 0.2 kPa, hexane vapors are added directly into the heating chamber, and the mixture is pumped past the surface cooled with liquid nitrogen. The samples obtained are defrosted in a nitrogen flow, which contains surfactants added for stabilization. The sizes of synthesized silver and copper particles were equal to 3 mm and could be controlled by varying the pressure of helium supplied to the chambers.

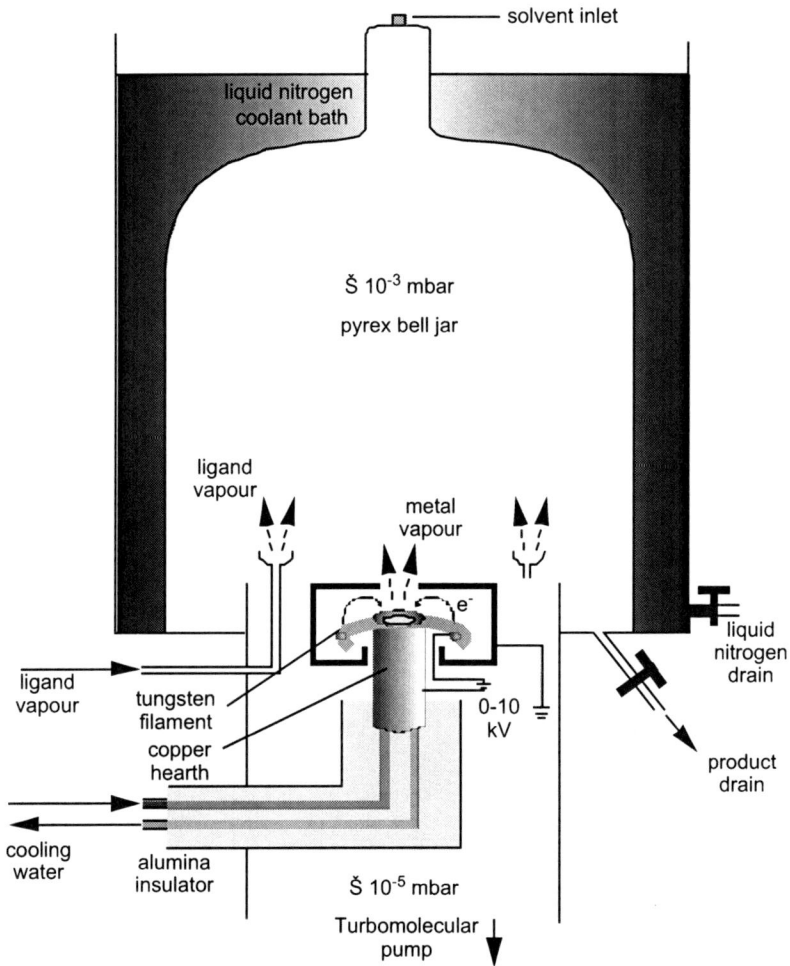

FIGURE 2.6 Unit for preparative synthesis of organometallic compounds by cryocondensation.[127]

The use of superfluid helium nanodroplets He_N ($N = 10_2$–10_5) opened up new possibilities in studying atoms, clusters, metastable molecules, and weakly bound complexes.[132] Nanodroplets were prepared by supersonic expansion of helium gas at high pressures and low temperatures. Surface evaporation cooled the droplets and maintained the temperature inside them at $T = 0.380$ K. Molecules, clusters, and weakly bound complexes were synthesized immediately upon incorporation of their atoms or molecules into helium nanodroplets. By applying this method, clusters of sodium[174] and silver[134,135] were obtained. In contrast to other atoms and molecules, alkali metal atoms form weakly bound states due to the Coulomb repulsion between the electron shells of helium and the valence electrons of a metal. van der Waals metal clusters formed on the

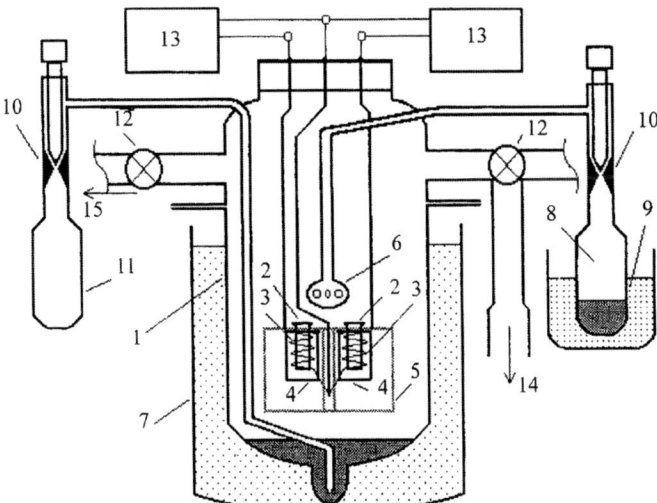

FIGURE 2.7 Scheme of semipreparative reactor for cryochemical reactions: (1) glass reactor (0.1 Pa), (2) quartz crucibles, (3) tungsten evaporators of metals, (4) ceramic tubes, (5) screening housing of metal evaporators, (6) inlet nozzle for organic component vapors, (7) Dewar vessel with liquid nitrogen, (8) glass ampoule with organic ligand, (9) thermostatically controlled bath, (10) Teflon valves, (11) collector of cryosynthesis products, (12) shutoff valves, (13) power supply units for evaporators of metal, (14) vacuum meter, and (15) vacuum line.

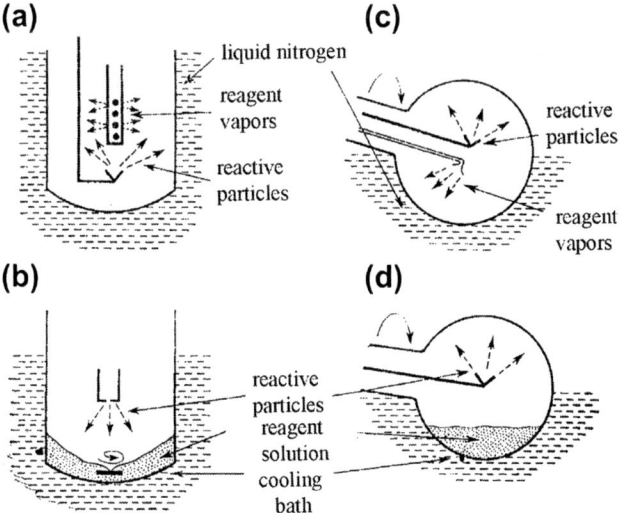

FIGURE 2.8 Scheme of reactors for preparative cryonanochemistry: (a) metal vapor condensation on Dewar walls, (b) condensation of metal into cold liquids, and (c, d) rotating Green reactors.[129]

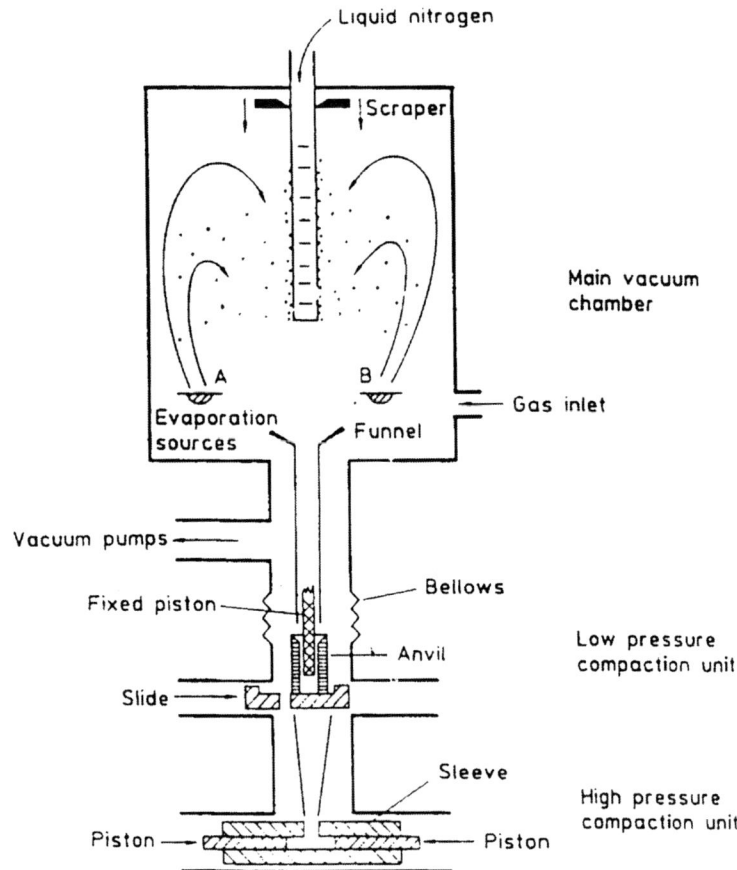

FIGURE 2.9 Scheme of a reactor for semipreparative synthesis of bulk composition materials.[130]

surface of nanodroplets were in the high-spin state with spins of all valence electrons parallel to one another. This phenomenon was studied by femtosecond multiphoton ionization spectroscopy. The observed effect of alternating of the intensity of peaks corresponding to different-size clusters was associated with peculiarities of the mechanism of spin relaxation.[133]

All methods that employ low temperatures and different versions of chemical vapor deposition (CVD) are discussed in Section 2.3.[136] The latter method is widely used in practice for preparation and application of various anticorrosion coatings.

Currently, nanoparticles of ammonium nitrate, hexogen $C_3H_6N_3(NO_2)_3$, and their mixtures are obtained by the method of low-temperature co-condensation.[137] According to atomic force microscopy results, the mixture contained ammonium nitrate particles measuring 50 nm and hexogen particles measuring 100 nm. The authors[65] did not specify how the properties of high-energy systems changed with the particle size.

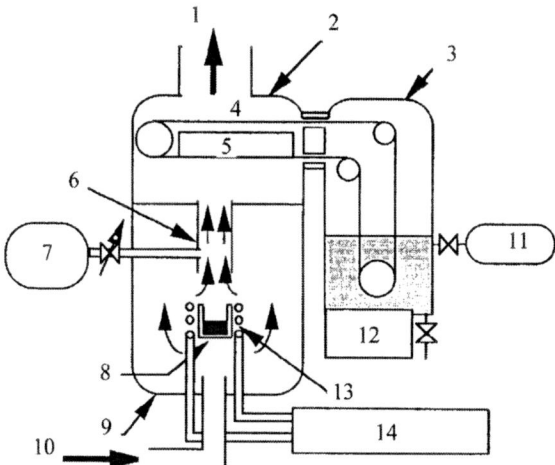

FIGURE 2.10 Scheme of a preparative setup for synthesizing nanoparticles of metals and their oxides: (1) pumping, (2, 3) volumes for collecting particles, (4) conveyor belt, (5) cooler, (6) particles inlet tube, (7) inlet of organic solvent, (8) evaporator, (9) vacuum chamber, (10) inlet of inert gas, (11) surfactant inlet, (12) ultrasonic mixer, (13) induction heating coil, and (14) power supply unit of induction heater.

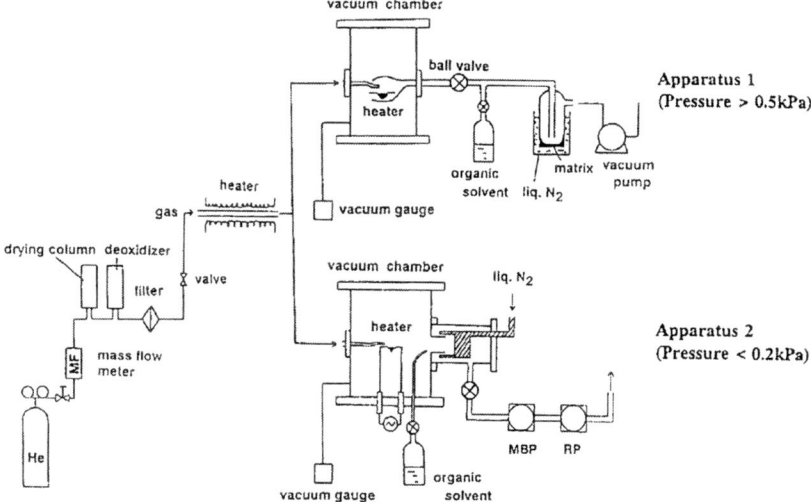

FIGURE 2.11 Scheme of an aerosol unit which that combines the jet and cryochemical methods:[131] (1) at P., 0.5 KPa, and (2) at P., 0.2 KPa.

When dealing with cryochemical and other methods of preparation of metal nanoparticles, scientists often come up with a problem that can be conditionally named "macro–micro." In essence, this problem is explained by a simple fact that dealing with samples, e.g., those prepared in reactors designed for spectral studies, we have particles of certain sizes and definite chemical reactions. However, attempts to reproduce the results by using larger amounts of substances often yield particles of different sizes and different chemical reactions. Naturally, opposite situations occur, when an experiment carried out at the macrolevel cannot be reproduced at the nanolevel. To resolve this contradiction is one of the challenging problems of chemistry and nanochemistry in particular.

2.5 PHYSICAL METHODS

There are many different physical methods for the preparation of metal nanoparticles. Among the major methods is the process based on combining metal evaporation into an inert gas flow with subsequent condensation in a chamber maintained at a certain temperature. Various versions of this method were analyzed in detail.[66] Physical methods of fabrication of nanoparticles traditionally involve those which that employ low-temperature plasma, molecular beams, gas evaporation,[138] cathode sputtering, shock waves, electroexplosion,[139] laser-induced electrodispersion,[140] supersonic jets, and different versions of mechanical dispersion.

Detailed descriptions of each method listed above are given elsewhere. Here, we only show only the schemes of several units proposed in the end of the 20th twentieth century for producing nanoparticles by means of different physical methods.

Figure 2.12 shows an original setup for preparation of highly porous nanoparticles of metals.[141] Operation of this setup is based on a closed gas cycle; particles of metals, e.g., silver, are deposited onto a filter, from which

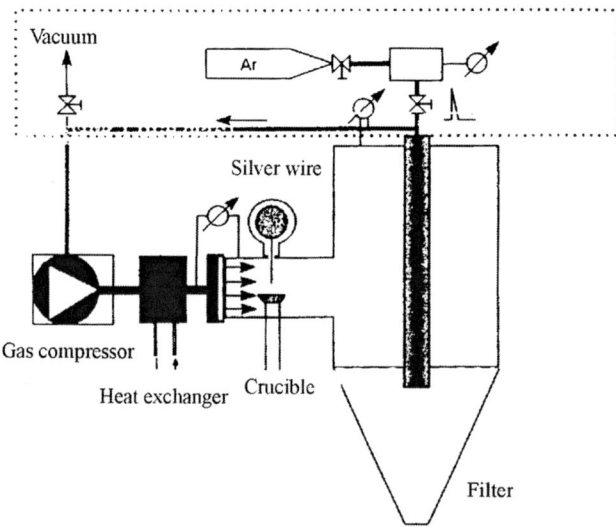

FIGURE 2.12 Scheme of a setup for continuous production of highly porous metals as a result of aggregation of metal nanoparticles.[141]

they are shaken off by gas pulses. This makes it possible to perform a virtually continuous process and obtain sufficiently large porous particles as a result of aggregation of nanoparticles.

Figure 2.13 shows a setup for plasma jet synthesis of metal–polymer compositions.[142] The setup has separate zones for plasma-induced preparation of nanoparticles and for coating them with monomers. Precursors of chlorides, carbonyls, and organometallic compounds are introduced into the discharge zone together with helium. The formed particles carry a charge that prevents them from collisions and the formation of clusters. This unit allowed polymer-coated particles of metal oxides, nitrides, sulfides, and carbides, measuring 5–20 nm, to be prepared.

Setups that employ laser evaporation for applying coatings on various particles and a hybrid combustion/chemical deposition procedure were developed.[143] Figure 2.14 illustrates one of such setups. A mixture of hydrogen and oxygen

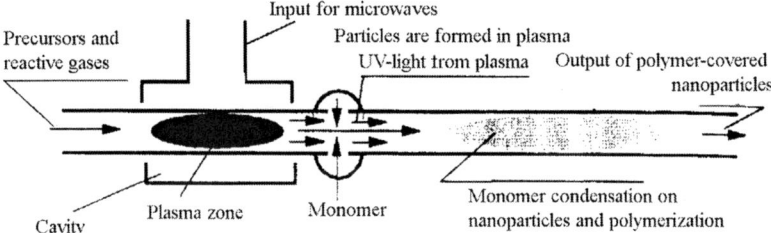

FIGURE 2.13 Unit for continuous production of nanoparticles encapsulated in polymers.[142]

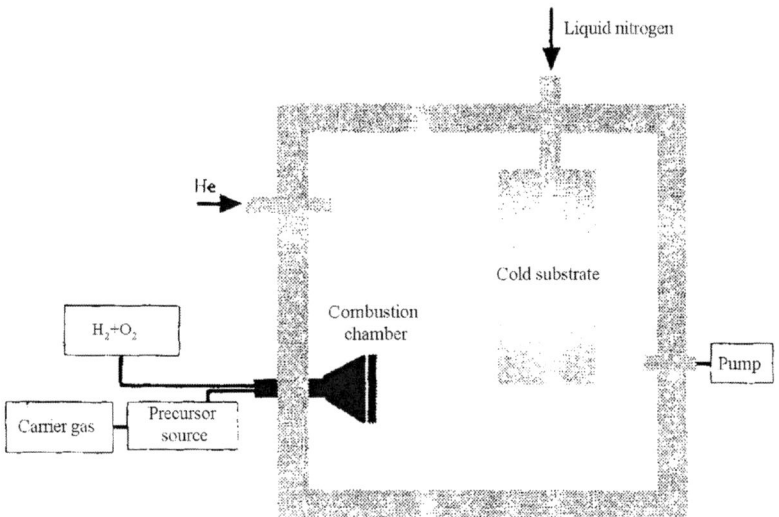

FIGURE 2.14 Scheme of a unit, that which combines the processes of combustion and chemical condensation in vacuum.[143]

serves as a flame source. In the synthesis of titanium oxide, titanium tetraethoxide was used as the precursor. Particles of TiO_2 formed were carried away from the formation zone with a helium flow to form a powder (20–70 nm) deposit on a cooled surface.

Three additional techniques also deserve mention.

The Pulsed Cluster Beam (PCB) utilizes a laser pulse to vaporize any desired element. The vapor plume is ejected into a flow tube where a pulse of cold helium (He) is simultaneously injected. This supersonic beam of atoms/inert gas finds itself in a relatively high pressure of the inert gas. The atoms begin to aggregate and are cooled to about 1–20 K as they form.[144–146]

The cluster growth can be moderated by He pressure, flow rate, and laser pulse power. The flowing clusters and inert gas enter a skimmer where a small amount is differentially pumped so that a portion is led into a chamber where the clusters are ionized by a second laser and then mass analyzed.

In more elaborate setups, certain ionized clusters can be magnetically separated and held in a vacuum cell for subsequent further study.

Figure 2.15 shows a schematic cross section of an improved, miniaturized version of the CB apparatus, especially constructed for generation of cluster ions that can be trapped and studied by Fourier transform-ion cyclotron resonance (FT-ICR). The laser target rod is rotated and translated under computer

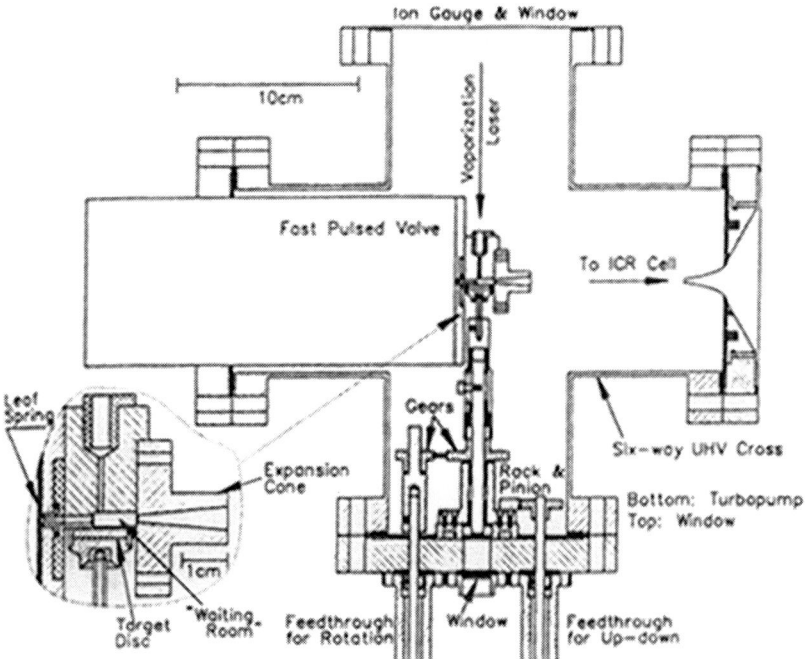

FIGURE 2.15 Pulsed cluster beam apparatus for production of gas-phase clusters.[144–146,157]

control so that fresh surface is always available for vaporization. The vaporization laser (second harmonic of a Nd–YAG, 10–30 mJ/pulse, 5 nsec pulse length focused on a 0.07-cm diameter spot) is fired on the leading edge of the rising carrier gas pulse. This allows the vapor plume to expand unimpeded for a short while before it is entrained in the rising density of the carrier gas pulse.

Operating with a He backing pressure of 10 atm, the pulsed valve is capable of putting out 0.05 Torr liter in a 125 μsec pulse. In a 3-liter chamber, such a fast pulse temporarily raises the pressure to 2×10^{-2} Torr.

In this design, the "waiting room" is the zone in the nozzle where clusters are formed and thermalized. The main flow of the carrier gas then passes through a 2.0-cm-long conical expansion zone. The gas can then undergo a free supersonic expansion with the central 0.2-cm diameter section of the jet being skimmed by about 8.4 cm downstream. After passing through the skimmer, the clusters can be ionized by a second laser and trapped or directly analyzed by MS.

Continuous Flow Cluster Beam (CFCB). A variation on this theme due to Riley and co-workers[147,148] utilizes a continuous flow of He or Ar. This apparatus allows more control of pressure and temperature and thus more meaningful kinetic analyses. The main disadvantage is the need for large pumping capacity in order to move the large volume of He gas rapidly enough.

A cross section of the central part of the apparatus is shown in Figure 2.16. An aluminum block with three inserting channels allows a pulsed laser beam to hit the sample rod ejecting vapor into the continuous main flow. (The target rod is continually rotated and translated automatically so that a fresh surface is available as vaporization continues.) The ejected vapor is rapidly cooled and nucleation and cluster growth occurs quickly. (An additional flow of gas over the laser window is needed to prevent metal film formation on the window.)

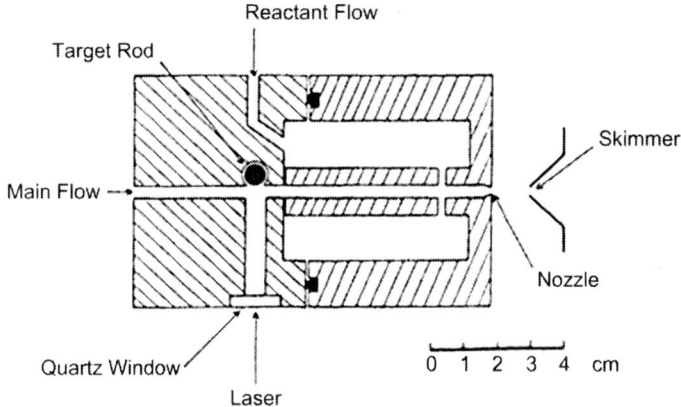

FIGURE 2.16 Continuous flow cluster beam aparatus.[148,157]

Reagent gas can be let into the metal cluster/carrier gas mixture through four inlets equally spaced down the circumference of the main channel. Any reaction that is taking place downstream of these inlets is quenched where the mixture expands into vacuum through a 0.1-cm diameter nozzle at the end of the main channels.

A portion of the expanded sample is collected by a skimmer, and this passes through several states of differential pumping to arrive at a time-of-flight mass spectrometer (40 cm downstream from nozzle). Ionization of a portion of the sample is achieved by a second-pulsed laser.

The vaporization laser used is a XeCl excimer typically operating with a pulse energy of 50 mJ and a repetition rate of 20 Hz. Carrier gas flows used were 500 std cm^3/min (sccm) He in the main channel, yielding a pressure of about 30 Torr in the reaction zone. The reagents are usually added as 1–10% diluted in He or as pure gases. Flows ranged from 5 to 310 sccm. Thus, partial pressures of reagent gases were varied from 3 μm to 12 Torr. Flow velocities were typically 5×10^3 cm/sec. Total pressure of cluster species were small, about 0.01 to 0.1 μm. The ionization laser was a collimated ArF excimer laser, and fluence was kept low (< mJ/pulse).

Gas-phase aerosol synthesis techniques offer an attractive route for the production of particulates as precursors for ceramic materials. One such method, aerosol spray pyrolysis, has seen considerable effort in recent years.[149–152] This method, simple in concept, involves dissolution of precursor salts, nebulizing the solution, drying and reacting the droplets in a heated reactor, and then collection of the particles. The technique has a number of positive attributes. Relatively pure particles in the submicrometer range can be produced. A wide range of chemical compositions can be created, including complex, multicomponent systems. Each droplet acts as a microreactor in which the constituents are mixed in the atomic level; hence, particle homogeneity is expected. The method has the potential for continuous creation of particles in one step.

Two major disadvantages of aerosol spray pyrolysis involve particle morphology and phase homogeneity. Very often, hollow or porous particles are created. Such nondense particulates are not suitable for further ceramic processing. This problem has been a major issue, of late.[153,154] It is due to surface precipitation in the droplet during the drying and/or reactor step.[155] Slow drying or use of high solubility salts (so the relative saturation can be low)[153] or high-temperature aerosol phase densification has been advocated to alleviate this problem. Porosity can occur regardless of whether the particles are uniform or hollow. This may result during heating of the dried salts as water of hydration and/or volatile anions leave the particle. Another common problem involves phase inhomogeneity or phase segregation—hence, the appearance of unwanted phases in the particles, despite the initial atomic level mixing of the precursor solution. This is an important problem since it degrades one of the major attributes of aerosol spray pyrolysis, namely, the ability to create complex, multicomponent systems. Phase segregation can occur during the drying step owing to a number of causes including differential precipitation of the components due to solubility differences, lack of isomorphic crystallization, differential reactions during drying, and a process that we will develop in this article, chemical segregation.[156,157]

The aforementioned physical methods of preparation of nanoparticles pertain to the group of condensation methods. Along with them, different versions of mechanical dispersion have gained wide application. In certain aspects, the latter methods resemble chemical ones. Mechanochemical dispersion for preparation of nanoparticles was discussed in detail in a number of special publications, which were reviewed.[6]

At present, the problem of preparation of nanoparticles of different elements is reduced to the development of methods that would allow synthesis and stabilization of particles with sizes of 1 nm and smaller. It is such particles that are of prime interest in chemistry.

2.6 PARTICLES OF VARIOUS SHAPES AND FILMS

Nowadays, great attention is paid to the problem of exercising control not only over the size but also over the shape of metal nanoparticles. Both the size and the shape of a particle are defined by the synthetic method; however, the ratio of nucleation to growth rates of particles is also of great importance. Each of these processes depends in turn on variations in the reaction conditions such as the temperature, the nature and concentrations of metal and ligand, and the nature of stabilizer and reducer. The problems of nucleation and particle growth are surveyed in detail elsewhere.[158]

For controlling shapes and sizes of nanoparticles, microemulsions are widely used.[159,160] Copper nanoparticles were obtained by using a functional surfactant, namely $Cu(AOT)_2$, which served as the source of copper particles and a stabilizer of water droplets. Droplets of $Cu(AOT)_2$ microemulsion in water and $Na(AOT)_2$ microemulsion in isooctane were mixed with a microemulsion of $NaBH_4$ stabilized by $Na(AOT)$. On mixing, the copper salt was reduced with sodium borohydride to give copper nanoparticles. The size and shape of nanoparticles were determined by the ratio $w = H_2O/AOT$ that defined the structure of micelles formed. For $w < 4$, copper particles measuring 1–12 nm were formed, while for $5 < w < 11$, spherical nanoparticles with diameters of 6.7 and 9.5 nm and rod-like nanoparticles with a diameter of 9.5 and a length of 22.6 nm were obtained. With the increase in water content ($w > 11$), only rods with lengths ranging from 300 to 1500 nm and diameters ranging from 10 to 30 nm were formed.

Salts of tetra-n-octylammonium with carboxylic acids of the general formula $(n\text{-}C_8H_{17})_4N^+(RCO_2)^-$ were proposed for use as reducers and stabilizers for exercising control over shapes of metal nanoparticles formed by the reduction of metal salts.[161] Palladium particles measuring 1.9–6.2 nm were formed in the following reaction:

$$Pd(NO_3)_2 + (n\text{-}C_8H_{17})_4N^+(RCO_2)^- \xrightarrow{66°C,\ THF} Pd\ particles$$

Their sizes and shapes were estimated by means of an electron microscope. It was shown that when acetate, dichloroacetate, pivalate, or pyruvate ions were used as anions, the formed particles were shaped largely as spheres. However, if palladium nitrate was treated under the same conditions with an excess of $(n\text{-}C_8H_{17})_4N^+(HOCH_2CO_2)^-$, triangular particles with an average size of 3.6 nm were formed together with spherical ones. The shape changes observed were attributed to the presence of a hydroxyl group in the anion.[161]

A nickel compound $Ni(COD)_2$ (COD stands for cycloocta-1,5-diene) was used for elucidating the role (reducer or stabilizer) of glycolic acid residue in the reduction of the glycolate group. The reaction is as follows:

$$Ni(COD)_2 + (n\text{-}C_8H_{17})_4N^+(HOCH_2CO_2)^- \xrightarrow{H_2,\ 66\,^\circ C} \text{Ni particles}$$

In as much as nickel in $Ni(COD)_2$ is in its zero-valence state, glycolate can act only as a stabilizer, whereas it is hydrogen that reduces COD to cyclooctane. Using electron microscopy, it was found that the reaction produces crystalline nickel particles with the average size of 4.5 nm, preferentially shaped as triangles. In testing experiments using $(n\text{-}C_8H_{17})_4N^+Br^-$ or as stabilizers, spherical particles were formed. The presence of triangular particles was additionally confirmed by STM. The glycolate effect on the morphology of particles seems to be associated with the selective adsorption of anions on the growing nanocrystals, which can be detected by the changes in absorption spectra. Upon synthesis of nanoparticles, a band at 1621 cm^{-1}, which belongs to dissolved glycolate, disappeared, and a new band at 1604 cm^{-1}, which corresponds to adsorbed glycolate, appeared. From the viewpoint of the authors,[161] the changes observed in in situ IR spectra agree with the mechanism put forward.

The formation of spherical and cylindrical silver nanoparticles was observed as a result of photochemical reduction of silver salts in the presence of polyacrylic acid.[101,162] Polyacrylic acid and Ag^+ ions form a complex which, being photolyzed, produces silver nanoparticles. According to the results obtained by electron microscopy and sedimentation analysis, photoreduction of this complex produced spherical silver nanoparticles measuring 1–2 nm. In the presence of a modified (e.g., partially decarboxylated) acid, in addition to spherical particles, prolonged particles (nanorods) up to 80 nm long and with characteristic light absorption in the 500–800 nm range were formed. Apparently, decarboxylation disturbs the cooperation of polyacrylic acid with silver cations, makes stabilization of spherical particles less effective, and favors the growth of nanorods.

The sizes of metal particles formed in the presence of macromolecular stabilizers depend on the conditions of formation of polymeric-protective coatings. If a polymer used is an insufficiently effective stabilizer, a particle may continue to grow after being bound to a macromolecule. By changing the nature of a monomer and the corresponding polymer and varying the polymer concentration in solution, one can control the sizes and shapes of particles formed.

An interesting method of changing the stabilizing ability of a polymer was put forward in Ref. 190. There, the authors studied how the conformation of poly-N-isopropylacrylamide affects the shape of platinum nanoparticles formed in the reduction of K_2PtCl_4 with hydrogen. This polymer can change its conformation with temperature. At $T<306$ K, polymer molecules are hydrophilic and represent swelled globules; in this case, up to 60% of platinum nanoparticles formed had irregular shapes. At $T>306$ K, polymer molecules are hydrophobic and begin to collapse. The stabilizing ability of such molecules decreases. At $T=313$ K, the reduction of platinum ions proceeded on the most active facet of a growing nanocrystal, and cubic nanoparticles were preferentially (with 68% efficiency) formed. The morphology of particles also depended on the concentration ratio of platinum salt and polymer in solution (however, this effect is weaker as compared with the temperature dependence).

To control the shapes and sizes of silver nanoparticles, methods of pulsed sonoelectrochemistry based on the use of ultrasonics in electrochemical conditions were applied.[163,164] Ultrasonics makes it possible to clean and degasify the electrode surface, accelerate mass transfer, and increase the reaction rate. Silver particles shaped as spheres, rods, and dendrites were obtained by electrolysis of aqueous $AgNO_3$ solutions in the presence of $N(CH_2COOH)_3$. These particles were characterized using electron microscopy, XRD, and electron spectroscopy. The shapes of particles were found to depend on the ultrasonic pulse duration and the reagent concentration. Spherical particles had diameters of ca. 20 nm. The diameters of rods were 10–20 nm. In certain cases, their surfaces exhibited protrusions that could develop into dendrites.

An interesting method of successive layer-by-layer deposition of thin (100–300 nm) films incorporating magnetic nanoparticles was discussed.[165] Alternating layers of magnetic nanoparticles, e.g., Fe_3O_4, and polydimethyldiallylammonium bromide were first deposited on a glass plate covered with paraffin and acetyl cellulose. Upon reaching the necessary film thickness, the cellulose layer was detached, and the whole sample was dissolved in acetone. A suspension thus obtained could be applied on any porous or dense support. It was noted that films incorporating uniform magnetic nanoparticles measuring ~10 nm can be used in memory devices.[166]

At present, exercising control over the shapes and sizes of nanoparticles seems to be among the most important problems of nanochemistry.[167] A synthesis of unishape spherical or rod-like particles of metal iron was described.[168] Nanoparticles were prepared by thermal decomposition of iron pentacarbonyl in the presence of stabilizing compounds. Spherical particles measuring 2 nm could be uniformly dispersed in solution and transformed into rods of 2 nm diameters and 11 nm lengths. Spherical particles were amorphous, whereas the rods had the face-centered cubic structure of α-iron.

Much attention was given to nanotubes based on inorganic materials.[169,170] Magnetic nanotubes of FePb and Fe_3O_4 were synthesized by reduction with hydrogen at 560 and 250 °C in nanochannels of porous alumina templates.[171] A new approach to synthesizing Fe_3O_4 nanotubes, based on wet etching of the MgO inner cores of MgO/Fe_3O_4 core–shell nanowires, was developed.[172] The procedure of preparation of crystalline nanowires of various oxides and subsequent deposition of different coatings on them to obtain core–shell structures was discussed in detail.[173,174] The aforementioned synthesis of Fe_3O_4 nanotubes involved pulsed-laser deposition of Fe_3O_4 on preliminarily prepared MgO nanowires. The resulting MgO/Fe_3O_4 structures were treated with 10% $(NH_4)_2SO_4$ solution at 80 °C to etch away the MgO cores. As evidenced by TEM results, the synthesized nanotubes had an outer diameter of 30 nm and a wall thickness of ~7 nm. The nanotube dimensions depended on experimental conditions and the sizes of original MgO nanowires.[172]

Various core–shell nanoparticles are widely used, particularly, in optoelectronics. Metal–semiconductor,[175] metal–carbon,[176,177] metal–dielectric,[178,179] dielectric–metal[180] systems, and gold shells with inner hollows[181,182] were obtained.

The devices used in electronics and optics usually employ solid-surface-supported core–shell structures with controlled properties. Ordinary chemical or wet methods fail to provide such structures with controlled properties. Highly ordered $In-In_2O_3$ structures on silica substrates were obtained by three-step oxidation at elevated temperatures.[183] First, a finely dispersed alumina mask was applied on the Si/SiO_2 substrate, then indium was deposited, and the mask was removed. The average size of indium nanoparticles could be varied from 10 to 100 nm. To obtain In_2O_3 shells, the deposit was first oxidized in an oxygen flow under atmospheric pressure at 146 °C. In the second stage, the temperature was varied between 146 and 800 °C. In the last stage, the sample was exposed at high temperature for 2 h for complete oxidation. The oxide shell thickness could be regulated, which allowed changing the photoluminescence properties of the samples.

The discovery of carbon nanotubes gave impetus to an active search for similar particles based on different elements. Recently, nanotubes were prepared from nonlayered compounds such as AlN^{184} and $GaN.^{184}$ Nanobelts and nanoribbons were synthesized based on oxides,[185] carbides (Al_4C_3),[186] and aluminum nitrides.[187] Sharp nanocones were prepared based on silicon carbide,[188] carbon,[189] zinc oxide,[190] and aluminum nitride.[191]

Different supports covered with films of 4-mercaptobenzoic acid, which were deposited in high vacuum, were used in the preparation of thin heterogeneous films with incorporated nanoparticles of gold, silver, and sulfate of cadmium. These heterogeneous films were synthesized by successive immersion of supports in the corresponding solutions.[192]

Quantum dots were synthesized based on ZnO-doped Mn^{2+} colloidal particles, and ferromagnetic nanocrystalline thin films were fabricated at room temperature.[193] According to TEM studies, the nanocrystals has an average diameter of 6.1 ± 0.7 nm.

Results of high-resolution electron microscopy showed that, being encapsulated into micelles, rod-like gold nanoparticles fabricated by an electrochemical method tend to grow along the {001} axis. In the process, the {100} face of particles remains stable, whereas {110} face is unstable and can transform into the stable face.[194]

Chapter | 2 Synthesis and Stabilization of Nanoparticles

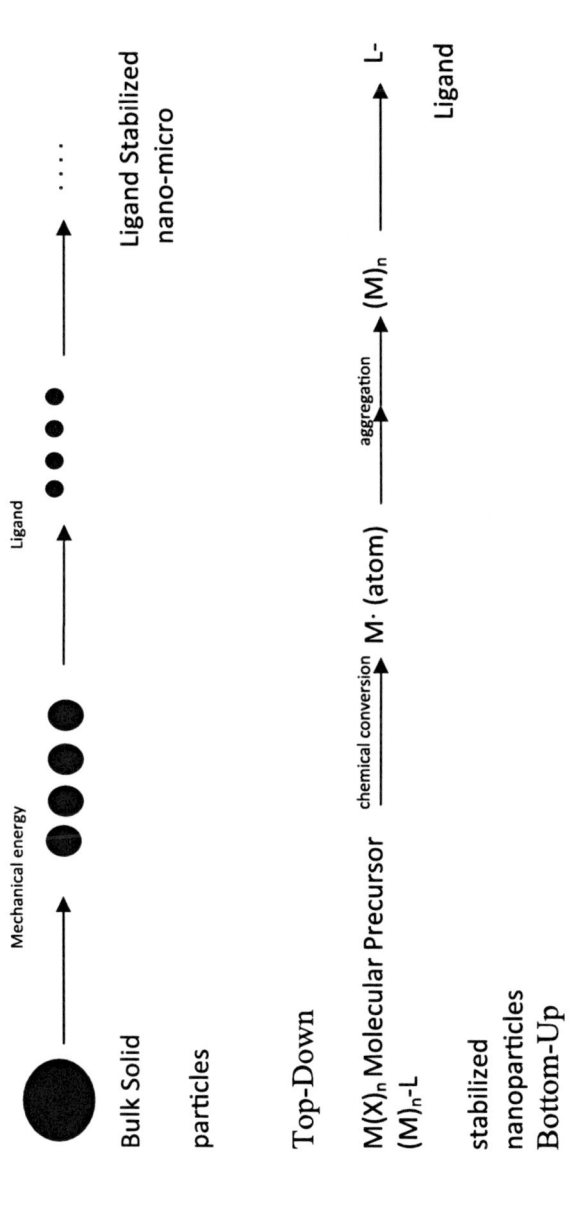

SCHEME 2.1 Representation of top-down method (grinding/pulverizing bulk solids) and bottom-up method (creating a reactive atomic or molecular precursor that aggregates to clusters and eventually nanoscale particles).

The possibilities of structural changes in silica-supported platinum nanoparticles and the microcrystalline structure of individual nanoparticles were examined by electron microscopy before and after heating in vacuum and atmospheres of hydrogen and oxygen.[195] In all the cases, the particle mass remained unchanged, but the sizes of particle-constituent crystals increased with temperature, which was assumed to be a consequence of surface fusion and self-diffusion of platinum particles. Adhesion of platinum particles to silica was studied by atomic force microscopy. Platinum nanoparticles were shown to be stable both under oxidizing and reducing conditions. However, heating was demonstrated to change the crystalline nature of particles and strengthen their adhesion to silica.

Controlled deposition of silver particles, and, on the platinum surface at different temperatures was studied.[196] As demonstrated by STM studies, a three-dimensional structure of becomes two-dimensional when heated from 60 to 140 K.

The most popular method of preparation of nanofilms consists in co-deposition of metal atoms from the gas phase at surfaces of different nature. The formation of supported films begins from nonuniform islets. The process depends on the surface temperature and the intensity and velocity of a flow of a substance to be deposited. At low temperatures, when diffusion of atoms is slow, small particles are formed, which, however, have a high density. The temperature and the relevant mobility of particles are the major factors that determine the self-assembling and formation of surface nanostructures from individual clusters.

Scheme 2.1 summarizes these concepts, and later chapters go into more detail on synthetic methods.

REFERENCES

1. Ozin, G. *Adv. Mater.* **1992**, *4*, 612–649.
2. Sviridov, V. V.; Vorob'eva, T. N.; Gaevskaya, T. V.; Stepanova, L. I. *Chemical Precipitation of Metals in Aqueous Solutions (Russ.), Izd*; Universitetskoe: Minsk, 1987.
3. Yonezawa, T.; Onoue, S.; Kimizuka, N. *Langmuir* **2000**, *16*, 5218–5220.
4. Sergeev, B. M.; Lopatina, L. I.; Prusov, A. N.; Sergeev, G. B. *Colloid J.* **2005**, *67*, 213–216.
5. Sergeev, B. M.; Lopatina, L. I.; Prusov, A. N.; Sergeev, G. B. *Colloid J.* **2005**, *67*, 72–78.
6. Bronstein, L. M.; Chernyshov, D. M.; Valetsky, P. M.; Wilder, E. A.; Spontak, R. J. *Langmuir* **2000**, *16*, 8221–8225.
7. Svergun, D. I.; Shtykova, E. V.; Kozin, M. B.; Volkov, V. V.; Dembo, A. T.; Shtykova, E. V., Jr., et al. *J. Phys. Chem. B* **2000**, *104*, 5242–5250.
8. Ershov, B. G.; Sukhov, N. L.; Janata, E. *J. Phys. Chem. B* **2000**, *104*, 6138–6142.
9. Ershov, B. G. *Russ. Chem. Bull. (Russ.)* **2000**, *49*, 1733–1739.
10. Henglein, A. *J. Phys. Chem. B* **2000**, *104*, 1206–1211.
11. Wang, S.; Xin, H. *J. Phys. Chem. B* **2000**, *104*, 5681–5685.
12. Wang, Z. L. *J. Phys. Chem. B* **2000**, *104*, 1153–1175.
13. Jeong, U.; Herricks, T.; Shahar, E.; Xia, Y. *J. Am. Chem. Soc.* **2005**, *127*, 1098–1099.
14. Rodrigues-Sanchez, L.; Blanko, M. L.; Lopez-Quintela, M. A. *J. Phys. Chem. B* **2000**, *104*, 9683–9688.
15. Petrii, O. A.; Tsirlina, G. A. *Russ. Chem. Rev.* **2001**, *70*, 285–298.

Chapter 2 Synthesis and Stabilization of Nanoparticles

16. Hyeon, T.; Lee, S. S.; Park, J.; Chung, Y.; Na, H. B. *J. Am. Chem. Soc.* **2001**, *123*, 12798–12801.
17. Sun, S.; Zeng, H.; Robinson, D. B.; Raoux, S.; Rice, P. M.; Wang, S. X.; et al. *J. Am. Chem. Soc.* **2004**, *126*, 273–279.
18. Shevchenko, E. V.; Talapin, D. V.; Rogach, A. L.; Kornowski, A.; Haase, M.; Weller, H. *J. Am. Chem. Soc.* **2002**, *124*, 11480–11485.
19. Shevchenko, E. V.; Talapin, D. V.; Schnablegger, H.; Kornowski, A.; Festin, O.; Svedlindh, P.; et al. *J. Am. Chem. Soc.* **2003**, *125*, 9090–9101.
20. Yu, W. W.; Peng, X. *Angew. Chem.* **2002**, *114*, 2474–2477.
21. Battaglia, D.; Peng, X. *Nano Lett.* **2002**, *2*, 1027–1030.
22. Gerion, D.; Zaitseva, N.; Saw, C.; Casula, M. F.; Fakra, S.; Van Buuren, T.; et al. *Nano Lett.* **2004**, *4*, 597–602.
23. Jun, Y.; Casula, M. F.; Sim, J. -H.; Kim, S. Y.; Cheon, J.; Alivisatos, A. P. *J. Am. Chem. Soc.* **2003**, *125*, 15981–15985.
24. Puntes, V. F.; Zanchet, D.; Erdonmez, C. K.; Alivisatos, A. P. *J. Am. Chem. Soc.* **2002**, *124*, 12874–12880.
25. Cordente, N.; Respaud, M.; Senocq, F.; Casanove, M. -J.; Amiens, C.; Chaudret, B. *Nano Lett.* **2001**, *1*, 565–568.
26. Manna, L.; Milliron, D. J.; Meisel, A.; Scher, E. C.; Alivisatos, A. P. *Nat. Mater.* **2003**, *2*, 382–385.
27. Jun, Y.; Choi, C. -S.; Cheon, J. *Chem. Commun.* **2001**, *1*, 101–102.
28. Mekis, I.; Talapin, D. V.; Kornowski, A.; Haase, M.; Weller, H. *J. Phys. Chem. B* **2003**, *107*, 7454–7462.
29. Kim, S.; Fisher, B.; Eisler, H. -J.; Bawendi, M. *J. Am. Chem. Soc.* **2003**, *125*, 11466–11467.
30. Zeng, H.; Li, J.; Wang, Z. L.; Liu, J. P.; Sun, S. *Nano Lett.* **2004**, *4*, 187–190.
31. Sobal, N. S.; Ebels, U.; Moehwald, H.; Giersig, M. *J. Phys. Chem. B* **2003**, *107*, 7351–7354.
32. Sobal, N. S.; Hilgendorff, M.; Moewald, H.; Giersig, M.; Spasova, M.; Radetic, T.; et al. *Nano Lett.* **2002**, *2*, 621–624.
33. Talapin, D. V.; Koeppe, R.; Goetzinger, S.; Kornowski, A.; Lupton, J. M.; Rogach, A. L.; et al. *Nano Lett.* **2003**, *3*, 1677–1681.
34. Gu, H.; Zheng, R.; Zhang, X.; Xu, B. *J. Am. Chem. Soc.* **2004**, *126*, 5664–5665.
35. Mokari, T.; Rothenberg, E.; Popov, I.; Costi, R.; Banin, U. *Science* **2004**, *304*, 1790.
36. Cushing, B. L.; Kolesnichenko, V. L.; O'Connor, C. J. *Chem. Rev.* **2004**, *104*, 3893–3946.
37. Fernandez-Garcia, M.; Martinez-Arias, A.; Hanson, J. C.; Rodriguez, J. A. *Chem. Rev.* **2004**, *104*, 4063–4104.
38. Foos, E. E.; Stroud, R. M.; Berry, A. D.; Snow, A. W.; Armistead, J. P. *J. Am. Chem. Soc.* **2000**, *122*, 7114–7115.
39. Fang, J.; Stokes, K. L.; Wiemann, J.; Zhou, W. *Mater. Lett.* **2000**, *42*, 113–120.
40. Bronstein, L. M.; Chernyshov, D. M.; Timofeeva, G. I.; Dubrovina, L. V.; Valetsky, P. M.; Khokhlov, A. R. *J. Colloid Interface Sci.* **2000**, *230*, 140–149.
41. Lee, M.; Oh, S. -G.; Yi, S. -C. *J. Colloid Interface Sci.* **2000**, *226*, 65–70.
42. Litmanovich, O. E.; Litmanovich, A. A.; Papisov, I. M. *Polym. Sci.* **2000**, *42*, 442–446.
43. Gubin, S. P. *Colloids Surf. A* **2002**, *202*, 155–163.
44. Bronshtein, L. M.; Sidorov, S. N.; Valetskii, P. M. *Russ. Chem. Rev.* **2004**, *73*, 501–516.
45. Haes, A. J.; Hall, W. P.; Chang, L.; Klein, W. L.; Van Duyune, R. P. *Nano Lett.* **2004**, *4*, 1029–1034.
46. Valsesia, A.; Colpo, P.; Silvan, M. M.; Meziani, T.; Ceccone, G.; Rossi, F. *Nano Lett.* **2004**, *4*, 1047–1050.

47. Yi, G. -R.; Moon, J. H.; Manoharan, V. N.; Pine, D. J.; Yang, S. -M. *J. Am. Chem. Soc.* **2002**, *124*, 13354–13355.
48. Kosiorek, A.; Kandulski, W.; Chudzinski, P.; Kempa, K.; Giersig, M. *Nano Lett.* **2004**, *4*, 1359–1363.
49. Yan, F.; Goedel, W. A. *Adv. Mater.* **2004**, *16*, 911–915.
50. Han, M. G.; Foulger, S. H. *Adv. Mater.* **2004**, *16*, 231–234.
51. Choi, D. -G.; Kim, S.; Lee, E.; Yang, S. -M. *J. Am. Chem. Soc.* **2005**, *127*, 1636–1637.
52. Choi, D. -G.; Yu, H. K.; Jangm, S. G.; Yang, S. -M. *J. Am. Chem. Soc.* **2004**, *126*, 7019–7025.
53. Mertz, E.; Zimmerman, S. C. *J. Am. Chem. Soc.* **2003**, *125*, 3424–3425.
54. Kim, H.; Spivak, D. A. *J. Am. Chem. Soc.* **2003**, *125*, 11269–11275.
55. Morikawa, M.; Yoshihara, M.; Endo, T.; Kimizuka, N. *J. Am. Chem. Soc.* **2005**, *127*, 1358–1359.
56. Keki, S.; Torok, J.; Deak, G.; Daroczi, L.; Zsuga, M. *J. Colloid Interface Sci.* **2000**, *229*, 550–553.
57. Muzafarov, A. M.; Rebrov, E. A. *Polymer Sci. C* **2000**, *42*, 55–77.
58. Grohn, F.; Bauer, B. J.; Akpalu, Y. A.; Jackson, C. L.; Amis, E. J. *Macromolecules* **2000**, *33*, 6042–6050.
59. Esumi, K.; Hosoya, T.; Suzuki, A.; Torigoe, K. *J. Colloid Interface Sci.* **2000**, *226*, 346–352.
60. Schmid, G.; Meyer-Zaika, W.; Pugin, R.; Sawitowski, T.; Majoral, J. -P.; Caminade, A. -M.; et al. *Chem.–Eur. J.* **2000**, *6*, 1693–1697.
61. Kim, Y. -G.; Oh, S. -K.; Crooks, R. M. *Chem. Mater.* **2004**, *16*, 167–172.
62. Zheng, J.; Dickson, R. M. *J. Am. Chem. Soc.* **2002**, *124*, 13982–13983.
63. Niu, Y.; Yeung, L. K.; Crooks, R. M. *J. Am. Chem. Soc.* **2001**, *123*, 6840–6846.
64. Ooe, M.; Murata, M.; Mizugaki, T.; Ebitani, K.; Kaneda, K. *Nano Lett.* **2002**, *2*, 999–1002.
65. Ye, H.; Scott, R. W.; Crooks, R. M. *Langmuir* **2004**, *20*, 2915–2920.
66. Chung, Y. -M.; Rhee, H. -K. *Catal. Lett.* **2003**, *85*, 159–164.
67. Chung, Y. -M.; Rhee, H. -K. *J. Mol. Catal. A* **2003**, *206*, 291–298.
68. Mallin, M. P.; Murphy, C. J. *Nano Lett.* **2002**, *2*, 1235–1237.
69. Shibata, T.; Bunker, B. A.; Zhang, Z.; Meisel, D.; Vardeman, C. F., II; Gezelter, J. D. *J. Am. Chem. Soc.* **2002**, *124*, 11989–11996.
70. Cao, Y. W.; Jin, R.; Mirkin, C. A. *J. Am. Chem. Soc.* **2001**, *123*, 7961–7962.
71. Selvakannan, P.; Swami, A.; Srisathiyanarayanan, D.; Shirude, P. S.; Pasricha, R.; Mandale, A. B.; et al. *Langmuir* **2004**, *20*, 7825–7836.
72. Scott, R. W.; Wilson, O. M.; Oh, S. -K.; Kenik, E. A.; Crooks, R. M. *J. Am. Chem. Soc.* **2004**, *126*, 15583–15591.
73. Scott, R. W. J.; Datye, A. K.; Crooks, R. M. *J. Am. Chem. Soc.* **2003**, *125*, 3708–3709.
74. Wilson, O. M.; Scott, R. W.; Garcia-Martinez, J. C.; Crooks, R. M. *Chem. Mater.* **2004**, *16*, 4202–4204.
75. Garcia-Martinez, J. C.; Scott, R. W.; Crooks, R. M. *J. Am. Chem. Soc.* **2003**, *125*, 11190–11191.
76. Garcia-Martinez, J. C.; Crooks, R. M. *J. Am. Chem. Soc.* **2004**, *126*, 16170–16178.
77. Wilson, O. M.; Scott, R. W.; Garcia-Martinez, J. C.; Crooks, R. M. *J. Am. Chem. Soc.* **2005**, *127*, 1015–1024.
78. Lang, H.; May, R. A.; Iversen, B. L.; Chandler, B. D. *J. Am. Chem. Soc.* **2003**, *125*, 14832–14836.
79. Scott, R. W. J.; Wilson, O. M.; Crooks, R. M. *Chem. Mater.* **2004**, *16*, 5682–5688.
80. Velarde-Ortiz, R.; Larsen, G. *Chem. Mater.* **2002**, *14*, 858–866.
81. Choi, H. C.; Kim, W.; Wang, D.; Dai, H. *J. Phys. Chem. B* **2002**, *106*, 12361–12635.

82. Lang, H.; Maldonaldo, S.; Stephenson, K.; Chandler, B. D. *J. Am. Chem. Soc.* **2004**, *126*, 12949–12956.
83. Zhou, W.; Heiney, P. A.; Fan, H.; Smalley, R. E.; Fischer, J. E. *J. Am. Chem. Soc.* **2005**, *127*, 1640–1641.
84. Deng, S.; Locklin, J.; Patton, D.; Baba, A.; Advincula, R. C. *J. Am. Chem. Soc.* **2005**, *127*, 1744–1751.
85. Liu, M.; Yan, X.; Liu, H.; Yu, W. *React. Funct. Polym.* **2000**, *44*, 55–64.
86. Rollins, H. W.; Lin, F.; Johnson, J.; Ma, J. -J.; Liu, J. -T.; Tu, M. -H.; et al. *Langmuir* **2000**, *16*, 8031–8036.
87. Siiman, O.; Burshteyn, A. *J. Phys. Chem. B* **2002**, *104*, 9795–9810.
88. Romanovsky, B. V.; Gabrielov, A. G. *Mendeleev Commun.* **1991**, 14–15.
89. Roggenbuck, J.; Tiemann, M. *J. Am. Chem. Soc.* **2005**, *127*, 1096–1097.
90. Krivovichev, S. V.; Kahlenberg, V.; Tananaev, I. G.; Kaindl, R.; Mersdorf, E.; Myasoedov, B. F. *J. Am. Chem. Soc.* **2005**, *127*, 1072–1073.
91. Choi, M.; Kleitz, F.; Liu, D.; Lee, H. Y.; Ahn, W. -S.; Ryoo, R. *J. Am. Chem. Soc.* **2005**, *127*, 1924–1932.
92. Liu, B.; Zeng, H. C. *J. Am. Chem. Soc.* **2003**, *125*, 4430–4431.
93. Feng, X.; Feng, L.; Jin, M.; Zhai, J.; Jiang, L.; Zhu, D. *J. Am. Chem. Soc.* **2004**, *126*, 62–63.
94. Bugaenko, L. T.; Kuzmin, M. G.; Polak, L. S. *High Energy Chemistry (Russ.)*; Khimiya: Moscow, 1988.
95. Kiryukhin, M. V.; Sergeev, B. M.; Sergeev, V. G. Physical chemistry of ultra-dispersed systems (Russ.). *Conference Proceeding* **2001**, 133–136.
96. Kimura, K. *J. Phys. Chem.* **1994**, *98*, 11997–12002.
97. Satoh, N.; Hasegawa, H.; Tsujii, K.; Kimura, K. *J. Phys. Chem.* **1994**, *98*, 2143–2147.
98. Takeuchi, Y.; Ida, T.; Kimura, K. *J. Phys. Chem. B* **1997**, *101*, 1322–1327.
99. Karpov, S. V.; Popov, A. K.; Slabko, V. V. *Russ. Phys. Bull (Russ.)* **1996**, *60*, 43.
100. Karpov, S. V.; Bas'ko, A. P.; Koshelev, S. V.; Popov, A. K.; Slabko, V. V. *Colloid J.* **1997**, *59*, 708–716.
101. Kiryukhin, M. V.; Sergeev, B. M.; Prusov, A. N.; Sergeev, V. G. *Polym. Sci. B* **2000**, *42*, 158–162.
102. Dokuchaev, A. G.; Myasoedova, T. G.; Revina, A. A. *Khim. Vys. Energ.* **1997**, *31*, 353.
103. Revina, D. A.; Egorova, E. M.; Karataeva, A. D. *Russ. J. Phys. Chem.* **1999**, *73*, 1708–1715.
104. Ershov, B. G. *Russ. Chem. Rev.* **1997**, *66*, 93–106.
105. Ershov, B. G. *Russ. Chem. Bull. (Russ.)* **1994**, *1*, 25–29.
106. Ershov, B. G.; Sukhov, N. L.; Troitskii, D. A. *J. Phys. Chem. (Russ.)* **1994**, *68*, 820–824.
107. Henglein, A.; Giersig, M. *J. Phys. Chem. B* **2000**, *104*, 5056–5060.
108. Henglein, A. *J. Phys. Chem. B* **2000**, *104*, 6683–6685.
109. Hodak, J. H.; Henglein, A.; Hartland, G. V. *J. Phys. Chem. B* **2000**, *104*, 9954–9965.
110. Hodak, J. H.; Henglein, A.; Hartland, G. V. *Pure Appl. Chem.* **2000**, *72*, 189–197.
111. D'Souza, L.; Sampath, S. *Langmuir* **2000**, *16*, 8510–8517.
112. Mizukoshi, Y.; Fujimoto, T.; Nagata, Y.; Oshima, R.; Maeda, Y. *J. Phys. Chem. B* **2000**, *104*, 6028–6032.
113. Rousset, J. L.; Aires, F. J. C. S.; Sekhar, B. R.; Melinon, P.; Prevel, B.; Pellarin, M. *J. Phys. Chem. B* **2000**, *104*, 5430–5435.
114. Novakova, A. A.; Agladze, O. V.; Tarasov, B. P. *Russ. J. Inorg. Chem.* **2000**, *45*, 1168–1172.
115. Novakova, A. A.; Kiseleva, T. Yu.; Levina, V. V. *Russ. J. Inorg. Chem.* **2000**, *45*, 1267–1272.
116. Bharathi, S.; Fishelson, N.; Lev, O. *Langmuir* **1999**, *15*, 1929–1937.
117. Galkin, A. A.; Kostyuk, B. G.; Lunin, V. V.; Poliakoff, M. *Angew. Chem. Int. Ed.* **2000**, *39*, 2738–2740.

118. Galkin, A. A.; Turakulova, A. O.; Kuznetsova, N. N.; Lunin, V. V. *Vestn. Mosk. Univ., Ser. 2: Khim.* **2002,** *42,* 305–308.
119. Galkin, A. A.; Kostyuk, B. G.; Kuznetsova, N. N.; Turakulova, A. O.; Lunin, V. V.; Polyakov, M. *Kinet. Catal.* **2001,** *42,* 172–181.
120. Gorichev, I. G.; Izotov, A. D.; Il'yukhin, O. V.; Gorichev, A. I.; Kutepov, A. M. *Russ. J. Phys. Chem.* **1999,** *73,* 1619–1624.
121. Meziani, M.; Sun, Y. -P. *J. Am. Chem. Soc.* **2003,** *125,* 8015–8018.
122. Meziani, M. J.; Pathak, P.; Hurezeanu, R.; Thies, M. C.; Enick, R. M.; Sun, Y. -P. *Angew. Chem. Int. Ed.* **2004,** *43,* 704–707.
123. Pathak, P.; Meziani, M. J.; Desai, T.; Sun, Y. -P. *J. Am. Chem. Soc.* **2004,** *126,* 10842–10843.
124. Moskovits, M.; Ozin, G. Eds.; *Cryochemistry (Russ.)*; Wiley: New York, 1976, 594.
125. Sergeev, G. B.; Batyuk, V. A. *Cryochemistry*; Khimiya: Moscow, 1978, 296.
126. Mochida, K.; Manrishi, M. *Chem. Lett.* **1984,** *13,* 1077–1080.
127. Arnold, P. Ph.D. Thesis, Univ. of Sussex, 1997, 209.
128. Arnold, P. L.; Petrukhina, M. A.; Bochenkov, V. E.; Shabatina, T. I.; Zagorskii, V. V.; Sergeev, G. B.; et al. *J. Organomet. Chem.* **2003,** *688,* 49–55.
129. Benfield, F. W. S.; Green, M. L. H.; Ogden, J. S.; Young, D. *J. Chem. Soc., Chem. Commun.* **1973,** *22,* 866–867.
130. Siegel, R. W. In *Material Science and Technology. Processing of Metals and Alloys*; Cahn, R. W., Ed.; VCH: Weinheim, 1991; pp 583–614.
131. Tohno, S.; Itoh, M.; Aono, S.; Takano, H. *J. Colloid Interface Sci.* **1996,** *180,* 574–577.
132. Northby, J. A. *J. Chem. Phys.* **2001,** *115,* 10065–10077.
133. Vongehr, S.; Scheidemann, A. A.; Wittig, C.; Kresin, V. V. *Chem. Phys. Lett.* **2002,** *353,* 89–94.
134. Doeppner, T.; Diederich, Th.; Tiggesbaeumker, J.; Meiwes-Broer, K. -H. *Eur. Phys. J. D* **2001,** *16,* 13–16.
135. Diederich, Th; Tiggesbaeumker, J.; Meiwes-Broer, K. -H. *J. Chem. Phys.* **2002,** *116,* 3263–3269.
136. Sergeev, G. B.; Batyuk, V. A. *Cryochemistry, Mir* **1986,** 326.
137. Frolov, Yu. V.; Pivkina, A. N.; Zav'yalov, S. A. *Doklady Phys. Chem.* **2002,** *383,* 81–83.
138. Petrov, Yu. I. *Clusters and Small Particles (Russ.)*; Nauka: Moscow, 1986, 387.
139. Kotov, N. A.; Samatov, O. M. *Nanostruct. Mater.* **1999,** *12,* 119–122.
140. Kozhevin, V. M.; Vavsin, D. A.; Kouznetsov, V. M.; Busov, V. M.; Mikushkin, V. M.; Nikonov, S. Yu.; et al. *J. Vac. Sci. Technol. B* **2000,** *18,* 1402–1405.
141. Busman, H. -G.; Guenter, B.; Meyer, U. *Nanostruct. Mater.* **1999,** *12,* 531–537.
142. Kim, S. Y.; Yu, J. H.; Lee, J. S. *Nanostr. Mater.* **1999,** *12,* 471–474.
143. Kim, B. K.; Lee, G. G.; Park, H. M.; Kim, N. J. *Nanostr. Mater.* **1999,** *12,* 637–640.
144. Dietz, T. G.; Ducan, M. A.; Powers, D. E.; Smalley, R. E. *J. Chem. Phys.* **1981,** *74,* 6511.
145. Morse, M. D.; Geusic, M. E.; Heath, J. R.; Smalley, R. E. *J. Chem. Phys.* **1985,** *83,* 2293.
146. Maruyama, S.; Anderson, L. R.; Smalley, R. E. *Rev. Sci. Instrum.* **1990,** *61,* 3686.
147. Parks, E. K.; Liu, K.; Richtsmeier, S. C.; Pobo, L. G.; Riley, S. J. *J. Chem. Phys.* **1985,** *82,* 5470.
148. Richtsmeier, S. C.; Parks, E. K.; Liu, K.; Pobo, L. G.; Riley, S. J. *J. Chem. Phys.* **1985,** *82,* 3659.
149. Roy, D. M.; Neurgaonkar, R. R.; O'Holleran, T. P.; Roy, R. *Ceram. Bull.* **1997,** *56,* 1023–1024.
150. Ruthner, M. J. *Ceram. Powders* **1983,** 515–531.
151. Sproson, D. W.; Messing, G. L.; Gardner, T. J. *Ceram. Int.* **1986,** *12,* 3–7.
152. Kodas, T. T. *Adv. Mater.* **1989,** *6,* 180–192.
153. Zhang, S. -C.; Messing, G. L.; Borden, M. *J. Am. Ceram. Soc.* **1990,** *73,* 61–67.

154. Ortega, J.; Kodas, T. T. *J. Aerosol Sci.* **1992**, *23*, 5253–5256.
155. Charlesworth, D. H.; Marshall, W. R., Jr. *AIChE J.* **1960**, *6*, 9–23.
156. Li, Q.; Sorensen, C. M.; Klabunde, K. J.; Hadjipanayis, G. C. *Aerosol Sci. Technol.* **1993**, *19*, 453–467.
157. Klabunde, K. J. *Free Atoms, Clusters, and Nanoscale Particles*; Academic Press: San Diego, 1994, pp 8–9.
158. Suzdalev, I. P.; Suzdalev, P. I. *Russ. Chem. Rev.* **2001**, *70*, 177–210.
159. Pileni, M. P. *Langmuir* **1997**, *13*, 3266–3276.
160. Pileni, M. P.; Gulik-Krzywicki, T.; Tanori, J.; Filankembo, A.; Dedily, J. C. *Langmuir* **1998**, *14*, 7359–7363.
161. Bradly, J. S.; Tesche, B.; Busser, W.; Maase, M.; Reetz, M. T. *J. Am. Chem. Soc.* **2000**, *122*, 4631–4636.
162. Kiryukhin, M. V.; Sergeev, B. M.; Prusov, A. N.; Sergeev, V. G. *Polym. Sci. B* **2000**, *42*, 324–328.
163. Zhu, J.; Aruna, S. T.; Koltypin, Y.; Gedanken, A. *Chem. Mater.* **2000**, *12*, 143–147.
164. Zhu, J.; Liu, S.; Palchik, O.; Koltypin, Y.; Gedanken, A. *Langmuir* **2000**, *16*, 6396–6399.
165. Mamedov, A. A.; Kotov, N. A. *Langmuir* **2000**, *16*, 5530–5533.
166. Sun, S.; Murray, C. B.; Weller, D.; Folks, L.; Moser, A. *Science* **2000**, *287*, 1989–1992.
167. Peng, X.; Manna, L.; Yang, W.; Wickham, J.; Scher, E.; Kadavanich, A.; et al. *Nature* **2000**, *404*, 59–61.
168. Park, S. -J.; Kim, S.; Lee, S.; Khim, Z. G.; Char, K.; Hyeon, T. *J. Am. Chem. Soc.* **2000**, *122*, 8581–8582.
169. Fan, R.; Wu, Y.; Li, D.; Yue, M.; Majumdar, A.; Yang, P. *J. Am. Chem. Soc.* **2003**, *125*, 5254–5255.
170. Li, D.; Xia, Y. *Nano Lett.* **2004**, *4*, 933–938.
171. Sui, Y. C.; Skomski, R.; Sorge, K. D.; Sellmyer, D. J. *J. Appl. Phys.* **2004**, *95*, 7151–7153.
172. Liu, Z.; Zhang, D.; Han, S.; Li, C.; Lei, B.; Lu, W.; et al. *J. Am. Chem. Soc.* **2005**, *127*, 6–7.
173. Han, S.; Li, C.; Lei, B.; Zhang, D.; Jin, W.; Liu, X.; et al. *Nano Lett.* **2004**, *4*, 1241–1246.
174. Zhang, D.; Liu, Z.; Han, S.; Li, C.; Lei, B.; Stewart, M. P.; et al. *Nano Lett.* **2004**, *4*, 2151–2155.
175. Yang, Y.; Nogami, M.; Shi, J.; Chen, H.; Liu, Y.; Quian, S. *J. Mater. Chem.* **2003**, *13*, 3026–3032.
176. Kim, M.; Sohn, K.; Na, H. B.; Hyeon, T. *Nano Lett.* **2002**, *2*, 1383–1387.
177. Nikitenko, S. I.; Koltypin, Yu.; Felner, I.; Yeshurun, I.; Shames, A. I.; Jiang, J. Z.; et al. *J. Phys. Chem. B* **2004**, *108*, 7620–7626.
178. Lu, Y.; Yin, Y.; Li, Z. -Y.; Xia, Y. *Nano Lett.* **2002**, *2*, 785–788.
179. Aoki, K.; Chen, J.; Yang, N.; Nagasawa, H. *Langmuir* **2003**, *19*, 9904–9909.
180. Graf, C.; Blaaderen, A. *Langmuir* **2002**, *18*, 524–534.
181. Sun, Y.; Xia, Y. *Anal. Chem.* **2002**, *74*, 5297–5305.
182. Hao, E.; Li, S.; Bailey, C.; Zou, S.; Schatz, G. C.; Hupp, J. T. *J. Phys. Chem. B* **2004**, *108*, 1224–1229.
183. Lei, Y.; Chim, W. -K. *J. Am. Chem. Soc.* **2005**, *127*, 1487–1492.
184. Wu, Q.; Hu, Z.; Wang, X.; Lu, Y.; Chen, X.; Xu, H.; et al. *J. Am. Chem. Soc.* **2003**, *125*, 10176–10177.
185. Pan, Z. W.; Dai, Z. R.; Wang, Z. L. *Science* **2001**, *291*, 1947–1949.
186. Zhang, H. -F.; Dohnalkova, A. C.; Wang, C. -M.; Young, J. S.; Buck, E. C.; Wang, L. -S. *Nano Lett.* **2002**, *2*, 105–108.
187. Wu, Q.; Hu, Z.; Wang, X.; Chen, Y. *J. Phys. Chem. B* **2003**, *107*, 9726–9729.

188. Wu, Z. S.; Deng, S. Z.; Xu, N. S.; Chen, J.; Zhou, J.; Chen, J. *Appl. Phys. Lett.* **2002,** *80,* 3829–3831.
189. Zhang, G.; Jiang, X.; Wang, E. *Science* **2003,** *300,* 474.
190. Li, Y. B.; Bando, Y.; Golberg, D. *Appl. Phys. Lett.* **2004,** *84,* 3603–3605.
191. Liu, C.; Hu, Z.; Wu, Q.; Wang, X.; Chen, Y.; Sang, H.; et al. *J. Am. Chem. Soc.* **2005,** *127,* 1318–1322.
192. Sastry, M.; Gole, A.; Sainkar, S. R. *Langmuir* **2000,** *16,* 3553–3556.
193. Norberg, N. S.; Kittilstved, K. R.; Amonette, J. E.; Kukkadapu, R. K.; Schwartz, D. A.; Gamelin, D. R. *J. Am. Chem. Soc.* **2004,** *126,* 9387–9398.
194. Wang, Z. L.; Gao, R. P.; Nikoobakht, B.; El-Sayed, M. A. *J. Phys. Chem. B* **2000,** *104,* 5417–5420.
195. Eppler, A. S.; Rupprechter, G.; Anderson, E. A.; Somorjai, G. A. *J. Phys. Chem. B* **2000,** *104,* 7286–7292.
196. Schaub, R. Controlled deposition of mass selected silver clusters on Pt(1 1 1) surface. Ph.D. Thesis, 2000, 149.

Chapter 3

Solvated Metal Atom Dispersion (SMAD) for Making Metal Nanoparticles

Chapter Outline

3.1 Experimental Techniques 55
3.2 Aggregation of Metal Atoms or Reactive Molecules in Low-Temperature Matrices/Solvents 56
   3.2.1 Control of the Gold–Tin (Au–Sn) Bimetallic System 57
      3.2.1.1 Experimental Results on Au Atom–Sn Atom Clusters in Cold Solvents 58
3.2.2 Reactivity of Aggregates (Nanoparticles or Nanocrystals) 61
3.2.3 Trapping and Stabilization 61
3.3 Examples of Useful Synthesis 61
   3.3.1 Gold Nanoparticles 61
   3.3.2 Silver and Copper 63
   3.3.3 Other Metals 63
   3.3.4 Binuclear Compounds 63
3.4 Digestive Ripening or "Nanomachining" 64
3.5 Rods, Wires, and Stars 69

3.1 EXPERIMENTAL TECHNIQUES

In the last chapter (Chapter 2), the general methods of preparing metal nanoparticles by "bottom-up" methods were summarized. These involve ways of producing a reactive species such as a metal atom in an environment under which it is kinetically and thermodynamically favorable for these metal atoms to quickly aggregate, forming metal particles or films.

Initially, this technique was referred to as "metal atom chemistry" and was very useful for preparing many new organometallic compounds.[1,2] As shown in Figure 3.1, the vapors of a reactant can readily be cocondensed with the vapors

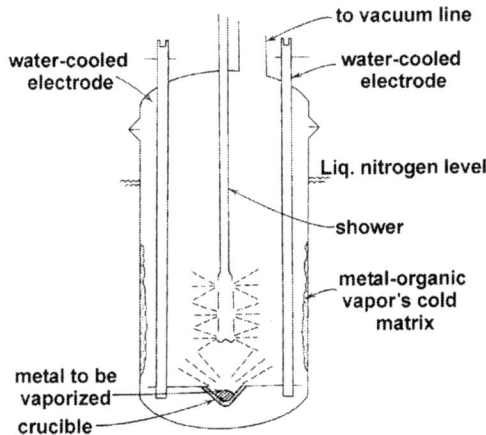

FIGURE 3.1 Schematic representation of a SMAD reactor.[1]

of a metal (atoms). Reactions then proceeded with very low activation energies, for example, chromium atoms cocondensed with benzene yield the stable bis(benzene)chromium [$Cr(C_6H_6)_2$].[3]

Next, this experimental method lent itself to preparing "solvated metal atoms." Thus, normally unstable metal-organic complexes could be prepared and serve as sources of highly reactive metal. An early example was published where nickel atoms were cocondensed with different solvents such as pentane, toluene, and tetrahydrofuran (THF), or other solvents. A solvate of Ni atoms with pentane would be expected to be very unstable, and it is.[4–7]

In fact, upon warming, these metal atom-solvates aggregated to form nanoparticles of different shapes (Figure 3.2) These different shaped nanoparticles possessed vastly different reactivities and different reaction paths, with phosphine, and as catalysts for hydrogenation of 1,3-butadiene.[8]

These initial examples showed the way for this experimental method to be used to prepare many nanoparticles of metals,[9] as well as semiconductors, such as CdS, CdSe, and CdTe.[10,11]

3.2 AGGREGATION OF METAL ATOMS OR REACTIVE MOLECULES IN LOW-TEMPERATURE MATRICES/SOLVENTS

As reviewed in Chapter 5, and in several earlier reports, metal atom aggregation can be controlled by slow temperature increase in matrices of frozen argon, xenon, perfluorocyclobutane, THF, acetone, toluene, alkanes, and others.[12,13,14,15] In this way, spectra of various metal atom dimers, trimers, tetramers, and sometimes larger clusters have been obtained.

Herein, our main interest is in preparing nanoscale particles that contain many more atoms, generally thousands, or tens of thousands, of atoms. Indeed, the SMAD method has been employed successfully for preparing highly

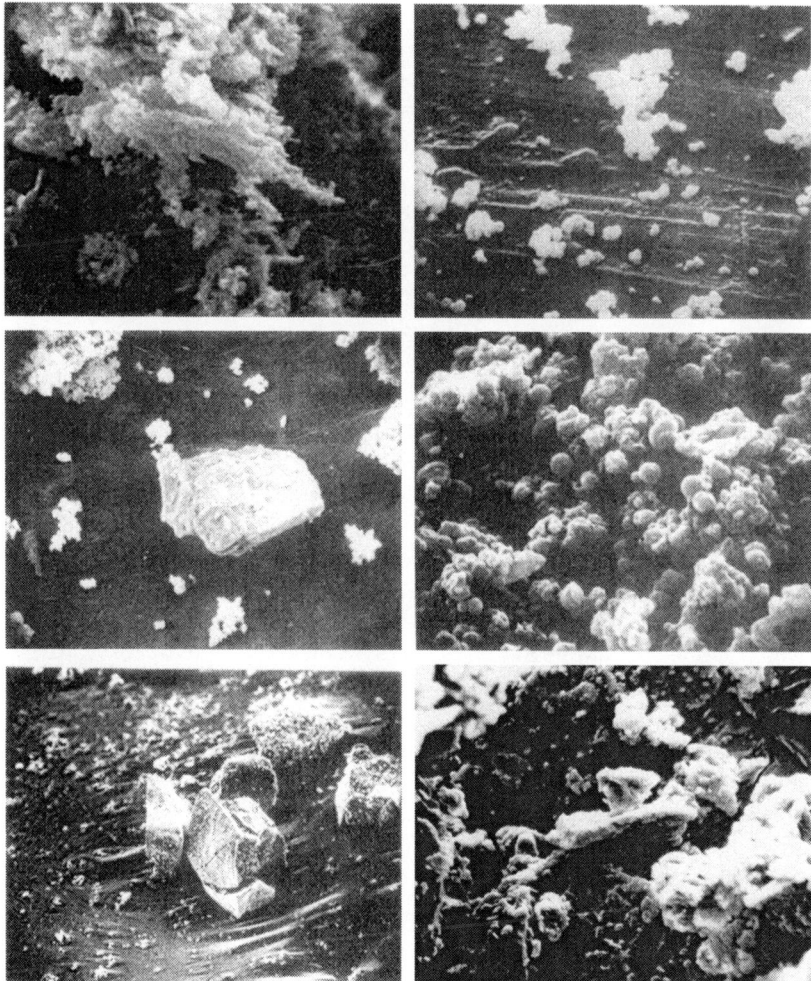

FIGURE 3.2 Nickel nanoparticles formed by warming Ni-(Solvent)$_x$ complexes; aggregation took place forming different shapes from different solvents.[4]

dispersed supported catalysts[16] and bimetallic particles[17] where crystallite sizes are in the 4- to 20-nm range. This size range required that thousands and tens of thousands of metal atoms migrate in cool matrices of solvents and form particles/crystallites that precipitate and are the final products.

3.2.1 Control of the Gold–Tin (Au–Sn) Bimetallic System[17]

This rationale for study of this bimetallic system is based on the goal of learning if nanocrystal growth could be controlled. The choice of Au/Sn was based on several things: (1) relatively low reactivity of Au and Sn atoms, (2) ease of evaporation of gold and tin and similarity in vapor pressures, (3) the fact that gold and tin form several well-characterized intermetallic compounds through the entire composition range of the two substituents, and (4) the ability to study both gold and tin by Mössbauer techniques (although this proved less useful than anticipated).

Experimental parameters such as the evaporation method, solvent polarity and viscosity, and warming rate during cluster formation were varied. Cluster/crystallite size and particle surface area were monitored. Additional information was gleaned from Mössbauer, Differential Scanning Calorimetry (DSC), and X-ray Photoelectron Spectroscopy (XPS).

An introduction to the Au–Sn system is appropriate before our experimental results are summarized. Considerable work on the interdiffusion of thin layers of gold on tin has been carried out—rapid diffusion occurs even at room temperature. Thus, the formation of AuSn is a very favorable process.[18,19] By this diffusion process, crystallite sizes of 20–60 nm for the AuSn have been observed.[20] Other stoichiometries have been observed if excess Au or Sn was present, although formation of such species as Au_3Sn, $AuSn_2$, and $AuSn_4$ is usually slower than formation of AuSn.[21]

In fact, many Au–Sn Stoichiometric ratios are possible. However, the favored species, according to film diffusion studies, appear to be AuSn, $AuSn_2$, $AuSn_4$, and Au_3Sn.

Electronic behavior of Au–Sn alloys has been studied by XPS.[22] Positive shifts in core level-binding energies for both Au $4f_{7/2}$ and Sn $3d_{5/2}$ were observed for $Au_{0.96}Sn_{0.04}$ and $AuSn_4$. The average shifts for AuSn were 0.9 eV for Au and 0.3 eV for Sn. Similarly, the valence band of this alloy was narrowed and shifted to higher binding energy.

These intermetallic species in molecular form have been examined by Knudson Cell-Mass Spectroscopy techniques. The dissociation energy D_0^0 of AuSn(g) was reported as 252.6 ± 7.2 kJ/mol and its heat of formation as 414.6 ± 7.2 kJ/mol, a rather substantial value.[23] Other molecular species such as $AuSn_2$, Au_2Sn, Au_2Sn_2, and $AuSn_3$ were also evaluated. These results further point out that the formation of these intermetallics is energetically quite favorable.

3.2.1.1 Experimental Results on Au Atom–Sn Atom Clusters in Cold Solvents

3.2.1.1.1 Cluster Growth

Would solvated atoms, allowed to slowly warm and nucleate to Au–Sn intermetallic alloy particles, show any selectivity toward growth of particular Au_xSn_y species? What properties of the solvent or other experimental parameters affect this selectivity? These are the questions we attempted to answer with the following experiments.

Indeed, with mono-metallic studies, we have found that solvent polarity and warm-up procedure can drastically affect resultant crystallite sizes,[5] magnetic properties,[2,24] and ability to form stable colloidal solutions.[25] However, a bimetallic such as Au–Sn had not been examined in this way before.

In the present study, gold and tin were usually evaporated from two separate crucibles, and the atoms cocondensed simultaneously with a high excess of the vapor of the solvent of choice. This resulted in the almost complete matrix isolation of the atoms at −196 °C (the ratio of Au:Sn could be determined by weighing the used crucibles later). Upon warming, the atom clustering took place, mainly in the cold liquid solvent as it melted.

A typical X-ray powder diffraction (XRD) pattern for the isolated, dry Au–Sn powder from an experiment where the Au:Sn ratio was 1.3:1 was compared with known diffraction patterns of Sn, AuSn, and Au_5Sn, and it could be seen that AuSn was the major component with the remaining smaller peaks assignable to Sn and Au_5Sn. So, some selectivity was exhibited, considering all the possibilities for other compositions, and statistical/random clustering does not appear to dominate.

Other experiments bear on this point more strongly. Layering experiments where gold and tin were evaporated at different times were carried out. In this way, a layer of frozen Sn atoms/acetone was covered by a layer of frozen Au/acetone with about the same molar ratio of Au:Sn overall. Upon warming and clustering, the same product distribution was obtained. These results show that the clustering process takes place mostly after solvent liquification and mixing of solvated atom solutions. If this were not true, large amounts of Sn and Au particles should have been formed due to the proximity of these species with each other. In addition, such results prove that the clustering process does not take place in the gas phase prior to codeposition and that pyrolysis of trace of amounts of solvent on the hot crucibles does not affect the clustering process.

The results of the layering experiments allowed our experimental design to be much simplified, if we so desired. That is, we placed gold and tin metals in the same crucible and simply evaporated all the metal. Tin evaporates at a lower temperature, and so, this is really another type of layering experiment. Again, the same products were obtained, further supporting that cluster formation of the intermetallic compounds occurs in the matrix during and after solvent liquification and complete mixing.

The next series of experiments dealt with solvent variation. It soon became apparent that the product mix of AuSn, Sn, and Au_5Sn did not change significantly with variation in solvent, if warm-up procedures were kept constant. Thus, selectivity was not changing, according to XRD. However, product morphology did change. For example, with ethanol, the largest crystallite size and lowest surface area was encountered. However, warm-up procedure had the most striking effect in that, in every case, a slow warm-up yielded smaller average AuSn crystallite sizes and larger surface areas. Moreover, this effect was minimized for solvents of higher viscosity but was marked for solvents of lowest viscosity.

These results tell us that the cluster growth process in this case does exhibit same selectivity toward the formation of the more stable phases. Also, for similar functional groups in the solvents, viscosity has a limited effect on the crystallinity or particle size. And although the warm-up rate had the most striking effect, it was noted that for the lowest viscosity solvents, the largest changes were caused by warm-up variation. These points suggest that the growth process occurs within a rather narrow temperature and viscosity range.

As an illustration, one could imagine a competition between processes: solvated atoms nucleate and begin to grow vs. the reaction of the growing particles with the solvent, and surface ligation occurs and particle growth is slowed or stopped. With earlier work on Ni-pentane nucleation and growth, this was expressed as a competition between particle growth and reaction with the pentane host.[2,26]

With bimetallic systems, it is more complex, and we express these concepts as shown in Figure 3.3. It is likely that reactions 1–3 are quite rapid when the solvent just liquifies, but these rates can be affected by large viscosity changes as would be encountered when a solvent melts to a very nonviscous liquid over a small temperature range. As the particle grows, and mobility decreases, (with simultaneous increase in temperature), reaction 3 would become slower. However, the rate of reaction 4 would increase rapidly once the minimum temperature for reaction with the solvent is reached. In other words, reactions 1, 2, and 3 occur below the temperature where reaction 4 becomes competitive.

The slower warming rate and minimization of large viscosity gradients tend to allow reaction 4 to compete better with reaction 3. Thus, solvent fragmentation/ligation (reaction 4) competes better over the slow warm-up period.

One additional interesting point is that the most polar solvents allow the growth of the largest crystallites. This, again, can be explained by the logical assumption that reactions 1, 2, and 3 become slower and more selective since displacement of more strongly ligating solvent ligands (due to higher polarity)

$$Au(solv) + Sn(solv) \longrightarrow AuSn(solv) \quad (1)$$

$$2\,Au(solv) + AuSn(solv) \longrightarrow Au_2Sn_2(solv) \quad (2)$$

$$Au_2Sn_2(solv) + AuSn(solv) \longrightarrow Au_nSn_n(solv) \quad (3)$$

$$Au_nSn_n(solv) + solv \longrightarrow Au_nSn_n(R)_x(solv) \quad (4)$$
$$\text{stabilized toward further growth}$$

•Mobility decreases with size. Therefore k_3 decreases.

•Mobility is low at high viscosities. Therefore, competes with k_3 more effectively.

•Interdiffusion of Au and Sn does not occur from particle to particle, because each particle is a separate entity due to the solvation shell.

FIGURE 3.3 A generalization of rate processes for cluster growth and cluster reactions with host solvent, solv = solvent molecule, R = fragment of solvent that serves as a ligand.

would be more difficult. Thus, these growth steps become more selective and crystal growth more favorable. In other words, fewer nucleation sites would be formed, and larger crystallites would grow.

These findings might be summed up by saying that slow warm-up causes a "milder" clustering process to occur, while fast warm-up causes a "wilder" process to occur.

3.2.2 Reactivity of Aggregates (Nanoparticles or Nanocrystals)

Initially, the idea of synthesizing solvated metal atoms, and allowing the low temperature solvates to melt and decompose, led to partial reactions with the solvents,[1,4,26,27,28] eventually leading to "highly active metal slurries." (The term "nanoparticle" or "nanocrystal" was not in use in 1979 or earlier). The metal particles were protected from further aggregation by solvent fragments.

The thrust of the work at that time was to show what chemical reactions occurred with (a) metal atoms, (b) solvated atoms, and (c) very small metal particles. Table 3.1 shows few examples regarding these reactivities.

Very few direct comparisons have been possible, but it is generally the case that atoms are most reactive, followed by solvated atoms, followed by their respective nanoparticles, and followed by larger particles. Sometimes, exceptions occur such as when at very low temperatures, Mg_2, Mg_3, and Mg_4 clusters were more reactive than Mg atoms with alkyl halides.[31]

3.2.3 Trapping and Stabilization

In more recent years, the SMAD process has been dedicated to nanoparticle synthesis and has moved away from "metal atom chemistry." For example, solvents were selected for gold atom (vapor) codeposition such that the warming, melting matrix would allow very small gold nanoparticles to form and would stabilize these particles. Although many solvents have been tested (pentane, toluene, ethanol, acetone), volatile ketones were found to be the best for gold and silver, while for copper, ethers such as THF and diglyme were the best.[9,25,32]

So, solvents behave as moderators of nanoparticle growth from solvated atoms, and solvent viscosities and rate of warming have effects on the final nanomaterials formed, and these results have been discussed in Section 3.2.1 with the gold–tin bimetallic system.

3.3 EXAMPLES OF USEFUL SYNTHESIS

3.3.1 Gold Nanoparticles

Stable colloidal solutions of gold nanoparticles in acetone, butanone, or pentanone, have been prepared and examined in the electron microscope, and the images are shown in Figures 3.4–3.6.

TABLE 3.1 Metal Reactions as Atoms, Solvated Atoms, or Metal Nanoparticles

Metal	Solvent	Reagent	Product	Reference
Cd nanoparticles	Several solvents	C_2H_5I	C_2H_5CdI	Klabunde[6]
Sn nanoparticles	Toluene	CH_3I	CH_3SnI_3 72%	Klabunde[6]
	Dioxane		42	Klabunde[6]
	THF		16	Klabunde[6]
	Hexane		0	Klabunde[6]
In nanoparticles	Diglyme	C_2H_5I	$EtInI_2$ 64%	Klabunde[6]
	Dioxane		41	Klabunde[6]
	Xylene		10	Klabunde[6]
	THF		9	Klabunde[6]
Ni atoms	PF_3	PF_3	$Ni(PF_3)_4$ 100%	Timms[29]
Ni atom solvate	Toluene	$P(OEt)_3$	$Ni[P(OEt)_3]_4$ 40%	Davis and Klabunde[26]
Ni Nanoparticles		$P(OEt)_3$	1%	Davis and Klabunde[26]
Ni Atoms	1,3-butadiene	1,3-butadiene	$(1,3\text{-butadiene})_3$Ni 39%	Klabunde et.al.[1]
Cu Nanoparticles	Toluene	C_6H_5I	$C_6H_5\text{-}C_6H_5$ (89%) + CuI	Ponce and Klabunde[30]
	Diglyme		52	Ponce and Klabunde[30]
	THF		80	Ponce and Klabunde[30]
	Pentane		70	Ponce and Klabunde[30]

As seen from the TEM images, the particle size and the degree of aggregation of the particles increase as the solvent is changed from the more polar acetone to the less polar pentanone. The colloidal solutions of these exhibit UV-vis absorption λ_{max} at ~550 nm with a broad tail out to 1000 nm. In each case, the particles look as though they have "necked" together, suggesting high reactivity and a tendency to bind together and grow, as expected for reactive particles attempting to lower surface energies.

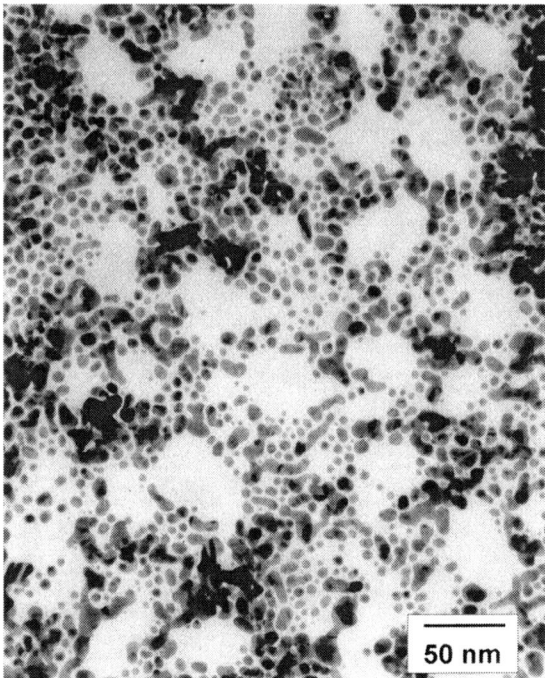

FIGURE 3.4 TEM micrograph of Au nanoparticles from acetone, prepared by the SMAD method.[33]

3.3.2 Silver and Copper

For silver, a jungle of necked particles was observed for Ag-acetone (Figure 3.7). Similarly, the Cu-toluene (acetone was too reactive to codeposit with copper) final SMAD product showed a tortuous interlinked nanostructure (Figure 3.8).

3.3.3 Other Metals

Indeed, many metals can be prepared in nanostructured form as "very active" metals and catalysts, including Al, Fe, Co, Ni, Cu, Zn, Ge, Pd, Ag, Cd, In, Sn, Au, and Pb.[2,12]

3.3.4 Binuclear Compounds

Even binuclear solids such as ZnS, CdS, CdSe, CdTe, PbS can be prepared as nanoparticles by the SMAD method. As an example, CdTe codeposited with pentane yielded a high surface area sample of $172\,m^2/g$, with toluene $57\,m^2/g$, and with THF $27\,m^2/g$. Figure 3.9 shows a TEM image of the CdSe-pentane SMAD sample. Note the similarities to the metal nanoparticle SMAD samples, although the fundamental building block particles are rather small, 5–10 nm.

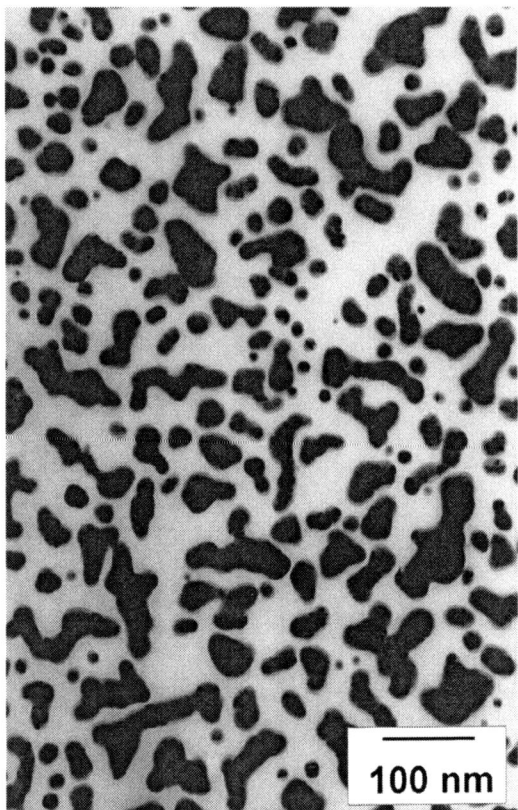

FIGURE 3.5 TEM micrograph of Au nanoparticles from 2-butanone, prepared by the SMAD method.[33]

It should be emphasized again that all of these SMAD-prepared samples are quite chemically reactive, especially with oxygen, water, or certain chemicals. For example, the Au-butanone sample proved to be an excellent catalyst for hydrolysis of alkylsilanes ($RSiH_3$), producing many siloxane nanowires under mild conditions.[37] Another example is Ag-acetone particles, when placed in water, are very biocidal toward bacteria and viruses.[38]

3.4 DIGESTIVE RIPENING OR "NANOMACHINING"

An important discovery coming out of the experimental work on the chemistry of nanoparticles, especially SMAD particles, is that certain particle sizes seem to be thermodynamically preferred. Indeed, the reactivity of SMAD particles allows them to be manipulated by added ligands. The best example is an alkyl thiol such as $C_{12}H_{25}SH$ attacking Au-acetone SMAD particles, which were allowed to contact the thiol immediately upon meltdown of the acetone to a

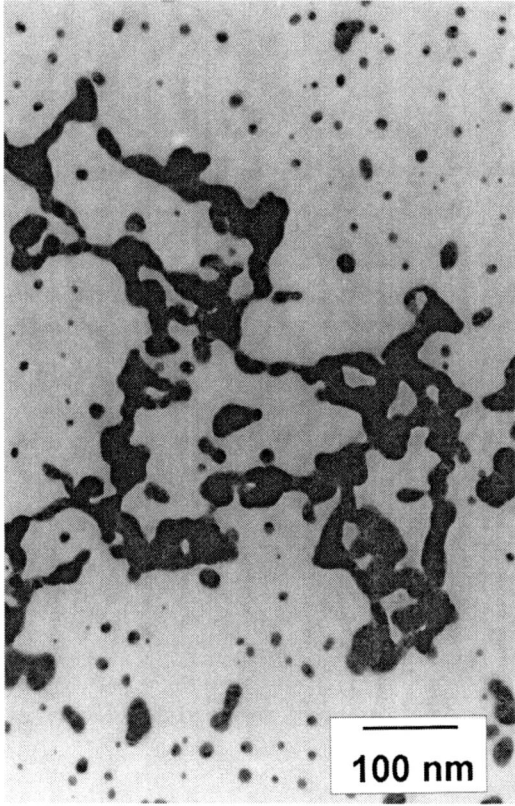

FIGURE 3.6 TEM micrograph of Au nanoparticles from 3-pentanone, prepared by the SMAD method.[33]

toluene-thiol mixture. In this way, the thiol displaced the acetone solvent forming a crude gold nanoparticle system that had some solubility in toluene. From TEM images, it can be seen that a field of large and small particles is present (Figure 3.10).

The long-chained thiol ($C_{12}H_{25}SH$) brings the gold particles up as a red-purple colored colloidal solution. Upon refluxing this solution, a remarkable thing happens: all of the particles eventually become the same size.

Interesting results are obtained by TEM from the digestive ripened colloids after they cool down from reflux temperature. The TEM micrographs of colloids cooled down for a different amount of time are shown in Figure 3.11. The amazing result is that the particles predominantly organize the TEM grid in large 3-D structures in only about 15 min. after the "digestive ripening" process is finished. Even larger 3-D superlattices (>3 µm) are observed after 1 day and after ~2 months (Figure 3.11). Noteworthy is the near-perfect organization of the Au particles in Figure 3.11c. The high degree of ordering in the superlattice

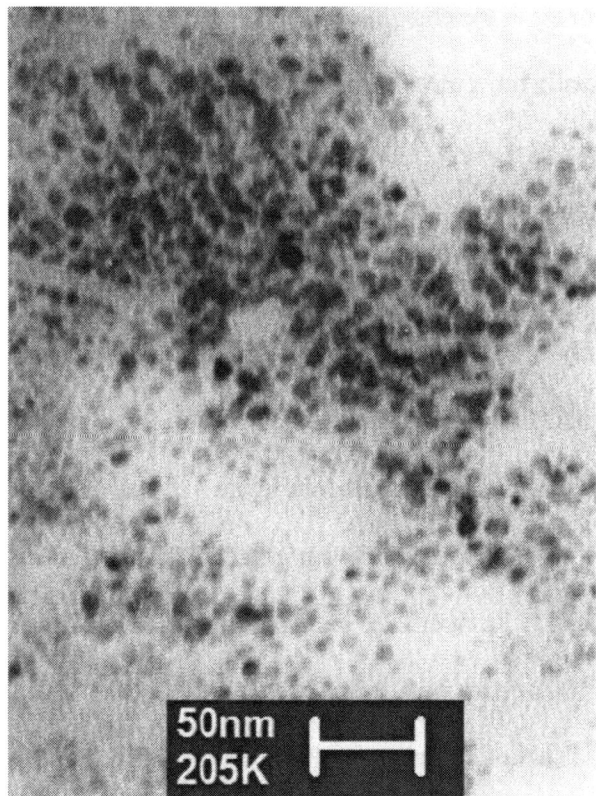

FIGURE 3.7 TEM micrograph of Ag nanoparticles from acetone, prepared by the SMAD method.[34]

structures is proven by the small-angle diffraction pattern given in the inset of Figure 3.12.

One of the most interesting features of the synthetic sequence reported herein is the "digestive ripening" step, and the mechanism for this remarkable process is not entirely clear. Only a few useful facts are known. First of all, nanoparticles are the necessary starting material, that is, normal gold powder is not susceptible to digestive ripening, showing again that nano-sized particles are intrinsically and more chemically reactive than bulk samples. The ripening process probably involves the dissolution of surface atoms or clusters of atoms by the ligand molecules, which are in excess in the solution. This phenomenon is best known for gold nanoparticles, but other metals can behave in a similar way with the appropriate ligands. So, a "dissolving and re-precipitation" process is likely, and reactive sites (corners, edges) would be the first atoms susceptible. The most intriguing question, though, is "why all the particles become the same size, and why 4–5?" One possible rationale is that at 4–5 nm, the thiol

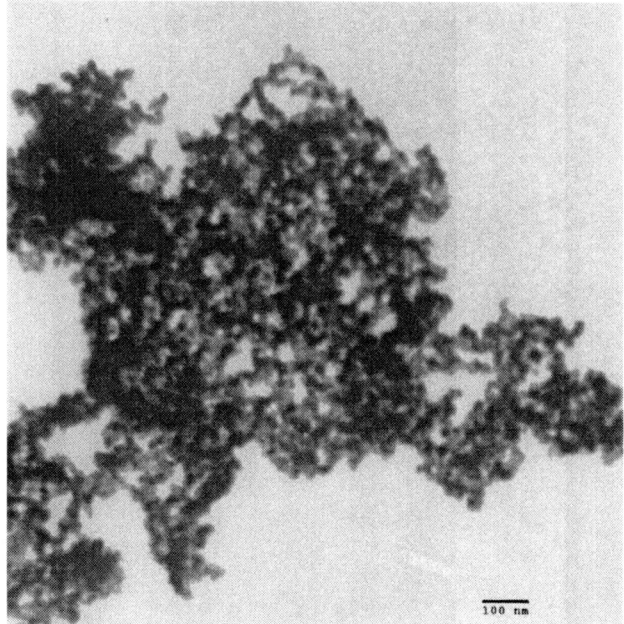

FIGURE 3.8 TEM micrograph of Cu nanoparticles from toluene, prepared by the SMAD method.[35]

head groups and the 12-membered carbon chains are close packed in the most ordered fashion (which would be akin to formation of a crystalline surface coating) for the curvature of a 4–5 nm gold particle.

This process of bringing a polydisperse gold colloid to a monodisperse one is different from "Ostwald ripening" where large particles grow at the expense of small ones. We coined the term "digestive ripening," and this process is controlled by thermodynamics rather than kinetics. Also, digestive ripening is broader in context than the SMAD method of synthesis since digestive ripening can occur with gold prepared by inverse micelle or other materials, as well as SMAD prepared gold nanoparticles.

What about other elements or compounds? Numerous other elements have been found to be susceptible to this thermodynamic processing. These include Cu, Ag, Pd, In, Co, and Mg. Even binuclear solids are susceptible, including Fe_2O_3, CdS, CdSe, and CdTe.[39–50] However, much remains to be understood.

One publication is devoted to a theoretical understanding of digestive ripening.[51] This report found that a minimum in the potential energy curve could be achieved for a certain sized particle if the particles possessed a charge. It is possible that charged particles are involved as is the case for many colloidal solutions. However, we know that the surface-stabilizing ligands are critically important, and this aspect was not dealt with, in the report discussed herein.[51]

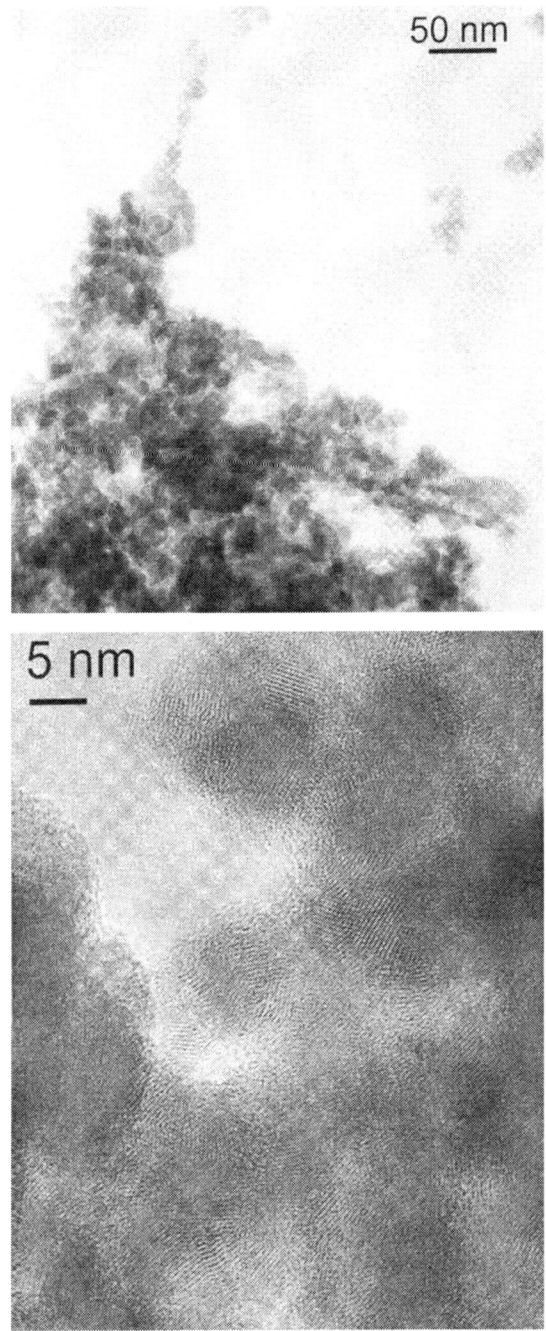

FIGURE 3.9 TEM micrograph of CdTe from pentane solvent, prepared by the SMAD method.[36]

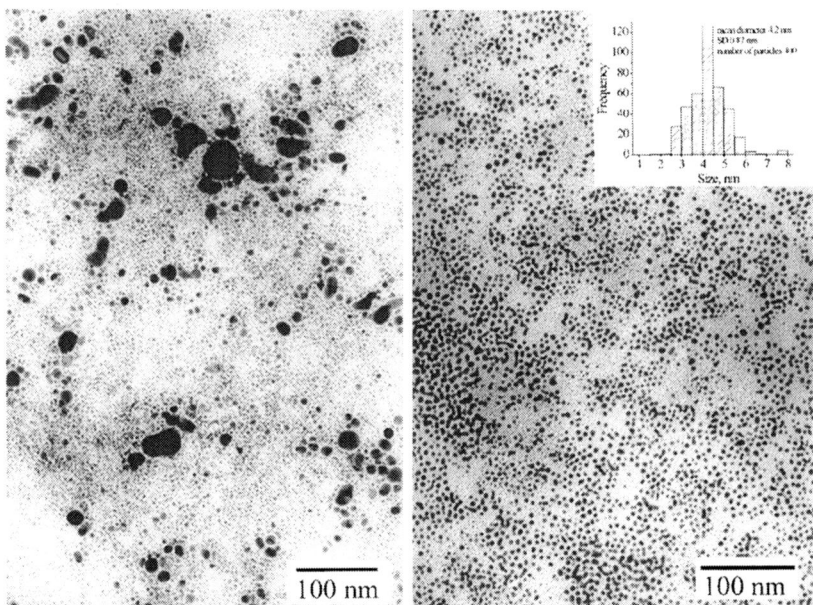

FIGURE 3.10 TEM of an Au-SMAD from acetone → toluene –RSH mixed at the bottom of the reactor.[33]

Clearly, much more needs to be done in order to understand this fascinating digestive ripening process, also called "Nanomachining."

3.5 RODS, WIRES, AND STARS

Nanorods of gold can be shaped by selective ligand mixtures when gold ions are reduced.[52–54]

Another interesting development is the variation in catalytic properties based on different nanoparticle shapes.[8] For example, there is an interesting report on a catalytically active platinum multiarmed nanostar single crystal,[55] and discussions of the importance of shape, catalytic activity, and recycling potential.[56]

Another recent finding is that the shape of CuO nanoparticles affects diffuse reflectance properties, greatly.[57] These nanorods of CuO about 2 μm in length and 0.5 μ thick were quite effective in diffuse reflectance of near infrared light, while spherical particles and nanorods of metallic Cu were not nearly as effective.

Indeed, nanorods and nanostars often exhibit quite unique optical properties, and further studies are ongoing in many laboratories, due to the potential to be used as IR and visible obscuration or "cloaking."

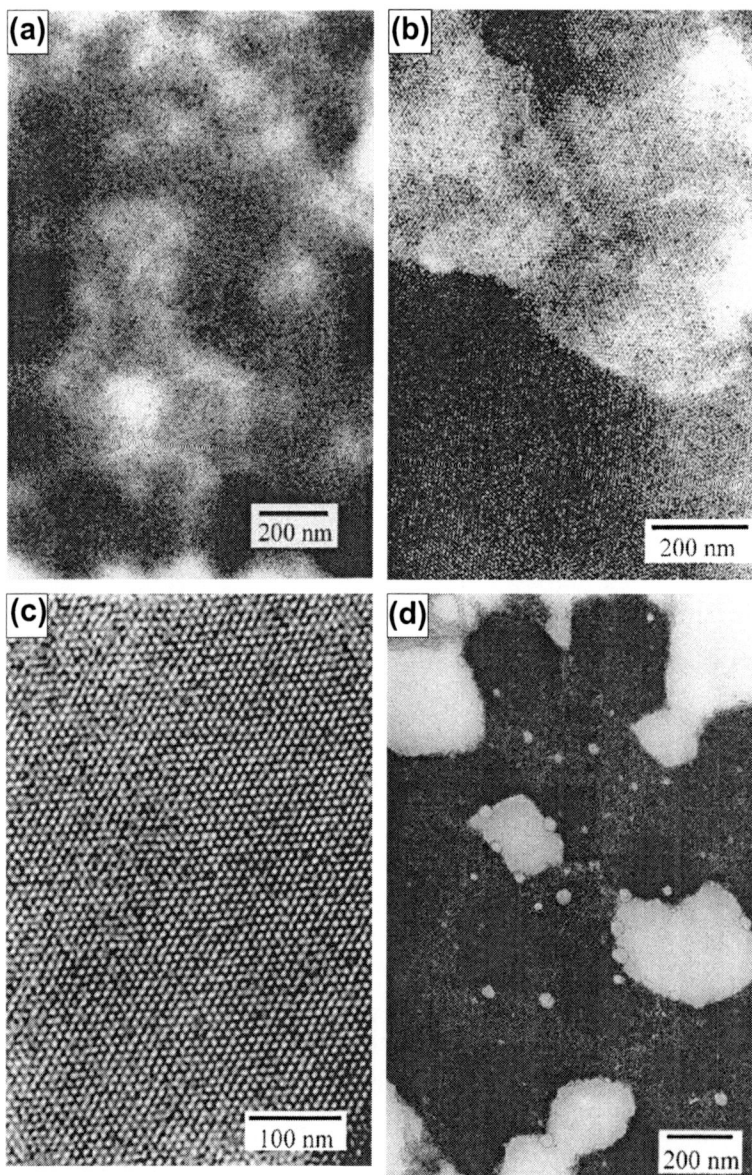

FIGURE 3.11 TEM micrographs of dodecanethiol-stabilized gold particles prepared by the SMAD process at different time intervals since the digestive ripening process (a) After 15 minutes, (b) After 1 day, (c) After 1 day, and (d) After 2 months.[33]

Chapter | 3 Solvated Metal Atom Dispersion (SMAD) for Making Metal   71

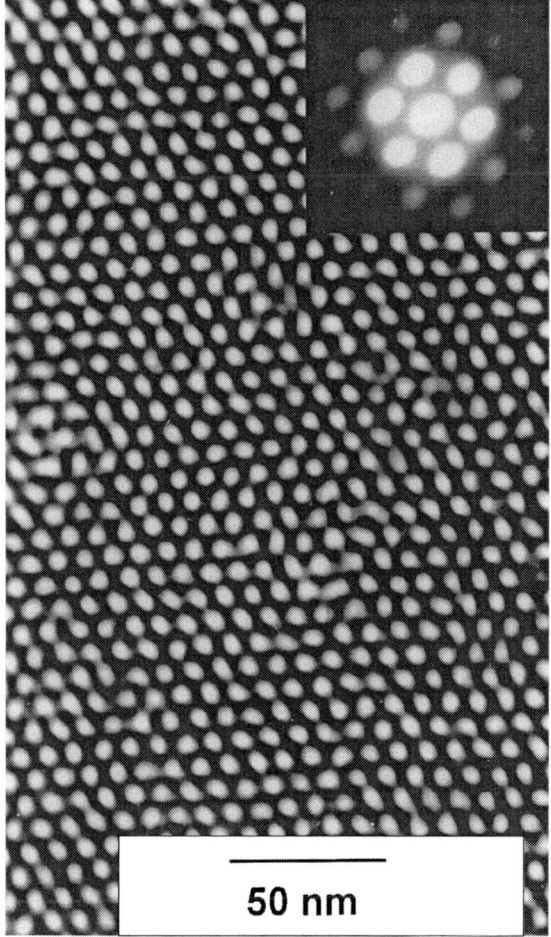

FIGURE 3.12 TEM micrograph of superlattice of dodecanethiol-stabilized gold nanoparticles, prepared by the SMAD process and the corresponding small-angle diffraction pattern.[33]

REFERENCES

1. Klabunde, K. J.; Timms, P. L.; Skell, P. S.; Ittel, S. *Inorg. Syn.* **1979,** *19,* 59.
2. Klabunde, K. J. *Chemistry of Free Atoms and Particles*; (238 page monograph); Academic Press: New York, N.Y, 1980.
3. Radonovich, L.; Zuerner, C.; Efner, H. F.; Klabunde, K. J. *Inorg. Chem.* **1976,** *15,* 2976.
4. Davis, S. C.; Severson, S. J.; Klabunde, K. J. *J. Amer. Chem. Soc.* **1981,** *103,* 3024.
5. Klabunde, K. J.; Efner, H. F.; Murdock, T. O.; Roppel, R. *J. Amer. Chem. Soc.* **1976,** *98,* 1021.
6. Klabunde, K. J. *Ann. New York Acad. Sci.* **1977,** *295,* 83.
7. Klabunde, K. J.; Murdock, T. O. *J. Org. Chem.* **1979,** *44,* 3901.
8. Klabunde, K. J.; Davis, S. C.; Hattori, H.; Tanaka, Y. *J. Cat.* **1978,** *54,* 254.
9. Stoeva, S.; Klabunde, K. J.; Sorensen, C.; Dragieva, I. *J. Am. Chem. Soc.* **2002,** *124,* 2305.

10. Cingarapu, S.; Yang, Z.; Sorensen, C. M.; Klabunde, K. J. *Chem. Mater.* **2009**, *21*, 1248.
11. Heroux, D.; Ponce, A.; Cingarapu, S.; Klabunde, K. J. *Adv. Funct. Mater.* **2007**, *17*, 3562.
12. Klabunde, K. J. *Free Atoms, Clusters, and Nanoscale Particles*; (311 page monograph); Academic Press: San Diego, CA, 1994.
13. Hauge, R. H.; Fredin, L.; Kafafi, Z. H.; Margrave, J. L. *Appl. Spectrosc.* **1986**, *40*, 588.
14. Godber, J.; Huber, H.; Ozin, G. H. *Inorg. Chem.* **1986**, *25*, 2909.
15. Segeev, G. B. *Nanochemistry*; Elzevier Pub.: Amsterdam, 2006.
16. Klabunde, K. J.; Li, Y. X.; Tan, B. J. *Chem. Mater.* **1991**, *3*, 30.
17. Wang, Y.; Li, Y. -X.; Klabunde, K. J. Platinum-Tin and Gold-Tin Bimetallic Particles Prepared from Solvated Metal Atoms. Structure and Catalysis. In *Selectivity in Catalysis. ACS Symposium Series 517, Chapter 10*; Davis, M. E.; Suib, S. L., Eds.; 1993; pp 136–155.
18. Chang, C.; Callcott, A. T.; Arkawa, E. T. *J. Appl. Phys.* **1982**, *53*, 7362.
19. Neel, S.; Arkawa, E. T.; Inagaki, T. *J. Appl. Phys.* **1984**, *55*, 4132.
20. Buene, L.; Falkenberg-Arell, H.; Tafto, J. *Thin Solid Films* **1980**, *67*, 95.
21. Simic, V.; Marinkovic, Z. J. *Less Common Metal* **1983**, *95*, 259.
22. Friedman, R. M.; Hudis, J.; Perlman, M. L.; Watson, R. E. *Phys. Rev. B* **1973**, *8*, 2433.
23. Kingcade, J. E., Jr.; Gingerich, K. A. *J. Chem. Phys.* **1986**, *84*, 3432.
24. Kernizan, C. F.; Klabunde, K. J.; Sorensen, C. M.; Hadjapanayis, G. C. *Chem. Mater.* **1990**, *2*, 70.
25. Lin, S. J.; Franklin, M. J.; Klabunde, K. J. *Langmuir* **1986**, *2*, 259.
26. Davis, S. C.; Klabunde, K. J. *J. Am. Chem. Soc.* **1978**, *100*, 5973.
27. Murdock, T. O.; Klabunde, K. J. *J. Org. Chem.* **1976**, *41*, 1076.
28. Klabunde, K. J.; Ralston, D.; Zoellner, R.; Hattori, H.; Tanaka, Y. *J. Cat.* **1978**, *55*, 213.
29. Timms, P. L. *J. Am. Chem. Soc.* **1970**, *A*, 2526.
30. Ponce, A. A.; Klabunde, K. J. *J. Mol. Catal. A: Chem.* **2005**, *225*, 1.
31. Imizu, Y.; Klabunde, K. J. *Inorg. Chem.* **1984**, *23*, 3602.
32. Smetana, A. B.; Klabunde, K. J.; Sorensen, C. M. *J. Colloid Interface Sci.* **2005**, *284*, 521.
33. Stoeva, S. Ph.D. Thesis, Kansas State University, 2003, 136–138.
34. Smetana, A. Ph.D. Thesis, Kansas State University, 2006, 3, 12.
35. Ponce, A. Ph.D. Thesis, Kansas State University, 2002, 67.
36. Heroux, D. Ph.D. Thesis, Kansas State University, 2004, 85.
37. Prasad, B. L.V.; Stoeva, S. I.; Sorensen, C. M.; Zaikovski, V.; Klabunde, K. J. *JACS* **2003**, *125*, 10488.
38. Smetana, A.; Klabunde, K. J.; Marchin, G. R.; Sorensen, C. M. *Langmuir* **2008**, *24*, 7457.
39. Prasad, B. L.V.; Stoeva, S. I.; Sorensen, C. M.; Klabunde, K. J. *Langmuir* **2002**, *18*, 7515.
40. Son, S. U.; Jang, Y.; Yoon, K. Y.; Kang, E.; Hyeon, T. *Nano Lett.* **2004**, *4*, 1147.
41. Prasad, B. L.V.; Stoeva, S. I.; Sorensen, C. M.; Klabunde, K. J. *Chem. Mater.* **2003**, *15*, 935.
42. Horinouchi, S.; Yamanoi, Y.; Yonezawa, T.; Mouri, T.; Nishihara, H. *Langmuir* **2006**, *22*, 1880.
43. Ganesan, M.; Freemantle, R. G.; Obare, S. O. *Chem. Mater.* **2007**, *19*, 3464.
44. Lin, X. M.; Sorensen, C. M.; Klabunde, K. J. *J. Nanopart. Res.* **2000**, *2*, 157.
45. Stoeva, S. I.; Klabunde, K. J.; Sorensen, C. M.; Dragieva, I. *JACS* **2002**, *124*, 2305.
46. Cingarapu, S.; Yang, Z.; Sorensen, C. M.; Klabunde, K. J. *J. Chem. Mater.* **2009**, *21*, 1248.
47. Heroux, D.; Ponce, A.; Cingarapu, S.; Klabunde, K. J. *J. Adv. Funct. Mater.* **2007**, *17*, 3562.
48. Smetana, A. B.; Klabunde, K. J.; Sorensen, C. M. *J. Colloid Interface Sci.* **2005**, *284*, 521.
49. Stoeva, S. I.; Smetana, A. B.; Sorensen, C. M.; Klabunde, K. J. *J. Colloid Interface Sci.* **2007**, *309*, 94.
50. Kalidindi, S. B.; Jagirdar, B. R. *Inorg. Chem.* **2009**, *48*, 4524.
51. Lee, D. K.; Park, S.; Lee, J. K.; Hwang, N. *Acta Materialia* **2007**, *55*, 5281.

52. Sau, T. K.; Murphy, C. J. *Langmuir* **2005,** *21,* 2923.
53. Orendorff, C. J.; Sau, T. K.; Murphy, C. J. *Small* **2006,** *2,* 636.
54. Sau, T. K.; Murphy, C. J. *Philosophical Magazine* **2007,** *87,* 2143.
55. El-Sayed, M. A. *J. Am. Chem. Soc.* **2008,** *130,* 4590.
56. Narayanan, R.; El-Sayed, M. *Top. Catalysis* **2008,** *47,* 15.
57. Shrestha, K. M.; Sorensen, C. M.; Klabunde, K. J. *J. Phys. Chem. C* **2010,** *114,* 14368.

Chapter 4

Experimental Techniques

Chapter Outline

4.1 Electron Microscopy 76	4.4.4 Photoelectron Spectroscopy 83
4.1.1 Transmission Electron Microscopy 77	4.4.5 Nuclear Magnetic Resonance (NMR) Spectroscopy 83
4.1.2 Scanning Electron Microscopy 77	4.4.6 Ultra Violet–Visible Spectrometry (200–800 nm) 84
4.2 Probe Microscopy 78	4.4.7 Dynamic Light Scattering 84
4.3 Diffraction Techniques 81	**4.5 Comparison of Spectral Techniques Used for Elemental Analysis** 85
4.3.1 X-ray Diffraction 81	
4.3.2 Neutron Diffraction 82	
4.4 Miscellaneous Techniques 82	
4.4.1 EXAFS 82	
4.4.2 X-ray Fluorescence Spectroscopy 82	
4.4.3 Mass Spectrometry 83	

This chapter is written based on Refs 1–14.

Sizes and physicochemical properties of the nanoparticles are closely interrelated and, moreover, are of paramount importance for studying their chemical transformations. Furthermore, there are different approaches to studying the properties of the particles on the surface and in the bulk.

The main techniques used for determining sizes and certain properties of the nanoparticles in the gas phase are as follows:

- ionization by photons and electrons, followed by an analysis of the obtained mass spectra by means of quadrupole and time-of-flight mass spectrometers;
- atomization and selection of neutral clusters with respect to masses; and
- electron transmission microscopy on grids (information on sizes and shapes of the particles).

To gain information on the particles located on the surface, the following techniques are used:

- transmission and scanning electron microscopies (TEM, SEM, information on the size/shape of the particles, their distribution, and topology);
- electron diffraction (information on size, phases (e.g. solid/liquid), structure, and bond lengths);
- STM (determination of size, shape, and the internal structure of the particles);
- adsorption of gases (information on the surface area);
- photoelectron spectroscopy (determination of the electronic structure);
- conductivity (information on the conduction band, percolation, and topology).

Miscellaneous techniques are also used for determining sizes and certain properties of the nanoparticles in the bulk or within a matrix.

The methods using TEM, SEM, conductivity measurements, and electron diffraction techniques provide information on the particles in the bulk, i.e. the data analogous to those obtained for the particles on the surface.

Several other techniques are used for studying the particles in the bulk. For example, X-ray diffraction can be used for determining the particle sizes and internal structures.

Extended X-ray absorption fine structure (EXAFS) technique makes it possible to measure the particle sizes. Electron paramagnetic(spin) resonance (EPR) and nuclear magnetic resonance (NMR) provide information on the electronic structure. Mössbauer spectroscopy, i.e. the resonance absorption of gamma quanta by atomic nuclei in solids (gamma resonance), is actively used for gaining insight into the internal structure of a number of elements, especially such important elements as iron. The energy of a gamma quantum is small (~150 keV), and its absorption excites a nucleus. The resonance condition is the equality of the nucleus excitation energy to the energy of a quantum transition, i.e. to the difference between the nucleus internal energy in the excited and ground states. The transition energy depends on the nature of a nucleus and gives insight into the microscopic structure of solids. The method cannot be applied to all elements; however, it provides valuable information on Fe^{57}, Sn^{119}, and Te^{125}.

4.1 ELECTRON MICROSCOPY

Here, we consider microscopy in sufficient detail, because it is the major technique for determining the nanoparticle size. As a rule, this concerns electron microscopy, which employs beams of accelerated electrons and also different versions of probe microscopes.

Electron microscopy, in turn, has the following two main directions:

- TEM, in which the high-resolution electron microscopy is currently a separate division;
- SEM.

4.1.1 Transmission Electron Microscopy

A sample shaped as a thin film is transilluminated by a beam of accelerated electrons with an energy of 50–200 keV in a vacuum of ca. 10^{-6} mmHg. Those electrons that were deflected at small angles by atoms in a sample and passed through the sample get into a system of magnetic lenses to form a bright-field image of the sample internal structure on a screen and a film. A resolution of 0.1 nm was achieved, which corresponds to a magnification factor of 10^6. The resolution depends on the nature of the sample and the method of its preparation. Usually, films of 0.01-μm thickness are studied; the contrast range can be extended using carbon replicas. Modern ultramicrotomes allow obtaining sections 10–100-nm thick. Metals are studied as thin foils. Transmission microscopes make it possible to obtain diffraction patterns, which provide information on the crystalline structure of a sample.

4.1.2 Scanning Electron Microscopy

This technique is largely used for studying the surface particles. An electron beam is constricted by magnetic lenses to give a thin (1–10 mm) probe, which travels point by point over a sample progressively, thus scanning the latter. The interaction of electrons with the surface generates several types of emission:

- secondary and reflected electrons;
- transmitted electrons;
- X-ray slowing-down radiation; and
- optic radiation.

Any of the radiation types listed above can be registered and converted into electrical signals. The signals are amplified and fed to a cathode-ray tube. A similar situation occurs in TV kinescopes. Images are formed on the screen and photographed. The major advantage of this technique is the great body of information it provides; its significant drawback concerns long scanning times. High resolution is possible only for low scanning rates. The method is usually employed for particles measuring more than 5 nm. A restriction on the sample thickness limits the method of application. For electrons with energies of 100 keV, the sample thickness should be about 50 nm. To prevent destruction of samples, special procedures are used for sample preparation. Moreover, the possible effect of electron emission on the samples should be taken into account—for instance, the electron-beam-induced aggregation of the particles.

One method used for preparation of samples consists in employing ultramicrotomes (their use is problematic for the cases of nonuniform deposition, particularly in islets). Chemical methods are also applied, especially matrix dissolution. The general view of a histogram obtained in microscopic studies often depends on the way the sample was prepared.

In the 1980s, a great breakthrough was observed in electron microscopy. Microscopes equipped for the computer analysis of the elemental composition were developed on the basis of energy-loss spectrometers. The energy-loss spectrometry was used in combination with TEM and SEM. A rearrangement of the magnetic-prism system allowed one to regulate the image contrast, which depends on the incidence angle, atomic number, and the reflection factor. Modern devices make it possible to obtain selective images of elements from boron to uranium with a resolution of 0.5 nm and sensitivity up to 10^{-20} g, which amounts to (for example) 150 atoms for calcium. High-resolution electron microscopy provides insight into such objects.

An important stage in the development of electron microscopy was associated with elaboration of computerized techniques for processing images, which allowed histograms over shapes, orientations, and sizes to be obtained. Now, it is also possible to separate details of the structure, statistically process information, estimate local microconcentrations, and determine lattice parameters. Built-in processors make it possible to exercise versatile control over microscopes.

4.2 PROBE MICROSCOPY

Another breakthrough in microscopy was associated with the development of scanning probes. In 1981, G. Binnig and H. Rohrer created STM, and in 1986, they were awarded the Nobel prize. The microscope allows the study of surfaces with nanoscale and subnanoscale resolution. The principle of gaining information on the properties of surfaces under study is general for all types of probe microscopes.

The main tool employed in these microscopes is a probe that is brought into either mechanical or tunneling contact with the surface. In doing so, the equilibrium between probe–sample interactions is established. This equilibrium can involve the attractive and repulsive forces (electrical, magnetic, Van der Waals) and the exchange of tunneling electrons and photons.

Upon establishment of the equilibrium, scanning is started. The probe moves line by line over a definite surface area determined by the number of lines scans, their length, and the interline spacing. The probe is driven by a piezomanipulator whose dimensions change under the effect of the applied potential difference, which allows one to shift a sample in three directions (Figure 4.1).

Now, we briefly discuss the general principles of probe microscopes. All the scanning probe microscopes are characterized by the presence of a certain selected type of interaction between the probe and a sample, which is used by a feedback system for fixing the probe–sample distance (d) in the course of scanning. To provide a high resolution, the intensity of this interaction should depend on the d. For example, in atomic force microscopes, this condition is satisfied by the repulsive forces between edge atoms on the probe and the sample; in tunneling microscopes, the exponential increase in the tunneling current

Chapter | 4 Experimental Techniques

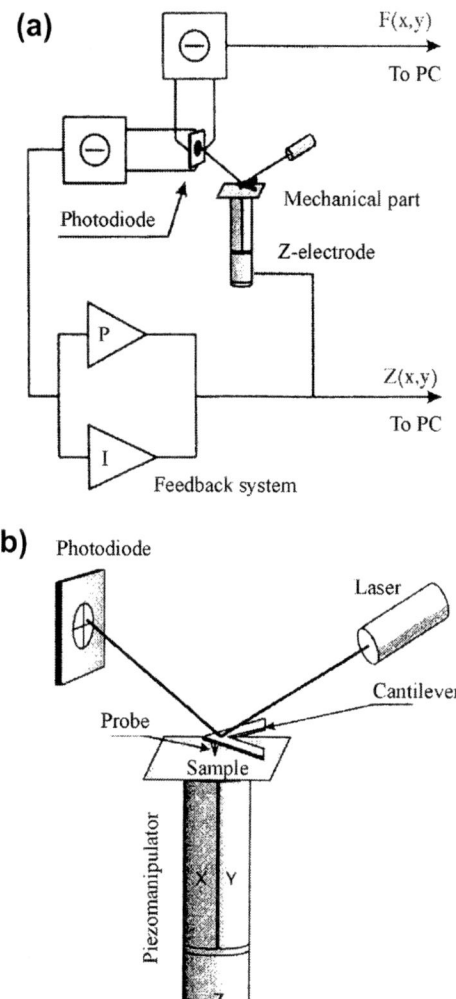

FIGURE 4.1 (a) General scheme of the operation of a probe microscope and (b) AFM mechanism.

with a decrease in the tunneling gap meets this condition, which allows one to achieve subnanoscale resolution (10^{-2} nm) for these devices.

In the course of scanning (the probe's motion in the *XY* plane), the feedback system shifts the probe in the direction *Z*, thus maintaining the signal at a given level corresponding to the working interaction amplitude. We designate the latter interaction as *A* (*X*, *Y*, *Z*). Signals at plates *X*, *Y*, and *Z* of the piezomanipulator are set by the computer. During scanning, the reproduction of the *A* (*X*, *Y*, *Z*) dependence by *Z* variations is equivalent to finding a dependence $Z|A = \text{const}$

(X, Y) that correlates with the local topographical features of the surface under study. Such dependence may be called the surface topography obtained in a mode of constant interaction A. If the intensity of interaction A (X, Y, Z) is different at different surface points, the detected picture is more complicated, representing a superposition of maps of the surface topography and the surface distribution of the A intensity.

In addition to the surface topography under constant-interaction conditions, scanning probe microscopy makes it possible to obtain a wide spectrum of other dependences $\Phi \mid A =$ const (X, Y), which provide valuable information on different surface properties. Here, the function Φ (X, Y) is measured under a condition of a constant interaction A, hence, under an approximation of a constant probe–sample distance, and can reflect the magnitude of some interactions different from A.

The principle of Φ (X, Y) variation under a condition $A =$ const. is used in various probe microscopes. The latter include different modifications such as the magnetic force microscope, operation of which is based on varying the forces between a magnetized probe and a surface with intrinsic magnetic properties; the near-field microscope which can detect the electromagnetic field that passes through a miniature diaphragm placed in the near-field zone of the source; and the electrostatic force microscope in which a conductive charged probe interacts with the sample. In all these microscopes, a corresponding interaction is analyzed when a constant probe–sample gap is maintained according to closed- and patched contact modes.

A circuit shown in Figure 4.1 is employed in atomic force microscopes for fixing the force interaction and maintaining it at a required level. The probe is adjusted to a free, unfixed end of a flexible arm-cantilever. When the probe approaches a sample or contacts it, the force interaction makes the cantilever bend, the magnitude of this bend being recorded by a precision transducer. The bend magnitude determines the contact force, and its maintenance at a required level in the course of scanning allows the surface profile to be reproduced. On the display, atoms look as hemispheres.

Most atomic force microscopies (AFMs) are equipped with optical sensors. A laser beam incident at an angle to the lever surface is reflected into the center of a four-section photodiode. A bend of the lever induces a difference between signals from the corresponding photodiode areas. The difference signal from the right and left segments, which corresponds to friction forces during scanning, is fed to the computer and reproduced on its display. The difference signal from the top and bottom segments, which passes through proportional and integration feedback circuits, is compared with the reference signal and fed to the Z-electrode of piezomanipulator. As a result, the sample is shifted in the vertical direction. The signal is also fed to the computer and the display, providing information on a surface under study.

AFM allows using a patched contact mode. For this purpose, an additional piezomanipulator is employed, which generates forced vibrations of the lever.

The forces of probe–sample interaction are complicated with regard to both their physical nature and performance. They are determined by the surface and geometrical properties of the materials that constitute the probe and the sample and also by the properties of the medium in which the study is carried out. For instance, studying polyparaxylylene films is complicated by the probe sticking in the films. The pressure of the probe is high, reaching 10^9 Pa, and can exceed the ultimate strength of many materials. Mica and graphite usually serve as supports. These materials easily scale and have smooth surfaces. AFM allows nondestructive measurements to be carried out. This is explained by the fact that the local pressure is distributed over three directions and the effective time is small, of an order of magnitude of 10^{-5} s. However, scientists are trying to reduce the probe–sample interaction forces, which is the central problem to date. The possible solutions may involve the choice of adequate media for the studies and the use of patched contact modes. At present, it is widely believed that probe microscopy has posed more problems than previously had been suspected.

4.3 DIFFRACTION TECHNIQUES

These techniques include diffraction of X-rays and neutrons and are less general when compared with electron microscopy. At the same time, the analysis of diffraction reflexes induced by atomic structures of the separate particles can be used for studying very small particles. The reflex angle width $\Delta\theta$ increases with an increase in the particle size ($\Delta\theta \sim 1/R$, i.e. the Scherrer effect). The smaller sizes correspond to smaller numbers of lattice planes that give rise to interference of the diffraction spot, while in larger clusters diffraction rings are usually observed.

4.3.1 X-ray Diffraction

When interacting with crystals, metal particles, and molecules, X-ray are scattered. An initial beam of rays with a wavelength $\lambda \sim 0.5 - 5$ Å gives rise to secondary rays with the same wavelength, the directions and intensities of which are related to the structure of scattering samples. The intensity of a diffracted ray also depends on the sizes and shapes of the particles. Polycrystalline particles give rise to secondary ray cones, each cone corresponding to a certain family of crystal planes. For small and abundant crystals, the cone is continuous, which results in a nonuniformly darkened ring.

A crystal represents a natural diffraction grating with strict periodicity. The crystals for studies should have sizes of ca. 0.1 mm and perfect structures. To elucidate a structure of average complexity, which contains 50–100 atoms in a unit cell, intensities of hundreds and even thousands of diffraction reflections are measured. This procedure is accomplished by means of microdensitometers and diffractometers controlled by computers. Earlier, these operations took months, but today they can be carried out in a single day.

When studying amorphous materials and incompletely ordered particles in polycrystals, X-ray diffraction allows one to determine the phase composition, size, and orientation of grains (texture).

The method of low-angle scattering, which allows studying spatial heterogeneities with sizes of 5–10 Å, is widely applied to date. It is used for studying porous, finely dispersed materials and alloys.

The determination of atomic structure, which involves size and shape estimation and assignment to a symmetry group, represents a complex analysis and cumbersome mathematical processing of intensities of all diffraction reflections. X-ray diffraction of materials embraces methods based on X-ray diffraction techniques for studying equilibrium and nonequilibrium states of materials, phase compositions, phase diagrams, residual stresses, etc.

4.3.2 Neutron Diffraction

A neutron is a particle the properties of which make it suitable to be used in the analysis of various materials. Nuclear reactors produce thermal neutrons with a maximum energy of 0.06 eV, which corresponds to the de Broglie wave with $\lambda \sim 1$ Å commensurable with interatomic distances. This forms the basis of the method of structural neutron diffraction. The commensurability of energies of thermal neutrons with those of thermal oscillations of atoms and groups of molecules is used for the analysis in neutron spectroscopy, while the presence of a magnetic moment lays the basis for magnetic neutron diffraction.

4.4 MISCELLANEOUS TECHNIQUES

4.4.1 EXAFS

The method is based on the measurements in the vicinity of the absorption edge, e.g. the K shell. In this case, the observed oscillations as a function of the photon energy are the result of interference of both primary waves and secondary ones scattered by neighboring atoms. The analysis of such oscillations allows one to find distances between neighboring atoms and to study deviations of "neighbors" in the particle surface layer as compared with compact metal lattices. The information on such deviations is important for understanding the optic properties of the metal particles. These deviations can affect the particle size, electron density, and optic properties. For example, atomic distances in Ag_2 and Au_2 are 0.210 and 0.253 nm, respectively, whereas in the compact metal this value is 0.325 nm.

4.4.2 X-ray Fluorescence Spectroscopy

The method is based on excitation of atoms in a substance under study by the emission of a low-power X-ray tube. This gives rise to secondary fluorescence emission, which falls on the crystal analyzer and, being reflected from the latter, is registered by a proportional detector. The crystal analyzer and the detector are driven by a goniometer. In doing so, each fixed position of the goniometer corresponds to a definite wavelength of secondary emission selected by the analyzer. The elemental composition of a sample is characterized by spectral lines with intensities unambiguously related to the quantitative contents of elements in a sample. Concentrations are usually computed by comparing with the values obtained for standard samples.

The X-ray fluorescence technique allows quick and high-precision analysis of practically all elements in the periodic table in solid, liquid, powder, and film samples.

4.4.3 Mass Spectrometry

This method is used for separating ionized species with respect to their masses, based on their interaction with magnetic and electrical fields. Dynamic devices analyze the time it takes for ions to fly a definite distance. In a quadrupole mass spectrometer, separation of ions is realized in a transverse electric field, which is created by a quadrupole capacitor constituted by four rods symmetric with respect to the central axis. In a time-of-flight mass spectrometer, an ion packet is let into the analyzer through a grid and "drifts" along the analyzer to a collector in such a way that heavy ions (m_1) lag behind light ions (m_2). The ion packet is separated, because all ions in the initial packet have the same energy, while their rates and, hence, the times of flight along the analyzer t are inversely proportional to the square root of their masses: $t = L(m/2n)^{1/2}$, where n is the accelerating potential and L the analyzer length.

4.4.4 Photoelectron Spectroscopy

The method is based on measuring the energies of electrons that escape solids under the action of photons. According to Einstein, the sum of the binding energy of an escaped electron (work function) and its kinetic energy is equal to the energy of the incident photon.

From the resulting spectra, the binding energies of electrons and their energy levels in a substance under study are determined. This method allows study of the electronic distribution in conduction bands and analysis of the substance composition and the chemical bond type.

Metals are characterized by intense photon reflections and a strong interaction with conduction-band electrons. The quantum yield is small ($\approx e/1$ photon), which complicates application of the method.

4.4.5 Nuclear Magnetic Resonance (NMR) Spectroscopy

NMR is becoming more powerful as an analytical tool as each year passes. Multiple nuclei are now accessible; high magnetic field strengths are becoming more common. Indeed, NMR is indispensable for organic and biomolecular identification/characterization.

For nanomaterials, NMR is less useful, but can be valuable for the study of ligand-stabilized nanoparticles, since the ligands are generally organic compounds, with NMR active nuclei ^{1}H, ^{13}C, and selected elements such as Pt or Si. Thus, if a nanomaterial is soluble, then NMR is quite valuable for characterization.

Solid-state NMR is also possible, and can yield useful results if the necessary spectrometer and sample spinners are available.

4.4.6 Ultra Violet–Visible Spectrometry (200–800 nm)

Many nanomaterials absorb UV and visible light. Solid samples can be studied by use of UV diffuse reflectance attachment, and even band gaps can be determined in this way. In solution, UV and visible absorption spectra are very useful. For example, ligand-stabilized gold and silver nanoparticles exhibit particular colors due to visible-light absorption known as plasmon bands. These absorptions are very characteristic; for example, 5 nm gold with thiol ligand shows absorption at about 520 nm, and this can shift slightly due to larger nanoparticle size (shift to longer wavelength), or smaller (shift to shorter wavelength), solvent employed, and ligand attached to the gold. For silver, the visible absorption is usually nearer 420 nm.

The particular advantage of UV-vis spectra is that characteristic groups can be recognized in molecules or nanomaterials of widely varying complexities.

4.4.7 Dynamic Light Scattering

Dynamic Light Scattering (DLS), also known as Quasi-Elastic Light Scattering (QELS), Photon Correlation Spectroscopy (PCS), and Light Beating Spectroscopy,[12–14] is a technique that relies upon temporal fluctuations in the light scattered from an ensemble of the particles to determine their motion. Usually the motion in any colloid or aerosol is a random Brownian diffusion that is quantified by a size-dependent diffusion coefficient. The DLS method measures the decay of the temporal fluctuations in the scattered light, which is related to their diffusion which, in turn, is related to their size.

Since the size is determined from diffusion, the viscosity of the suspending medium must be known. The size is an effective mobility size related to both size and shape. Size ranges from nearly molecular dimensions—a few nanometers to many microns can be determined. As with any light scattering measurement, larger particles scatter more than smaller, and hence they dominate the measurement in any polydisperse system.

DLS requires a coherent light source. Most common laboratory lasers such as HeNe, argon ion, Nd:YAG, etc. have enough longitudinal and transverse coherence to be useful for DLS. Good transverse coherence can be gained if the laser is operating in the TEM00 mode, which is characterized by a Gaussian beam profile. The donut profile TEM01* will work too but with some loss of signal to noise. The scattering volume should be small to ensure good transverse coherence on the detector. The second thing needed for DLS experimentation is a detector capable of detecting single photons such as a photomultiplier tube or avalanche photodiode. The detector output is typically amplified, discriminated, and shaped for the third thing needed, a digital correlator. These are available commercially and typically provide analysis routines to make use of the method straightforward.

4.5 COMPARISON OF SPECTRAL TECHNIQUES USED FOR ELEMENTAL ANALYSIS

Certain techniques used for analyzing the nanoparticles of different elements are beyond the scope of this section. The applications of electron spectroscopy, light scattering, and EPR are considered in other chapters of this book devoted to the studies of concrete reactions or applications of the metal clusters and nanoparticles. The methods for analyzing nonmetals are touched upon in a chapter that deals with fullerenes and carbon nanotubes.

Due to the wide diversity of methods used for analyzing elements, their comparison is almost impossible. Hence, we restrict ourselves to the most popular spectral techniques. The most important characteristics of any method are its detection limit and the size of samples it is capable of analyzing. Table 4.1 illustrates a comparative analysis of some techniques.

Summarizing, we mention certain problems associated with the analysis of the small clusters and metal nanoparticles. One of these problems is related to the size reproducibility of the particles obtained by different methods on micro- and macrolevels (this problem has already been touched upon above).

Studying the nanoparticles poses problems that arise when several analysis techniques are used simultaneously. Solving this problem is of vital importance in nanochemistry. For the majority of chemical reactions that involve the

TABLE 4.1 Comparative Analysis of Spectral Techniques

| Technique | Detection limits (solid state) | | Sample |
	Relative %	g	
Atomic emission spectral analysis	10^{-7}–10^{-4}	10^{-9}–10^{-7}	10–100 mg
Atomic absorption analysis	10^{-8}–10^{-5}	10^{-3}–10^{-11}	0.1–1 mg
Atomic fluorescence analysis	10^{-8}–10^{-6}	10^{-11}–10^{-9}	1–5 ml
X-ray fluorescence analysis	10^{-5}–10^{-4}	10^{-7}–10^{-6}	1–5 ml
Spectrophotometry	10^{-4}–10^{-3}	10^{-11}–10^{-8}	0.2–10 ml
Laser mass spectrography	10^{-8}–10^{-5}	10^{-12}–10^{-11}	5–100 mg

particles with sizes of ca. 1 nm, i.e. containing up to 10 atoms, a stoichiometric reaction equation is difficult to determine. This is due to the fact that most of such reactions occur under nonequilibrium conditions and do not allow one to follow the changes in concentrations and compositions of the starting and final products. The techniques used for studying the composition, size, and properties of the nanoparticles require further development and improvement.

To date, the complications associated with studying the highly active particles built of small numbers of atoms are overcome using various theoretical methods. The electronic structure of metal clusters is analyzed by two nonempirical methods: the density functional method and the Hartree–Fock method of configuration interactions. The former method provides information on the ground states of neutral and charged particles but fails to describe excited states. The Hartree–Fock method gives insight into both ground and excited states of neutral and charged particles. However, the optimal geometry of a cluster can be determined only for small particles, because calculations for multielectron clusters require much longer time.

For analyzing metals with strongly delocalized valence electrons, an electron shell model that assumes that valence electrons of each atom in a particle become free and are localized at the cluster boundaries is used. For the spherical particles, closed shells occur only for the electron number $n = 2, 8, 18, 34, 40, \ldots$. For clusters of alkali metals, the particles in such electron states are called "magic." In contrast to the "gel" model, the shell model considers the positive charges of atoms as spread over a homogeneous substrate and, hence, ignores the properties associated with the atomic structures of clusters. The particles with closed electron shells are spherical, whereas shells of open clusters are deformed, which is reflected in the energy of a shell and is taken into account in ab initio quantum-mechanical calculations.

Thus, a wide variety of techniques are used for studying the individual nanoparticles. However, to comprehensively investigate the metal nanoparticles, which have high reactivities and are able to change them depending on the kinetic and thermodynamic conditions, new methods should be developed. Moreover, these methods allow one to not only measure but also to follow in detail the changes in the properties of the nanoparticles during their formation and subsequent self-assembling and also should help to develop high-performance nanotechnological devices on their basis.

REFERENCES

1. Roco, M. C.; Williams, S.; Alivisatos, P., Eds. *Nanotechnology Research Directions: IWGN Workshop Report-vision for Nanotechnology in the Next Decade*; Kluwer: New York, 1999; pp 1–360.
2. Alferov, Zh. I. *Semiconductors* **1998**, *32*, 1–14.
3. Petrunin, V. F. Ekaterinburg, UrO RAN. Physical Chemistry of Ultradispersed Systems. In *Proceedings of Conference (Russ.)*, 2001, 5–11.

4. Andrievsky, R. A. *Russ. Chem. J.* **2002**, *46*, 50–56.
5. Gleiter, H. *Acta Mater.* **2000**, *48*, 1–29.
6. Pomogailo, A. D.; Rozenberg, V. I.; Uflyand, I. E. *Metal Nanoparticles in Polymers*; Khimiya: Moscow, 2000; pp. 1–672.
7. Roldugin, V. I. *Russ. Chem. Rev.* **2000**, *69*, 821–844.
8. Bukhtiyarov, V. I.; Slin'ko, M. G. *Russ. Chem. Rev.* **2001**, *70*, 147–160.
9. Sergeev, G. B. *Russ. Chem. Rev.* **2001**, *70*, 809–826.
10. Summ, B. D.; Ivanova, N. I. *Russ. Chem. Rev.* **2000**, *69*, 911–924.
11. Klabunde, K. J., Ed. *Nanoscale Materials in Chemistry*; Wiley: New York, 2001; pp 1–292.
12. Berne, B.; Pecora, R. *Dynamic Light Scattering*; John Wiley and Sons, Inc., 1976.
13. Dahneke, B. E. *Measurement of Suspended Particles by Quasi-Elastic Light Scattering*; John Wiley and Sons, 1983.
14. Sorensen, C. M. Private Discussions at Kansas State University, 2012.

Chapter 5

Cryochemistry of Metal Atoms and Nanoparticles

Chapter Outline

5.1	Reactions of Magnesium Particles	90	5.4.2 Reactions of Silver Particles of Various Sizes and Shapes	132
	5.1.1 Grignard Reactions	90	5.5 **Theoretical Methods**	**137**
	5.1.2 Activation of Small Molecules	93	5.5.1 General Remarks	137
	5.1.3 Explosive Reactions	96	5.5.2 Simulation of the Structure of Mixed Metallic Particles	138
5.2	Silver and Other Metals	100		
	5.2.1 Stabilization by Polymers	101	5.5.3 Simulation of Properties of Intercalation Compounds	143
	5.2.2 Stabilization by Mesogenes	110		
5.3	Reactions of Rare-earth Elements	115	5.5.4 Simulation of Structural Elements of Organometallic Co-condensates	145
5.4	Activity, Selectivity, and Size Effects	122		
	5.4.1 Reactions at Superlow Temperatures	122		

Low-temperature reactions of metal vapors date back to studies by N.N. Semenov in 1928. During condensation of cadmium and sulfur vapors on a surface cooled by liquid nitrogen, he observed a periodic reaction that propagated from the center of a condensed film.[1] The reacted substances formed zones shaped as concentric rings.

In the late 1950s, vapors of sodium, potassium, and magnesium were used for initiating low-temperature polymerization. Magnesium exhibited the highest activity. The joint condensates of magnesium vapors with acrylonitrile, methylacrylate (MA), acrylamide, and certain other monomers underwent rapid solid-phase polymerization at low temperatures. The results of these studies have been generalized.[2]

Our studies of low-temperature co-condensates of metal vapors with vapors of various ligands started in the late 1970s and, naturally, relied on the previous experience. The research carried out formed the basis of a new actively developing direction—cryochemistry of nanosize metal particles or cryonanochemistry.[3–9]

5.1 REACTIONS OF MAGNESIUM PARTICLES

The choice of magnesium was based on the previous experience in co-condensation of magnesium with acrylonitrile and their EPR investigation.[10]

New remarkable results were obtained for the reactions of atoms, clusters, and nanoparticles of magnesium with polyhalides of methane at low and superlow temperatures. The absence of magnesium reaction with carbon tetrachloride in solutions was usually cited as the evidence that polyhalide hydrocarbons do not form Grignard reagents. A radically different situation takes place at low temperatures. Carbon tetrachloride is "rigid" to a certain extent and at low temperatures can sometimes be used as a matrix.[11,12] In this connection, it is most likely that no large magnesium aggregates are formed in co-condensates of magnesium with excessive carbon tetrachloride at 77 K. Stabilization of small clusters is more probable.

5.1.1 Grignard Reactions

IR spectroscopic studies of low-temperature co-condensates of magnesium with carbon tetrachloride allowed the formation of a Grignard reagent to be detected at 77 K; in other words, it allowed the insertion of a magnesium atom into the carbon–chlorine bond to yield a trichloromethyl radical and dichlorocarbene.[13–16] As a result, the following scheme of parallel concurrent reactions was put forward:

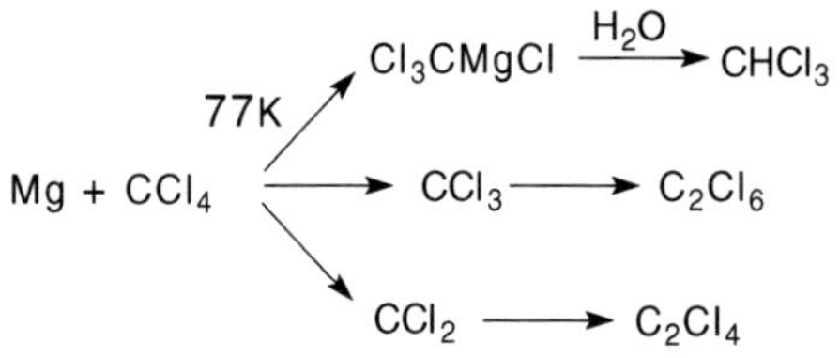

$$Mg + CCl_4 \xrightarrow{77K} \begin{cases} Cl_3CMgCl \xrightarrow{H_2O} CHCl_3 \\ CCl_3 \longrightarrow C_2Cl_6 \\ CCl_2 \longrightarrow C_2Cl_4 \end{cases}$$

The formation of trichloromethyl radicals and dichlorocarbene was confirmed by the presence of hexachloroethane and tetrachloroethylene in reaction products, which were identified in IR spectra and gas chromatograms. The reaction products of water with the low-temperature condensate were shown to contain chloroform, which additionally testifies the formation of a Grignard reagent.

A single-stage synthesis of a Grignard reagent with fluorobenzene was realized at liquid nitrogen temperature.[17] In solutions, such a process is hindered and proceeds in two stages.[18] EPR studies of the reaction of magnesium with benzyl halides C_6H_5X, where X = F, Cl, Br, or I helped to reveal radicals that appeared with the detachment of a halogen atom by magnesium. As seen from EPR spectra in Figure 5.1, the spectrum of benzyl bromide is sufficiently well resolved, whereas the benzyl fluoride spectrum demonstrates only a singlet that was assigned to the formation of a radical–ion pair.[19] Distinct acryl radicals were observed at the interaction of magnesium with alkyl chlorides.

The question of the participation of radicals or radical–ion pairs in the formation of Grignard reagents was solved by studying the effect of temperature on the reaction kinetics. Figure 5.2a shows the changes in EPR signal intensity for alkyl radicals (curve 1) and radical–ion pairs (curve 2) in a solid co-condensate of magnesium with n-octyl chloride as a function of temperature and time. As can be seen, the relative intensity of the alkyl radical signal gradually decreased with an increase in temperature, and the radicals disappeared at $T = 123$ K. The concentration of radical–ion pairs increased with an increase in temperature, reached a maximum at $T = 123$ K, and then decreased.

A correlation between the Grignard reagent yield (Figure 5.2b) and the concentration of radical–ion pairs (Figure 5.2a) was observed. The experimental results obtained suggest that the formation of a Grignard reagent involves radical–ion pairs rather than free radicals.[15,19] It is possible that radicals also take part in the Grignard reagent formation. Such mechanisms have been repeatedly

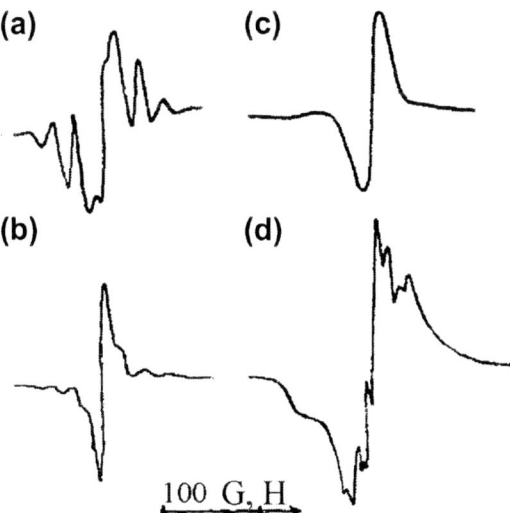

FIGURE 5.1 EPR spectra in the magnesium–halogen derivatives: (a) n-choropentane, (b) n-fluorooctane, (c) fluorobenzene, and (d) bromobenzene.

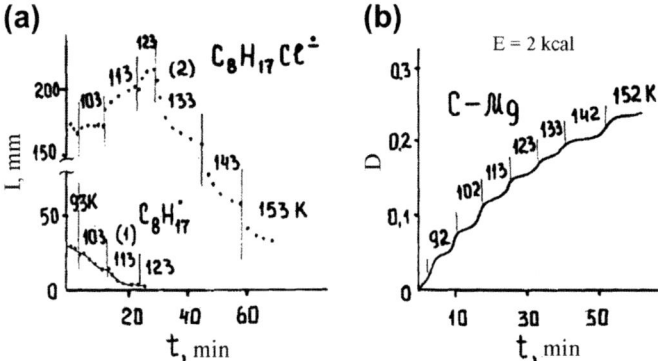

FIGURE 5.2 Kinetics of reactions in the magnesium–chlorooctane system: (a) changes in the EPR signals of (1) alkyl radicals and (2) radical–ion pairs; (b) changes in the optical density D ($v = 555$ cm^{-1}) corresponding to the C–Mg bond formation.

proposed in the literature. Under low-temperature conditions, the reaction of halogen abstraction by a metal atom competes with that of metal insertion into a carbon–halogen bond. For compounds with weak carbon–halogen bonds such as iodides and bromides, the formation of radicals and the products of their doubling prevail, i.e. the Würtz reaction is observed. The totality of results obtained for the low-temperature interaction of magnesium particles with alkyl and aryl halides allowed the proposal of the following scheme:

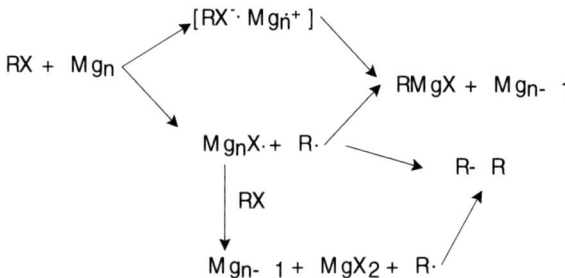

The formation of radical–ion pairs depended not only on the carbon–halogen bond strength but also on the metal–carbon ratio. An increase in the metal concentration assisted the charge transfer and the appearance of radical ions.

Considering the example of butyl halides (Würtz reaction), the dependence of the yield of radical-doubling products on the energy of carbon–halogen bond and the magnesium concentration was studied.[20] Figure 5.3 shows the results obtained. As seen, with an increase in the magnesium concentration in the n-C$_4$H$_9$Cl–Mg system, the octane yield decreases, whereas it remains virtually unchanged for the n-C$_4$H$_9$Br–Mg system and increases for the n-C$_4$H$_9$I–Mg system. These results agree with the reaction scheme proposed.

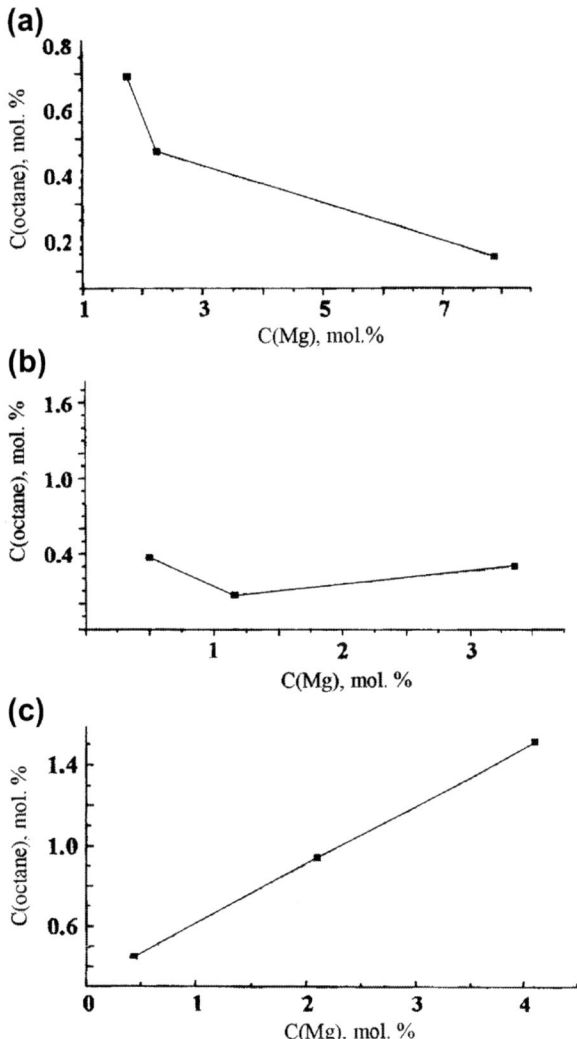

FIGURE 5.3 Dependence of the yield of *n*-octane C (mol%) on the magnesium concentration and the strength of C–X bond, where X = Cl, Br, I: (a) *n*-chlorobutane, (b) bromobutane, and (c) iodobutane.

5.1.2 Activation of Small Molecules

The easiness of carrying out unusual reactions such as the synthesis of Grignard reagents using tetrachlorocarbon and benzyl fluoride, which occur in low-temperature co-condensates and involve metal nanoparticles, pointed to the presence of stored energy in such systems. Such energy could be accumulated in the form of stabilized metal aggregates or in their metastable complexes with either organic or inorganic ligands. It seemed to be of interest to use the stored energy for activating low-active molecules, e.g. carbon dioxide. The choice of carbon dioxide is explained by the fact that comparatively few reactions occur with its participation. These are the syntheses of urea and carbonates of sodium and ammonium. To activate CO_2, binding it into complexes, especially with metal nanoparticles at low temperatures, holds a great promise.

Solid samples containing one to three components, namely, a metal, carbon dioxide, saturated or unsaturated hydrocarbons, and also, in special cases, argon as the diluter, were studied. In cryoreactions involving carbon dioxide, the following metals were used: lithium, potassium, sodium, silver, magnesium, calcium, zinc, cadmium, mercury, and samarium.

We were the first to carry out and study the low-temperature reactions of carbon dioxide with magnesium, calcium, and samarium particles in co-condensates at temperatures ranging from 4.2 to 293 K.[15,21,22]

EPR spectra recorded in co-condensates of lithium and sodium with carbon dioxide were assigned to the formation of $M^+CO_2^-$ complexes. This was additionally confirmed by analyzing IR spectra, which demonstrated heating-induced changes in the intensity of sample lines. For alkali metal–carbon dioxide systems, the formation of intermediates, which included CO_2 dimers and alkali metal atoms, was proposed.

Studies of the low-temperature interaction between magnesium and carbon dioxide carried out by IR and EPR spectroscopy techniques have shown that the first stage of this process is the electron transfer with the formation of anion and dianion radicals. It was assumed that magnesium–carbon dioxide complexes of various compositions and magnesium carbonyl were formed in the co-condensates. Apparently, at the instant of co-condensation, certain other products such as magnesium oxalates and carbonates can also be formed. Moreover, their formation occurs largely at the very moment of condensation.

The interaction of alkali and alkali-earth elements with carbon dioxide in low-temperature co-condensates involves the electron transfer and the formation of compounds of the $M^+CO_2^-$ type. Moreover, the ability of a metal atom to give away an electron, which is determined by its ionization potential, is of great importance. Indeed, there is a correlation between the functional ability of CO_2 and the ionization potential of a metal atom. Lithium ($I=5.39\,eV$), potassium ($I=4.34\,eV$), sodium ($I=5.14\,eV$), and samarium ($I=5.6\,eV$) easily enter into this reaction, whereas magnesium ($I=7.6\,eV$) and calcium ($I=6.1\,eV$) enter with difficulty. Cadmium, zinc, and mercury do not enter into this reaction, with ionization potentials being 8.49, 9.39, and 10.44 eV, respectively. For ionization potentials, reference data were used.[23]

Apparently, the aforementioned metal activity toward carbon dioxide and its relation with the ionization potential are not the only factors that determine the reactivity of the systems under study. There is yet no information on the exact number of metal atoms in an active particle; however, it is known that ionization potentials of metal particles vary within groups in the periodic system, strongly depend on the particle size, and, as a rule, decrease with an increase in the cluster size. The trend for a decrease in the ionization potential with an increase in the number of metal atoms was associated with possible delocalization of the positive charge that appears upon ionization of a cluster containing a large number of atoms.

Size-induced peculiarities of cryochemical reactions allow one to consider low-temperature co-condensates as systems that can accumulate and store energy. Moreover, of special interest as the accumulators of energy are multicomponent systems containing metal nanoparticles. In this case, along with the energy associated with the presence of metastable states and defects and the geometric sizes of a sample and particles involved in the reaction, the energy of the conjugated processes should be considered. The essence of such phenomena may be illustrated by the fact that among two- and three-component systems with a common reagent, the former system can be relatively stable, while the latter can have a greater activity.

The interaction of carbon dioxide with ethylene in the presence of magnesium was studied.[24] Samples were obtained by co-condensation of reagent vapors on a surface cooled by liquid nitrogen for a reagent ratio $Mg/CO_2/C_2H_4 = 1:5:50$. For a comparison, $Mg-CO_2$ and $Mg-C_2H_4$ co-condensates with compositions varying from 1:5 to 1:50 were studied. After the hydrolysis by water vapors, the reaction products were analyzed by chromatography–mass spectrometry technique. According to IR spectra, the co-condensate of magnesium with ethylene did not virtually differ from pure ethylene. The EPR spectrum revealed a signal with a g-factor of 1.9988 ± 0.0005 and a half-width of 7G. The signal intensity decreased with the heating of a sample up to 95 K and then disappeared.

A co-condensate of magnesium, carbon dioxide, and ethylene was stable at 77 K, and its EPR spectrum represented a superposition of signals of individual systems. Heating of a 100-μm-thick film to 100 K resulted in an explosion reaction accompanied by flashes.

According to chromatography data, the reaction products comprised at least 10 individual substances. Chromatography–mass spectrometry technique allowed identification of two compounds with the masses of molecular ions equal to 104 and 132. An interpretation of the spectra revealed the presence of diethylacetals of formic and propionic aldehydes. A possible reaction scheme is as follows:

The formation of compounds of magnesium with ethylene dimer was observed in low-temperature co-condensates.[25] This reaction could not be realized in films thinner than 10 mm under slow heating conditions owing to the evaporation of ethylene.

At 80 K, a reaction was shown to occur in a Li–CO_2–C_2H_4 system with the reactant ratio of 1:20:10. The presence of water traces was beneficial for the reaction. According to IR spectra, lithium propionate is the possible reaction product.[26] The substitution of sodium and potassium for lithium resulted in their reaction with carbon dioxide only. The authors failed to involve ethylene in this reaction. The reaction did not occur in the presence of water vapors. IR spectra of ternary co-condensates of carbon dioxide with samarium and silver, which were recorded after heating up to 293 K, demonstrated weak absorption of new products, which could not be identified.

Magnesium, zinc, and tin particles obtained in cryochemical synthesis were employed for the destruction of carbon tetrachloride in water.[27] The reaction products were analyzed by chromatography–mass spectrometry, IR spectroscopy, and chromatography. The activity of cryochemically synthesized metal particles was compared with that of the particles obtained by different methods. Cryoparticles exhibited the highest activity. It was assumed that the destruction of CCl_4 proceeds via the formation of intermediate compounds of metal insertion into the C–Cl bond, which easily reacted with water. Compounds CH_3Cl, CH_2Cl_2, and $CHCl_3$ also reacted via a similar scheme. The final products were hydrocarbons. A high activity of zinc cryoparticles was additionally confirmed by atomic force microscopy.[28]

On the one hand, the above examples show that low-temperature co-condensates accumulate large portions of energy and can sustain conjugated chemical reactions. On the other hand, these examples suggest that the chemical nature of a metal plays an important role in the realization of cryochemical reactions. Here, we have a problem of a balance between activity and selectivity.

5.1.3 Explosive Reactions

A phenomenon of fast, virtually explosive cryochemical reactions was studied most comprehensively by the example of magnesium–alkyl halide systems. For the first time, this phenomenon was revealed in the course of recording EPR spectra, when a slight shaking of the reactor resulted in an explosive reaction.[29] Filming of this process has shown that the reaction takes less than 0.01 s. Further studies with different halogen derivatives made it evident that the reaction depends on the thickness of magnesium–alkyl halide co-condensate films.[30,31] Table 5.1 shows the results obtained. The process was initiated by an impact of a needle.

A comprehensive study of co-condensate films formed by magnesium and dichloroethane allowed the mechanism of fast explosive reactions involving magnesium to be refined.[32] As seen from Figure 5.4, an explosive reaction between magnesium and dichloroethane occurs at a certain critical thickness L_{cr} (4 µm). The conversion reaches nearly 100%. The reaction is accompanied by heat liberation, gas evolution, and mechanical destruction of the film. The critical thickness depends on the reagent ratio and the temperature of a surface

Chapter | 5 Cryochemistry of Metal Atoms and Nanoparticles

TABLE 5.1 The Effect of the Alkyl Halide Nature and the Thickness of its Co-condensate with Magnesium on the Cryoexplosive Reactions[a]

System	Lower limit (μm)	Upper limit (μm)
$Mg-1,2,-C_2H_4Cl_2$	20	90–100
$Mg-C_6H_5Cl$	90–110	220–270
$Mg-C_6H_5J$	35–40	120–140

[a] The co-condensation rate was 2×10^{16} molecule/s cm^2, and the impact force was $5 \times 10^{-4} - 5 \times 10^{-3}$ J.

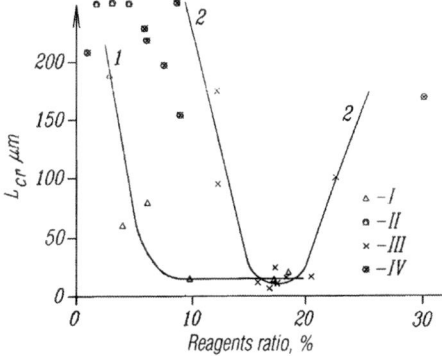

FIGURE 5.4 Dependence of the critical thickness of co-condensate film of magnesium and 1,2-dichloroethane on the composition and the support temperature: (1) at 80 K, (2) at 90 K; I and III—critical temperatures at 80 and 90 K, respectively; II and IV—limiting temperatures at which no explosive reactions were observed at 80 and 90 K, respectively.

onto which the reactants are condensed. The effect of magnesium content on the average critical thickness Lcr at temperatures lower than 80 K is illustrated as follows:

Mg content (mol%)	10–20	20–25	25–30	30–50
Critical thickness (mm)	9.5	3.6	2.7	1.5

As seen from these data, a threefold increase in magnesium content results in more than a sixfold decrease in L_{cr}. The critical value has a minimum corresponding to the equimolar magnesium/dichloroethane ratio of 1:1. An increase in the surface temperature narrows the explosion region. It is significant that the sample was amorphous in the initial state and crystalline in the final state; moreover, the reaction need not involve the whole film and could proceed on its part. A certain interaction between the reacting and the nonreacting parts of a sample was observed, self-propagating waves were formed, and the internal energy of mechanical stresses initiated the explosion. The probability of a fast reaction decreases with an increase in the surface temperature, which together with the absence of preexplosion heat up allows us to exclude the mechanism of thermal explosion from our consideration. The explosion is unlikely to occur at slower condensation rates. Explosions were also not observed at 110 K. To describe a phenomenon under consideration, a model was developed based on an assumption that the condensate film formation generates mechanical stresses that, in turn, cause plastic deformations and cracking as soon as the film grows to a certain critical thickness. Plastic deformations result in an increase in the mobility of molecules and in an acceleration of chemical interactions or crystallization.

The processes that occur in low-temperature film condensates can be studied by calorimetric techniques. An original thin-film differential scanning calorimeter that allows studying condensation of reagent vapors on a support with a temperature of 80–300 K has been designed.[33] This setup was used for recording calorimetric curves in condensates of butanol-1 and water. By the example of water condensates, a spontaneous crystallization of the amorphous water condensate that occurred upon a condensate film of a certain thickness was observed for the first time.[34,35] Amorphous water films of approximately 0.5-μm thickness were obtained by condensation of water vapors on a copper plate of the calorimeter under molecular beam conditions at 80 K. The condensation rate varied from 6×10^{14} to $5 \times 10^{16}\,cm^{-2}/s$ (0.01–1 μmol/min). After the film grew to a certain critical thickness, a spontaneous surge of heat liberation was observed. Figure 5.5 shows a typical calorimetric curve.

The critical film thickness at the instant of initiation of a fast process was 4 mm. The liberated heat amounted to 0.2–0.7 KJ/mol and depended on experimental conditions. Such substantial heat liberation was assumed based on the results on the crystallization of amorphous water, which was observed upon the formation of a film thinner than the critical value in the temperature range 163–167 K. The crystallization heat was 1.2 ± 0.1 KJ/mol at a scanning rate of 3 K/min.

As mentioned above, upon reaching the critical thickness, the fast, explosive reactions can be initiated by mechanical stresses that arise during the formation of a sample. A water condensate was studied in a cryotensometric setup,[30] which provided conditions of sample formation comparable with those in a calorimetric setup. After the film reached a certain critical thickness, the stresses were abruptly relieved from 11 to 1 mPa, and the formation of a visible net of cracks was observed.

On the basis of the results obtained, we can infer that upon the attainment of a critical thickness, internal mechanical stresses initiate film destruction and crystallization of a sample. It is reasonable to assume that, being interrelated via a positive feedback, the destruction and crystallization processes cause an autowave process of avalanche crystallization in the amorphous film. A more detailed study allowed the authors to relate the observed size effect with

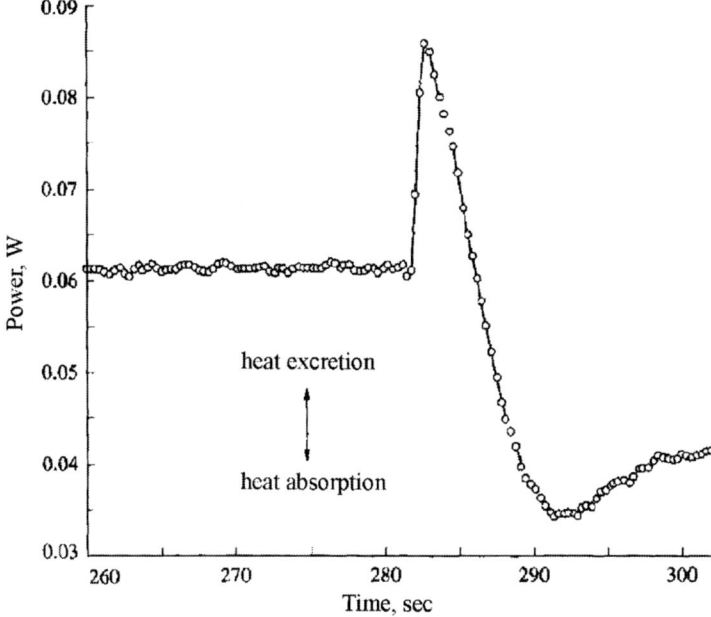

FIGURE 5.5 Dependence of the power of heat effects on time during the formation of water condensate (fast reaction domain, condensation rate, 0.9 μm/min; substrate temperature, 80 K; critical thickness, 4 μm).

fast heat liberation, generation of cracks, and a sharp decrease in the internal mechanical stresses.

The film destruction and the chemical reactions or the crystallization processes are interrelated by a positive feedback and generate autowave explosive chemical reactions or crystallization processes during the formation of co-condensate films. The same peculiarities as those observed above for the magnesium–dichloroethane system were also revealed for reactions of acetyl chloride with diethylamine and cyclopentadiene with $TiCL_4$ and also for polymerization reactions initiated by magnesium vapors.[36,37] A more sophisticated model of cryoexplosive reactions was developed,[32] and the dependence of the reactions on the conditions of sample formation was examined.[38,39] The observed extreme dependences on the condensation rate were explained on the basis of a kinetic model that took into account the nucleation processes during crystallization. Figure 5.6 illustrates the effect of mechanical energy on the processes occurring in growing films at low temperatures.[31,40] It should be stressed that explosive processes can involve only a part of a film and repeat if the film formation continues after the explosion.

An explosive reaction can be initiated by an external impact, either thermal or mechanical, or arise spontaneously after the end of film formation or during its exposure under isothermal conditions.

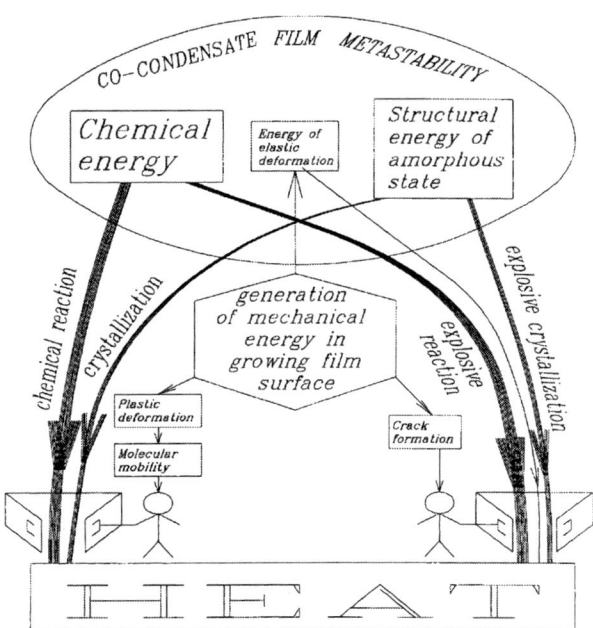

FIGURE 5.6 Generation of mechanical energy in a growing co-condensate and the processes accompanying this phenomenon.

Conditions of film formation determine the critical thickness at which explosive processes are observed. As already mentioned above, an increase in the support temperature and a decrease in the co-condensation rate inhibit spontaneous explosive processes. Hence, low-temperature processes associated with the formation of structures and their changes in co-condensates and also with the size of formed particles play the decisive role as regards concrete chemical mechanisms. The processes that occur in films at their formation can be initiated by cracking upon reaching an ultimate strength. Thus, in low-temperature reactions, physical and chemical processes are deeply intertwined.

5.2 SILVER AND OTHER METALS

Low-temperature co-condensation under kinetic control can involve the formation of metastable porous crystals, which are capable of incorporating other substances of certain sizes and shapes. Realization of chemical reactions in such systems opens up a possibility of controlling the reactivity under conditions of structural ordering. Thus, modifying the structure of succinic anhydride during low-temperature co-condensation changed its selectivity in the reaction with 1,2-diaminopropane.[40] A similar approach was applied for synthesizing metal-containing polymers. By the low-temperature condensation of acrylamine–potassium systems, a polymer with polymeric chains packed in layers (interlayer distance 11.2 Å) was obtained.[41] The polymer exhibited a sorption ability toward alcohols.

5.2.1 Stabilization by Polymers

Synthesis of materials, which either consist of metal nanoparticles or incorporate them in their compositions, is impeded by the high activity of these particles. Recently, a new method for stabilization of nanosize metal particles was proposed.[42–44] Essentially, the method consists in using monomers, which can be polymerized at low temperatures. Thus, polymeric films containing aggregates of metal atoms were formed when metal particles with vapors of *p*-xylylene obtained by pyrolysis of di-*p*-xylylene were co-condensed on a cooled surface and then heated up to 110–130 K or were illuminated with mercury-lamp light at 80 K. The polymerization process proceeded via the following scheme:

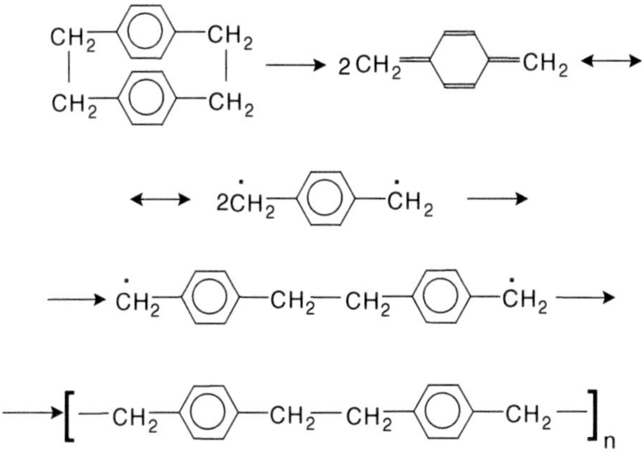

Poly-*p*-xylylene (PPX)

A polymer can incorporate and stabilize metal clusters. Films of poly-*p*-xylylene with incorporated metal particles could be withdrawn from the reaction vessel for further studies. Samples were studied using the electron microscopy techniques. Figure 5.7 shows one of the images obtained for lead clusters. As seen, the particles are globular.

A histogram in Figure 5.8 demonstrates that the particle size is distributed over the range 2–8 nm. Their average diameter was estimated to be 5.5 nm. Table 5.2 shows the synthetic conditions for poly-*p*-xylylene films and the average diameter of the particles. The tabulated data allow one to conclude that a rather wide variation of the lead content in a sample has virtually no effect on the average size of the particles. Along with lead, the nanosize particles of Zn, Cd, Ag, Mg, and Mn were stabilized in poly-p-xylylene films.

IR spectroscopy provides information on stabilization of globular particles of different metals in polymeric films. It was found that Zn, Pb, and Ag

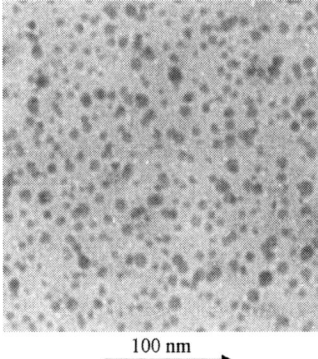

FIGURE 5.7 Electron microphotography of lead particles in poly-*p*-xylylene film.

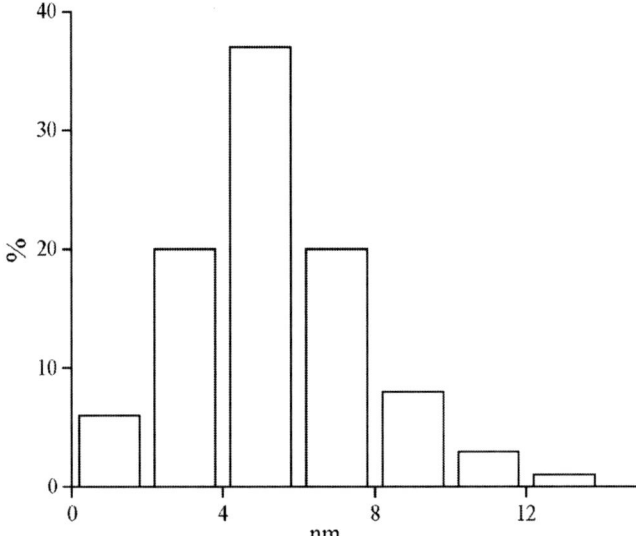

FIGURE 5.8 Size distribution of polymeric-film-isolated lead particles.

induce no noticeable changes in IR spectra of systems monomer–metal and polymer–metal. This points to the absence of strong interactions, e.g. those resulting in the formation of organometallic compounds (OMC).

A quite different situation is observed for magnesium. Magnesium–*p*-xylylene co-condensates revealed the formation of new bands at 1210 and 1483 cm^{-1}, which are probably associated with the electron transfer from magnesium to benzene rings in a magnesium–*p*-xylylene complex. Heating resulted in the appearance of new bands at 720 and 740 cm^{-1}, which indicated the transformation of a low-temperature π-complex to a σ-complex. Table 5.3

TABLE 5.2 Synthetic Conditions for Films and the Average Size of Lead Particles

Pb, T_b (°C)	p-xylylene, $T_{pyrolysis}$ (°C)	UV irradiation	Lead content (mass%)	Average diameter of particles (nm)
742	495	–	6.5	5.2
600	600	–	0.5	3.4
625	630		0.1	6.9
625	630	+	0.1	7.4
735	605	–	1.9	5.5
735	605	+	1.9	6.7

TABLE 5.3 Absorption Bands in Co-condensates of Metals with Poly-p-xylylene (PPX) at 80–300 K[266]

Metal	Absorption bands (cm^{-1})	Temperature (K)	Assumed assignment to a complex
Zn	855	80	π-complex of =CH$_2$ group with Zn
	1900 (ampl.)	120–300	π-complex of Zn with PPX
Cd	855	80	π-complex of =CH$_2$ group with Cd
	1900 (ampl.)	120–300	π-complex of Cd with PPX
Pb	865	80	π-complex of =CH$_2$ group with Pb
	1900 (ampl.)	120–300	π-complex of Pb with PPX
Ag	860	80	π-complex of =CH$_2$ group with Ag
	1900 (ampl.)	120–300	π-complex of Ag with PPX
	1790 (ampl.)	120–300	π-complex of Ag with PPX
Mg	1500	80	CTC of Mg with PPX
	1483	80–300	π-complex of Mg with PPX
	1210	80–300	π-complex of Mg with PPX
	740	140–300	σ-complex of Mg with PPX
	720	140–300	σ-complex of Mg with PPX
Mn	1592	80–300	Stable π-complex manganese–monomer
	1578	80–300	

shows the results of IR studies of joint condensates formed by metals and p-xylylene.[45]

Individual poly-p-xylylene films are good insulators. The introduction of 10 mass% lead did not affect their high insulation properties, and the film had a specific resistance of $10^{16}\,\Omega/cm^2$. The crystalline and metallic nature of particles incorporated into a poly-p-xylylene film was also detected by X-ray diffraction methods. The fact that such particles as ZnS and PbS could be incorporated into poly-xylylene films is also of certain interest.[43]

Along with incorporation of nanosize metal particles into poly-p-xylylene films, synthetic methods were developed for incorporating cryochemical particles into polyacrylamide gels. Moreover, in situ syntheses of silver nanoparticles in cross-linked polyacrylamide gels and inverse micelles based on sodium isooctylsulfosuccinate (AOT) were carried out. The potentialities of the various methods used for the formation of metal nanoparticles were compared.[44] Studies of the reactivity of metal nanoparticles have shown that the solvent nature and, in particular, its polarity are of great importance. A method for the synthesis of organic dispersions from metal clusters, which involves joint condensation of vapors of an organic solvent and a metal in vacuum on a support cooled to a low temperature, followed by resolution of metal particles, was proposed.[46]

Polymers with incorporated metal nanoparticles open up possibilities for synthesizing new materials. For instance, poly-p-xylylene films containing silver particles (1.5 mass%) exhibited catalytic activity in a model reaction of methanol oxidation. High sensitivity of lead-containing films to ammonium was observed. Such films were proposed for use as new ammonium sensors having a response enhanced by 4–5 orders of magnitude.[47,48] Figure 5.9 shows the results of AFM studies of poly-p-xylylene films with deposited lead particles. Such films can be used as sensors for wet ammonium.[49]

The use of monomers of the acryl series for stabilizing nanoparticles of various metals, particularly silver, was studied in detail. The results are summarized in Table 5.4.

A setup that allows one to obtain bimetallic cryoorganic dispersions by low-temperature vacuum condensation of vapors of two different metals and an organic substance was developed. A synthesis of bimetallic nanoparticles in the silver–lead–MA system was studied most thoroughly.[50]

In line with the elaborated systematic approach to studying bimetallic nanoparticles, the detailed studies of individual metals were also carried out. The choice of a silver–lead–MA system was based on the preliminary studies of the properties of Ag–MA and Pb–MA pairs, particularly their stability. The choice of the silver–lead system was also supported by quantum-chemical estimates of properties of mixed bimetallic nanoparticles.[51]

Bimetallic cryoorganodispersions were obtained using two independent evaporators mounted in a semicommercial glass reactor (Figure 2.7).

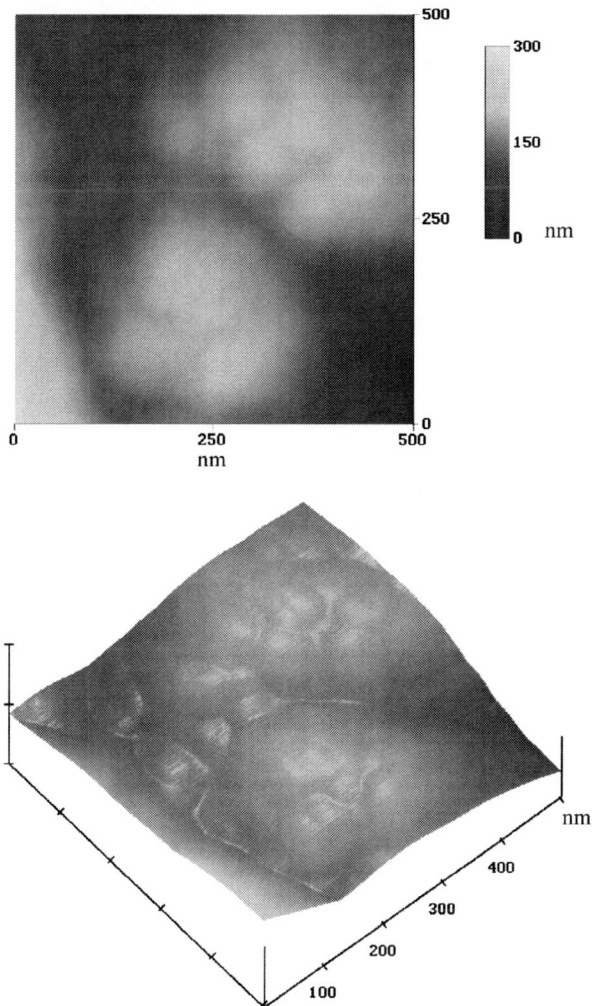

FIGURE 5.9 Lead nanoparticles on the surface of poly-p-xylylene film, synthesized by condensation.

The metal content in organosols was determined using atomic emission spectrometry with induction-confined plasma and also by analyzing X-ray fluorescence spectra.

Red-brown organosols Ag–Pb–MA formed during slow (~1 h) heating of low-temperature co-condensates retained stability in argon atmosphere for several days. A Pb–MA system that was studied in parallel behaved in a similar way. MA that was evaporated during the cryosynthesis could be quantitatively removed from the resulting Pb–MA and Ag–Pb–MA organosols, which points

TABLE 5.4 Low-temperature Co-condensation of Vapors of Metals and the Monomers of the Acryl Series

Co-condensation	Products	Polymer yield components
Acrylic acid (AA)	AA	0%
Ag–AA	Ag_n–poly-AA (solid film + AA)	≥30–50% (co-condensation)
		≤5–10% (layered condensation)
Mn–AA	Mn_n–poly-AA (solid film + AA)	>50%
Methylmethacrylate (MA)	MA	0%
Ag–MA	Ag_n–(MA + poly-MA) sol	1–3%
Particle size: 10–15 nm	Ag_n–poly-MA	Slow polymerization in Ar atmosphere
Mn–MA	Mn_{ok}–poly-MA	Slow polymerization in Ar atmosphere
Sm–MA	Sm_{ok}–poly-MA (solid film)	50%
Sn–MA	Sn_{ok}–poly-MA (solid film)	50%
Pb–MA	Pb_n–MA (suspension)	0%
Particle size: 5 nm		
Bimetallic systems		
(Pb–Ag)–MA	$(Pb_{ok}–Ag_n)$–MA (sol)	0%
Particle size: 5 nm	$(Pb_n–Ag_n)$–poly-MA	Slow polymerization in Ar atmosphere
(Ag–Mn)–MA	$(Ag_n–Mn_n + Mn_{ok})$–poly-MA	Slow polymerization in Ar atmosphere

to the absence of polymerization under experimental conditions. Thus, lead, in contrast to silver, does not initiate MA polymerization. Moreover, the behavior of the bimetallic system with respect to MA polymerization resembled the behavior of a Pb–MA system rather than of Ag–MA systems studied earlier. From our viewpoint, such a behavior is associated either with some nonadditive changes in nanoparticle properties during the transition from binary Pb–MA and Ag–MA systems to ternary Ag–Pb–MA systems or with the inhibition of silver-induced polymerization of MA with lead atoms.

MA is apparently a less-effective stabilizer of nanoparticles as compared with polymethylacrylate, which forms on their surface a polymeric coating that prevents their aggregation. Hence, lead nanoparticles and bimetallic Ag–Pb particles form aggregates in organosols, which is evident from their electron microscopic images. In both cases, the particle size did not exceed 5 nm, i.e. turned out to be smaller than the diameter of silver nanoparticles (7–15 nm) formed under similar conditions.

Interesting information follows from the absorption spectra of cryochemically synthesized organosols. The spectrum of Pb–MA organosols was characterized by the presence of the absorption band of lead plasmon with the maximum at ~220 nm. Oxidation of lead particles with air oxygen was accompanied by a fast decrease in organosol absorption in the visible region and by the appearance of opalescence. The absorption band of silver plasmon in an Ag–MA organosol had a maximum in the 416–420 nm range. In the Ag–Pb–MA organosol spectrum measured under argon atmosphere, this band was shifted to the red region ($\lambda_{max} = 438$ nm). Letting air into the system was accompanied by further shift of λ_{max} to 453 nm. During the next 1–1.5 h, λ_{max} gradually approached 466 nm, while the band intensity somewhat faded probably because of oxidation of lead contained in bimetallic particles. The interpretation of optical absorption spectra of colloidal dispersions of metal particles poses a multifactor problem. At the same time, the electron microscopy results allowed the aggregation of nanoparticles to be considered as the main reason for the red shift (416–420→438 nm) observed in an inert atmosphere. Lead oxidation by air oxygen can affect the stability of bimetallic particles. In this case, the observed long-wavelength shifts (438→453 and 453→465 nm) point to the evolution of aggregation processes and, probably, certain lead-oxidation-induced changes in the electronic state of bimetallic particles. Further studies would make it possible to elucidate how the composition of bimetallic particles and the structure and properties of the organic-ligand surface layer affect the optical and chemical properties of cryochemically synthesized nanoparticles and their organosols.

The conductivity of certain films containing two metals was studied.[52–54] Direct-current measurements were carried out in the range 10^{-11}–10^{-7} A by heating film samples from 80 K to room temperature. A system of co-condensates of silver and samarium vapors on poly-*p*-xylylene was studied most extensively. The results obtained were compared with the data on individual metals. The temperature dependence of conductivity of films containing two metals represented a superposition of corresponding dependences obtained for films containing individual metals. As in the case of individual samarium films, the exposure of a sample at 100 K followed by heating to 250 K resulted in an abrupt loss of conductivity for a system containing both silver and samarium. The further increase in the temperature entailed a conductivity rise by two orders of magnitude so that the further temperature dependence resembled that of a film containing silver only.

A system of silver–lead on poly-*p*-xylylene was studied in more detail. According to electron microscopy data, heating of a film containing only lead resulted in the formation of globular nanoparticles, which induced a decrease in the film conductivity. In contrast, heating of silver-containing systems induced an increase in conductivity. In our opinion, this fact is inconsistent with both the temperature dependence of conductivity of compact silver and the possibility of a heating-induced rupture of the islet silver film, owing to the higher temperature expansion coefficient of the polymeric sublayer. It can be assumed that the islet silver films form network or filamentary structures, which favors the increase in conductivity. In bimetallic systems, a second-metal effect that consisted in limiting the temperature-induced increase in the film conductivity was observed.

To elucidate whether the donor–acceptor interactions affect the conductivity of a metal in poly-*p*-xylylene, a series of experiments with naphthalene were carried out. The effect of naphthalene on the conductivity of islet films formed by sodium and silver was studied. As compared with films formed on both *p*-xylylene monomer and poly-*p*-xylylene, the conductivity of a metal layer on a napthalene-covered polymer underwent no changes at 100–300 K. When islet films formed by sodium and silver on naphthalene supports were used in place of poly-*p*-xylylene supports, the time and temperature dependences of conductivity were similar to those of films formed on poly-*p*-xylylene without naphthalene. Apparently, even if any charge–transfer complexes (CTC) with naphthalene were formed, they had no effect on the conductivity.

Dependences of conductivity on time and temperature were obtained in bimetallic systems of sodium–silver on poly-p-xylylene and sodium–lead on poly-p-xylydene. Preliminary studies were carried out with systems based on individual metals. "Overall" curves represented a superposition of curves for individual metals. Thus, the behavior of a sodium–silver system combined a sharp decrease in conductivity observed upon the completion of film deposition, which was typical of individual sodium on poly-*p*-xylylene, and a conductivity rise as a result of film heating above 250 K, which was typical of silver islet films. In a bimetallic Na–Ag system, the conductivity decreased more smoothly as compared with individual sodium, and its increase started at higher temperatures as compared with individual silver.

In the sodium–lead system, an abrupt drop in conductivity observed upon the completion of metal deposition, which is typical of sodium islet films, was followed by a smoother decrease, in contrast to islet films of individual lead, in which steady-state conductivity values were established only after the film was exposed to 100 K. Probably, the additivity of conductivity values observed by the example of Na–Ag and Na–Pb films may be attributed to the presence of a system of separate islets based on individual metals in a bimetallic system.

From our viewpoint, the conductivity measurements can be used as a control test that would allow one to elucidate the state of a system containing nanoparticles of one or two metals as well as the changes in its state induced by various chemicals.

Properties of polymer-stabilized silver nanoparticles were studied by visible and UV spectroscopy and by dynamic light scattering techniques.[55] Joint condensation of vapors of silver and 2-dimethylaminoethylmethacrylate was realized in vacuum on the walls of a glass vessel cooled by liquid nitrogen. As evidenced by electron microscopy, heating to room temperature produced polymer-stabilized silver nanoparticles measuring 5–12 nm. Dynamic light scattering studies have shown that the size distribution of silver particles is bimodal. Such a distribution is probably associated with the simultaneous existence of individual silver particles and their aggregates. The dependence of the radius of solvated particles on the solvent nature was studied. Of three solvents, namely, water, acetone, and toluene, the smallest silver particles were observed in acetone. Cryochemically formed silver particles were also stabilized in isopropanol, acetone, acetonitrile, and toluene.[56]

Considering the example of a silver–lead–2-dimethylaminoethylmethacrylate system, the effects of metals on one another were studied. Mono- and bimetallic particles in this system were synthesized by joint low-temperature condensation of vapors of components on a vacuum-reactor surface cooled to 77 K. Composition of co-condensates was varied by regulating the power of independent resistive evaporators of metals. By using optical spectroscopy techniques, the compositions of co-condensates containing different relative amounts of silver and lead and the processes, which occurred in both inert argon atmosphere and in the presence of air oxygen, were studied. It was shown that with an increase in the lead content, the absorption band of silver nanoparticles observed at 400 nm in the spectra of organodispersions (co-condensate melts) shifted in the short-wavelength direction. Simultaneously, the absorption in the UV range, which is typical of lead nanoparticles, increased. The results obtained make it possible to assume that the increase in the co-condensate temperature and its fusion induce the formation of bimetallic nanoparticles with uniformly distributed metals.

It was found that in the course of a low-temperature synthesis, 1–2% of 2-dimethylaminoethylmethacrylate is polymerized. The polymer formed stabilizes the organodispersion formed during fusion of co-condensates. In the absence of lead, a part of silver was oxidized in air in the course of polymerization of a sample and its storage in air. The presence of an oxidized silver form, probably its cations, was confirmed by the increase in the absorption intensity of silver nanoparticles upon the addition of a reducer, namely, hydroquinone to the melts and also by much greater resistance of organodispersions stored in argon atmosphere toward oxidation. Lead has a higher reduction potential as compared with silver. The presence of lead or some other reducing agent, e.g. hydroquinone in the co-condensate, prevents silver from oxidation. The introduction of silver nitrate into organodispersions containing either lead or lead with silver increased the absorption at 400 nm. From our viewpoint, this suggests that the volume fraction of silver nanoparticles increased when silver cations were reduced with metal lead.

The results obtained point to nonadditivity of properties of bimetallic nanoparticles formed by the cryochemical method. By the example of the silver–lead system, it was shown that the resistance of one of the components (silver) toward oxidation can be enhanced by varying the nanoparticle composition, i.e. by adding a more active metal (lead). By extending the conclusions drawn based on the results for silver–lead organodispersions to other bimetallic systems, we can extend the possibilities of controlling the stability and reactivity of dispersions involving nanoparticles of two metals.

The behavior of systems containing metal nanoparticles, which were synthesized by the cryochemical method, strongly depends on the low-temperature states and properties of compounds used either for stabilizing nanoparticles or for studying their chemical reactions. Detailed information on the phase composition of individual compounds that interact with metal nanoparticles at low temperatures can be obtained by using a newly developed low-temperature differential scanning calorimeter.[35,36]

The measurements of conductivity of highly active systems containing different metals and alkyl halides were modernized with the aim of enhancing their sensitivity. Comb-shaped supports with gaps between electrodes of 50×0.5 mm were fabricated from glass textolite sheets of 0.2 mm thickness, which allowed the sensitivity of measurements to be enhanced by a factor of 25 when using the same equipment. To eliminate the surface effects, the interelectrode gaps were filled with insulating varnish and then polished together with the conductive coating. By using a unit with enhanced sensitivity, the conductivity variations in low-temperature condensates were measured *in operando*, i.e. in the course of chemical processes. The results obtained in metals–alkyl halides systems, where metals were represented by magnesium and calcium and alkyl halides, i.e. by butyl chloride, butyl bromide, and butyl iodide—suggest that correlations can be drawn between conductivity variations, the metal activity, and the mechanism of cryochemical reactions involving metal particles of different sizes.

5.2.2 Stabilization by Mesogenes

Owing to its properties, the mesomorphic or liquid-crystal state of matter occupies an intermediate place between solid crystalline and isotropic liquid states. Liquid crystals or mesogenes are mobile as liquids and, at the same time, resemble crystals because they retain a long-range order as regards orientation and, in some cases, a translation. Studying chemical transformations in liquid crystals extends the possibilities of controlling the selectivity and the rates of chemical reactions.[57] Specific features of reactions in liquid crystals and, particularly, at low temperatures were considered by several authors.[58–60] Our analysis of the peculiarities in the structure and properties of mesogenic compounds, which were discussed in these studies, have led to an assumption that liquid crystals can be used for stabilizing nanosize particles of metals and, probably, for controlling their shapes.

Cyanobiphenyl (CB) films with silver particles were prepared by joint condensation of vapors of components on cooled surfaces of spectroscopic cryostats in vacuum.[61,62] In a number of cases, to improve the resolution of the spectra, co-condensation was carried out in the presence of excessive inert component, e.g. a saturated hydrocarbon—decane. Condensation rates of evaporated components were varied in the range 10^{14}–10^{16} molecules (sec/cm^2), and the film thickness was 2–100 mm. The metal-to-CB ratio was determined by special calibration of evaporation cells and varied in the range 1:1 to 1:100. The chemical analysis of obtained samples was carried out by an extraction-photometric detection of silver in its complex with dithizone. Ternary systems were prepared by joint condensation of vapors of silver and CB with a 100–1000-fold molar excess of decane.

Co-condensates Ag–CB were studied by IR, UV, optical, and EPR spectroscopies in a temperature range 80–350 K. The size of silver particles was determined by transmission electron microscopy at room temperature. The peculiarities of the observed spectra were compared with the results of quantum-chemical simulation of equilibrium structures and theoretical spectra. For the analysis, packs of programs GAMESS and ALCHEMY were used.

Spectroscopic studies of film co-condensates with a component ratio from 1:1 to 1:100 at 90 K and also of Ag–5CB–decane samples have revealed the presence of low-temperature metastable complexes, which were formed owing to the interaction of silver atoms with the π-electron system of CB molecules. IR spectra of Ag–5CB co-condensates (5CB is 4-phenyl-4'-CB) in the range of valence vibrations of CN groups revealed two new bands at 2080 and 2030 cm^{-1} as compared with a film of 5CB ligand (2230 cm^{-1}) containing no silver.[61] Thus, CN-group bands shifted by −150 and −200 cm^{-1} as compared with the 5CB spectrum. The observed decrease in the valence vibration frequency of the C≡N bond points to the formation of a π-complex of silver and CB. Indeed, the transfer of the electronic density from a binding orbital and the partial occupation of an antibinding orbital of the ligand should loosen the multiple bonds in the complex, i.e. reduce the vibrational frequency of the corresponding bond. A shift by 100–200 cm^{-1} to lower frequencies was observed for the formation of π-complexes of certain transition metals with unsaturated molecules.[63] Co-condensates Ag–5CB and Ag–5CB–$C_{10}H_{22}$ also demonstrated a new band in the region of 650–660 cm^{-1}, which was assigned to metal–ligand vibrations in the Ag–5CB π-complex. Similar results were obtained for silver co-condensates with different CBs. The fact that the intensities of IR spectrum bands corresponding to a π-complex varied in line with the co-condensate temperature variations allowed assigning the band to one and the same complex. The Ag–5CB complex was stable at low temperatures and decomposed at 200–300 K to give the starting compound and silver clusters. An IR spectrum of Ag–5CB co-condensate at room temperature resembled that of a film of individual condensate of 5CB molecules. The results obtained agreed with the low thermal stability of complexes of zero-valence metals with unsaturated compounds. The formation of π-complexes in the system under study was confirmed by the results of quantum-chemical calculations carried out for 5CB by the example of its fragment PhCN.[64]

Physicochemical evolution of metal-containing atomic–molecular systems of silver co-condensates with mesogenic CB 5CB was studied in a temperature range 80–300 K using the EPR method. As clearly seen from Figure 5.10, the co-condensate spectra measured at 80 K demonstrated signals with well-resolved hyperfine interactions (HFI) on metal atoms. This indicated the formation of complexes of silver atoms with mesogenic ligands under conditions of low-temperature co-condensation. Parameters of doublet signals in co-condensates of silver with 5CB and 4-pentyl-4′-cyanophenylpyridine (5Py), which were estimated by simulation of spectra, are typical of the formation of π complexes of Ag^{107} ($J=½$) and Ag^{109} ($J=½$) atoms. A comparison of HFI constants determined for complexes and isolated atoms A (Ag^{107})=611 G and A (Ag^{109})=705.4 G, made it possible to assess the unpaired electron density on the s-orbital of the metal: $\rho_s=0.89$ (Ag–5CB). The central signal "C" in the region of free-electron g-factor was associated with the absorption from the totality of silver clusters stabilized in the 5CB matrix. Such signals were observed for concentrated samples of silver co-condensates with inert gases and hydrocarbons. They were assigned to spin-resonance signals of conduction electrons in metallic nanoclusters.[65] An average silver cluster in a co-condensate with a ratio Ag/5CB = 1/10 at 90 K comprised several tens of atoms (1–2 nm) as assessed from an experimental spectrum of the sample. With an increase in the temperature in the range 80–150 K, the relative intensity of doublet components in the spectrum decreased, while the intensity of the central component increased, which indicated the thermal decomposition of the complex and the growth of silver nanoclusters.

EPR spectra of copper co-condensates with 5CB were measured.[66,67] The spectra demonstrated a strong anisotropic quartet signal. As in the A–5CB

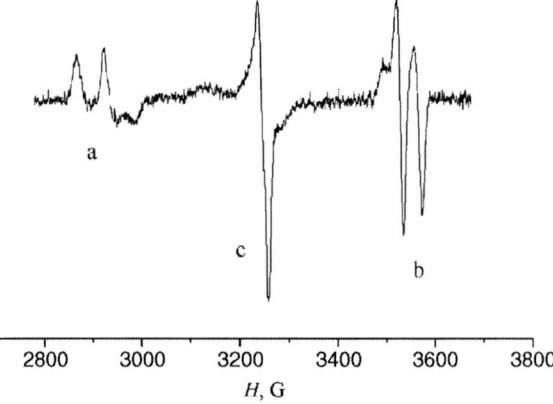

FIGURE 5.10 EPR spectrum in the system silver–4-pentyl-4′-CB: (a, b) hyperfine interaction on metal atoms and (c) absorption of silver clusters.

system, heating resulted in decomposition of the complex, aggregation of atoms, and formation of copper nanoparticles.

A possibility of photo-induced formation of silver nanoclusters in the temperature interval 80–90 K at UV irradiation of low-temperature samples was demonstrated. Furthermore, heating above 200 K resulted in a very fast decrease in the central line intensity and induced a very broad background absorption that can be attributed to the appearance of larger silver nanoparticles and their aggregates. As compared with the spectra measured in the absence of the metal, the optical spectra of Ag–5CB co-condensates at 90 K revealed a structured absorption band with a maximum at 360 nm, which corresponded to pale yellow co-condensate films.[62] A quantum-chemical simulation of the excited states for a complex of the proposed structure showed the presence of several intense charge–transfer transitions of metal–ligand and ligand–ligand kinds in this region.[68] It should be mentioned that in the range 390–420 nm, the absorption of small silver clusters can also be present. The structured band disappeared upon heating the co-condensate film to 200–300 K.[69] Thus, the heating of a sample up to room temperature initiated thermal degradation of the complex and aggregation of liberated silver atoms. A wide band with a maximum at 440 nm, which appeared at these temperatures, was attributed to the absorption of surface plasmons of nanosize silver particles[15,50] formed as a result of silver aggregation during decomposition of complexes. Instantaneous heating of a sample to 300 K transferred the sample to the nematic mesophase state, which manifested itself in an increase in the plasmon resonance absorption in the long-wavelength region of UV spectra and can be associated with both coarsening of silver particles as a result of further aggregation and the formation of nonspherical, anisotropic metal particles in the orientation-ordered matrix.

Thus, the samples obtained by low-temperature co-condensation of vapors of metal silver and CB and then heated to room temperature represented a nanocomposite material, which consisted of silver nanoparticles stabilized in a CB matrix.[70] The nematic properties of the material were retained. Thermograms of obtained samples and individual 5CB were identical. Their textures corresponded to the nematic phase.

Low-temperature layer-by-layer co-condensation of vapors of silver, 5CB, and *p*-xylylene monomer followed by heating of the obtained film sample resulted in encapsulation of the metal–mesogen system into a polymeric film. Electron microscopic studies of film samples, which were prepared by the encapsulation of a silver-containing 5CB sample into poly-*p*-xylylene under conditions of vacuum co-condensation of reactants, revealed the existence of two kinds of metal particles stabilized in the mesogenic matrix at room temperature.[71] These were globular silver particles with diameters of 15–30 nm and anisotropic rod-like metal–mesogen particles longer than 200 nm, which were stabilized in a CB matrix. UV–visible spectra of Ag and 5CB films formed in poly-*p*-xylylene demonstrated a wide absorption band at 440–600 nm at room temperature. An increase in the metal–ligand ratio in the sample resulted in the preferential growth of rod-like silver particles.[70]

Thus, in silver-containing films of mesogenic CBs obtained by low-temperature condensation of vapors of components under molecular beam conditions, metastable π complexes of metal atoms with CB dimers were formed. The formation of metastable π-complexes in the temperature interval of 90–200 K was confirmed by the results of IR, UV, and EPR spectroscopic studies in combination with the quantum-chemical simulations of the "silver–cyanophenyl" model system. With an increase in the temperature, thermally unstable complexes decomposed and silver atoms aggregated within an anisotropic liquid-crystal matrix to give nanoclusters and aggregates of silver nanoparticles. Heat- and photo-induced degradation of complexes in the temperature range 90–200 K induced the formation of nanosize silver particles and their further aggregation within an anisotropic matrix. Such systems are promising for catalytic applications and can exhibit valuable electrooptical properties.

Yet another approach to studying stabilization and reactivity of metal atoms, clusters, and nanoparticles was developed in our studies. This approach is based on an idea of "interception" of active particles by "third" molecules. Low-temperature interactions of atoms and small clusters give rise to the formation and stabilization of either molecular complexes or ligand-surrounded metal particles of various sizes. With further heating, such formations are stabilized by low temperatures and decompose with liberation of active metal particles, which, in turn, enter into reactions with other compounds. These compounds may either be present in the initial system or should be specially introduced during the heating. For such an approach to be realized, information on thermodynamic and kinetic peculiarities of the systems that involve metal particles and a stabilizing ligand should be gained.

Competitive interactions in ternary co-condensates prepared according to the principle "single metal—two organic reagents" were studied. By the example of a system 5CB–carbon tetrachloride, a possibility of using labile complexes the thermal decomposition of which produces highly active particles, was considered. The introduction of an electron–acceptor ligand into the system was realized for co-condensation of silver and 5CB with carbon tetrachloride. As shown earlier, the interaction of Mg with CCl_4 at low temperatures involves synthesis of a Grignard reagent, detachment of one or two chlorine atoms, and formation of C_2Cl_4 and C_2Cl_6 among the products. IR spectra of the Ag–5CB–CCl_4 system showed the absence of any products similar to those obtained in the reaction of Mg with CCl_4. At co-condensation of Ag, 5CB, and CCl_4, carbon tetrachloride took part in the formation of more stable complexes, which did not decompose throughout the temperature interval of the matrix existence. IR spectra measured at 90 K in the range of valence vibrations of CN groups revealed a new band with a maximum at 2264 cm^{-1}.

The band shift with respect to valence vibrations of CN groups in individual CB was 137 cm^{-1}. From our opinion, this points to the formation of a σ-complex, which is stable throughout the interval of matrix existence. The introduction to the Ag–5CB system of benzene or decane as the third component induced no changes in the metal–ligand interaction and entailed strengthening of the metal bonding, owing to the increase in the total solid-state mobility of molecules.

5.3 REACTIONS OF RARE-EARTH ELEMENTS

Nanochemistry of lanthanides or rare-earth elements (REE) has been insufficiently studied. In this section, we combined few literature data available and our own results. Lanthanides are multielectron systems, and their quantum-chemical consideration poses many problems, which complicates the comparison of calculations and experimental results. Our initial studies of low-temperature condensates including those of lanthanides date back to the early 1980s.[72]

At present, REE are actively studied. This is associated with at least two reasons. First, among the elements in the periodic table, REE are the least known and, second, their atoms, clusters, and relevant materials exhibit unique optical, magnetic, and catalytic properties. Vapors of REE such as Yb, Sm, and Eu when co-condensed with alkenes favor the insertion of a metal atom into the C–H bond, the rupture of the C–C bond, and oligomerization and dehydrogenation of C_2H_4, C_3H_6, and cyclopropane. However, attempts to extract individual OMC have failed.[73,74]

Relatively recently, cryochemical methods have allowed zero-valence OMC of lanthanum, ytterbium, and gadolinium with 1,3,5-tritretbutylbenzene to be obtained for the first time. Their yield approached 50%, and stability was retained up to (100 °C) 373.15 K or 373 K. The compounds had a sandwich structure. For gadolinium, the structure was established based on XRD data. Stable compounds were synthesized with Nd, Tb, Ho; labile compounds were prepared with La, Pr, and Sm.[75,76]

It was assumed that a metal atom should have an easily accessible d^2s^1 state. The instability can be due to a great covalent radius of a corresponding metal. The studies of magnetic properties of complexes confirmed the assumption drawn and a scheme, in which only three of the valence electrons of a lanthanide take part in binding benzene rings, while the other electrons remain in the f-shell.

In the periodic table, lanthanides are placed in a separate group because of their specific electronic configuration $4f^n6s^2$. In contrast to the d orbital, the population of a transition-metal $4f$ orbital usually has no effect on the chemical properties owing to the small size of this orbital and its strong screening by occupied $5s$ and $5p$ orbitals. All lanthanides readily form positive oxidation states. An oxidation degree +3 is typical for most members of this series, although Sm and Eu, for instance, can have an oxidation degree +2. For samarium nanoparticles, it was found that clusters containing less than 13 atoms have a valence 2, while for particles containing more than 13 atoms the valence state 3 predominates.[77]

When carrying out chemical reactions involving metal atoms or their clusters, it is of interest to compare their feasibility with similar reactions involving compact metals.

In low-temperature co-condensates formed by vapors of metals and different ligands, the high chemical activity of metal particles is combined with the high selectivity of the process and its dependence on the temperature. The totality of the mentioned factors allows realization of processes that never occurred with compact metals in the liquid phase at room temperature.

By studying reactions of lithium, sodium, magnesium, samarium, and ytterbium with acetone, it was shown that under cryosynthetic conditions clusters of lithium and sodium tend to form pinaconates, whereas compact metals form enolates.

Under conditions of cryosynthesis, samarium and ytterbium particles form pinaconates, whereas corresponding compact metals do not react with acetone at room temperature. Magnesium, which has the highest first-ionization potential, forms enolate at low temperatures and pinaconate at room temperature. Reactions of metals with acetone and their possible mechanisms were considered.[78,79]

In addition to reactions with acetone, reactions with acetylacetone were also considered for samarium and ytterbium.[38] Co-condensation of samarium or ytterbium on a surface at 80 K for a metal-to-ligand ratio equal to 1:(20–500) produced light-brown films. These films lost their color when heated to 130–135 K, which was accompanied by the formation of samarium tris-acetylacetonate as an adduct with one acetylacetone molecule.

$$Sm + 4\ CH_3\text{-}C(O)\text{-}CH_2\text{-}C(O)\text{-}CH_3 \longrightarrow \left[\begin{array}{c} H_3C \\ \diagup \\ \diagdown \\ H_3C \end{array}\right]_3 SmCH_3\text{-}C(O)\text{-}CH_2\text{-}C(O)\text{-}CH_3 + 3/2\ H_2$$

The composition of this product was found by its elemental analysis and from the IR spectrum. As compared with the liquid-phase synthesis involving compact samarium, the cryochemical method made it possible to obtain anhydrous sublimated acetylacetone. The sublimated product yield depended on the molar ratio of the reagents in the co-condensate.

An attempt to enhance the volatility of samarium acetylacetonate by carrying out its reaction with fluorine derivatives failed. In co-condensates of samarium and hexafluoroacetylacetone with a molar ratio of 1:50, a spontaneous explosive reaction accompanied by a bright flash was observed when a certain film thickness was reached. Analysis of IR spectra made it possible to conclude that concurrent reactions occur at two centers, namely, $C=O$ and $C-F$ bonds, which points to the high reactivity of these bonds at low temperatures.

A comparative study of the chemical reactivity of atoms and small clusters of sodium, magnesium, and samarium in low-temperature co-condensates was performed on the example in their co-condensates with alcohols.[80–82] Samarium reactions were studied more comprehensively.

A cryosynthesis involving a small excess of alcohol (5:1) with respect to samarium yielded samarium alcoholate; however, with a decrease in the metal ratio in the co-condensate (1:500 and lower), the alcoholate yield decreased. Reaction products contained hydrocarbons (e.g. n-pentane and traces of decane, for the case of n-pentanol-1). The pentane yield was 1.5 moles per mole of deposited samarium. Thus, in co-condensates strongly diluted with respect to metal, alcohols were reduced to hydrocarbons. Special experiments showed that hydrocarbons were formed in samarium–alcohol systems only in the course of condensation at 80 K. Alcoholates were formed in the course of heating of solid co-condensates.

The surface deposition of samarium and an alcohol was accompanied by completion of processes of atom–ligand interaction and aggregation of atoms. The following scheme was proposed:

$$Sm \xrightarrow[80K]{ROH} RSmOH \xrightarrow[80K]{ROH} 1.5\, RH + (RO)_{1.5}Sm(OH)_{1.5}$$

$$\downarrow Sm_{n-1}$$

$$Sm_n \xrightarrow[80 - 300K]{ROH} Sm(OR)_3 + H_2$$

Presumably, this scheme includes an OMC, RSmOH, which is similar to the Grignard reagent. The experimental yield of hydrocarbons coincided with an estimate obtained based on the kinetic analysis of the reaction scheme. Moreover, such an analysis makes it possible to assess the effective ratio of rate constants for reactions of samarium atoms, which can either enter into the dimerization process or get inserted into a C–O bond of an alcohol molecule.[82] It was found that the interaction between samarium atoms is more likely than the reaction of samarium atoms with alcohol molecules. Moreover, the process was observed as an alternative for alcohol molecules containing even/odd numbers of carbon atoms.

The study of the evolution of co-condensation has shown that the reduction of alcohols to hydrocarbons proceeds at the instance of co-condensation. Probably, at the same moment, the intermediate compound of the insertion of Sm atom into the C–O bond as well as samarium clusters is formed, which are transformed to a corresponding alcoholate during further heating.

The scheme shown above and the kinetic analysis give only a partial picture of the processes occurring in low-temperature co-condensates. The complicated nature of such reactions is evidenced by both the synthesis and catalytic properties of samarium co-condensates with hydrocarbons.[83,84] This was studied by the example of cyclohexane, hexane, hexane-1, and cyclohexene. Catalysts were prepared by the co-condensation of vapors at low temperatures. The catalytic activity was determined as a hydrogenation rate constant per mass unit of vaporized samarium. A catalytic system obtained by joint condensation of samarium and hexane turned out to be more active in the hydrogenation of hexane-1 as compared with cyclohexene hydrogenation. On the other hand, a catalytic system prepared by co-condensation of samarium with cyclohexan was more active in cyclohexene hydrogenation and, vice versa, less active toward hexane-1. This phenomenon was named "memory effect."

Thus, the chemical activity of a system depends on the size and shape of the particles that take part in its formation. During co-condensation, a certain linear or cyclic hydrocarbon provides the positions for samarium atoms and fixes the latter in these positions in a cluster, which predetermines subsequent hydrogenation reactions. In this case, we have an analogy with enzyme catalysis, where the enzyme–substrate interaction involves dynamic reconstruction and adjustment of an enzyme active center to the corresponding substrate.

For similar ratios of components and equal co-condensation rates, the catalytic activity is independent of the presence of double bonds in the hydrocarbon chain. For instance, the activity of co-condensates with hexane or hexene in hydrogenation of the same substrate was virtually the same. In contrast to catalytic activity, the specific surface area of catalysts was independent of the hydrocarbon/samarium ratio and amounted to $100\,m^2/g$. The results shown above and the fact that on heating co-condensates evolved hydrogen allowed us to conclude that the catalytic activity is related with the formation of an OMC, and, moreover, the nature of the latter is the same in co-condensates containing alkanes and alkenes. Such a compound was assumed to be $RC\equiv CSmH$.[84]

Below, we show a possible reaction scheme for a metal/hydrocarbon ratio equal to 1:1000.

$$RCH_2CH_3 \xrightarrow{Sm} RCH_2CH_2SmH \xrightarrow[T_m]{-2H_2} RC \equiv CSmH$$

$$RCH_2=CH_3 \xrightarrow{Sm} \underset{RCH\ \ SmH\ \ CH_3}{CH} \xrightarrow[T_m]{-H_2} RC \equiv CSmH$$

Yet another limiting case is possible for a metal/hydrocarbon ratio equal to 1:1. Such a condition favors the formation of samarium clusters. In hydrolysis products, a pronounced increase in the hexine fraction was observed. Inasmuch as no dehydrogenation was observed during condensation of hydrocarbons on pure samarium, it was assumed that samarium enters into the reaction in the form of clusters Sm_n by the scheme.

$$RCH_2CH_3 \xrightarrow{Sm_n} RC \equiv CH + Sm_nH_{2\alpha} + (2-\alpha)H_2 \quad 0<\alpha<2$$

$$2n\ RC \equiv CH \xrightarrow{Sm_n} \left[(RC \equiv C)_2Sm\right]_n + H_2$$

Intermediate dilution of the metal by an order of magnitude of 100 makes both schemes possible. An increase in the hydrocarbon–samarium ratio increases the fraction of metal atoms that produce a catalytically active compound RC≡CSmH. Samarium dihexinide formed at low hydrocarbon–samarium ratios proved to be catalytically inactive.

In the above examples, the involvement of atoms or larger particles in the reaction was judged from the metal/ligand ratio. It was assumed that, as shown by the example of samarium reactions with hydrocarbons, a 1000-fold ligand excess favors reactions of atoms, whereas a 1:1 ratio favors reactions of clusters and nanoparticles. Of great importance are the nature of metal and ligand, the temperature of the co-condensation surface, dilution by inert compounds, and certain other factors mentioned above. The direct determination of sizes of reacting particles currently remains the central problem.

Low-temperature joint condensates of samarium and mesogenic 5CB and 4-octyl-4'-CB (8CB) in the temperature range 6–300 K were studied by IR and UV spectroscopic techniques.[85–87] The formation of two labile complexes with metal/ligand ratios equal to 1:2 and 1:1 was observed.[88] At temperatures 170–210 K, a solid-phase transformation of the $Sm(CB)_2$ complex to $Sm_2(CB)_2$ took place. The kinetics of this process is multistep, which points to the wide distribution over both the reactivity of complexes and the activation energy of solid-phase transformation.[89] IR spectra of Sm–5CB co-condensates measured in the temperature range 95–273 K and also for different metal–ligand ratios at 95 K revealed two new bands with maximums at 2135 and 2085 cm^{-1} in the region of valence vibrations of CN groups, as compared with the IR spectrum of a film of pure 5CB. A shift by about 100 cm^{-1} in the low-frequency direction made it possible to assign these absorption bands to π-complexes. A co-condensate with a metal/ligand ratio of 1:1 had a single absorption band (2135 cm^{-1}) at 95 K. With heating from 95 to 213 K, the absorption increased at 2085 cm^{-1} in synchronism with its reduction at 2135 cm^{-1}. Thus, a transformation of one complex to another occurred. Moreover, the changes in the band ratio, which accompanied the metal–ligand ratio variations, allowed the presence of two complexes with different compositions to be assumed.[90] Spectra of Sm–5CB film co-condensates in UV and visible ranges demonstrated two new overlapping bands with maximums at 390 and 420 nm.[89] Absorption in this spectral range is typical of the CTC of transition metals with unsaturated organic molecules. The intensity of these absorption bands was also temperature-dependent. This trend provided an additional confirmation to the existence of two complexes. The proposed structures of these complexes are as follows:

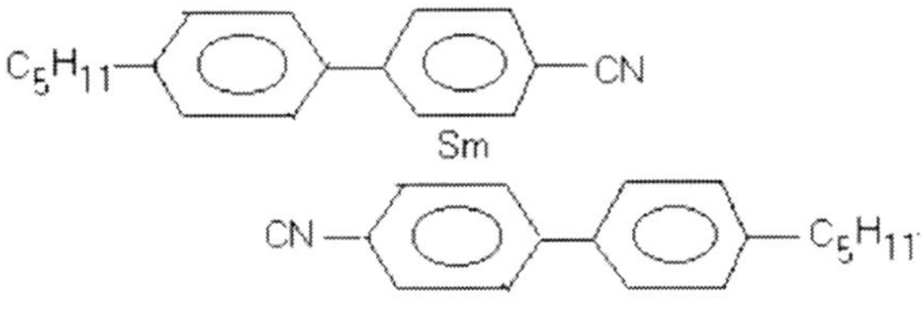

C_5H_{11}—〈◯〉—〈◯〉—CN
Sm
CN—〈◯〉—〈◯〉—C_5H_{11}

Complex 1/2 [Sm(5CB)$_2$]

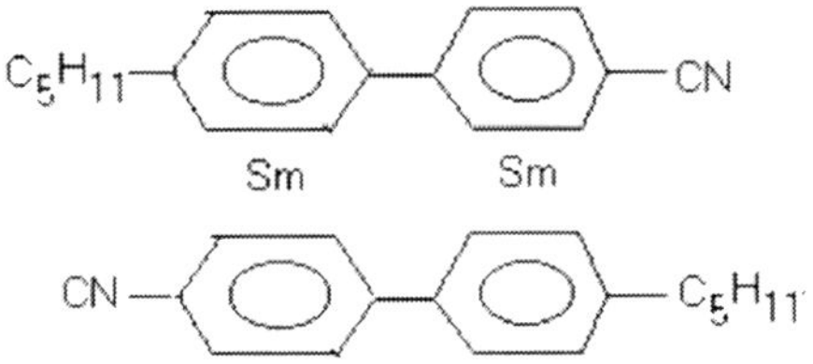

C_5H_{11}—〈◯〉—〈◯〉—CN
Sm   Sm
CN—〈◯〉—〈◯〉—C_5H_{11}

Complex 1/1 [Sm$_2$(5CB)$_2$]

These models reflect the equivalence of 5CB molecules in a complex and the possibility of formation of sandwich compounds for zero-valence lanthanides.

The time dependences of absorption by Sm (CB)$_2$ complex cannot be linearized in the coordinates corresponding to reactions of the first and second orders; hence, the solid-phase transformations of one complex to another are of the polychronic nature.[311] This fact was associated with the wide distribution of molecules over rate constants. From the kinetic dependence $dA/(d\ln t)$, we can obtain the function of distribution of molecules over reactivity $\phi(G)$ as

$$1/C_0 dC/d(\ln t) = 1/A_0 dA/d(\ln t) = -RT\phi(G).$$

Kinetic data on the transition of SmCB complex of a 1:2 composition to a complex with a 1:1 composition can be described by a linear dependence between the absorption of the 1:2 complex and ln t, which points to the rectangular distribution of particles over free energy of activation.

In this connection, the thermal stability of zero-valence samarium complexes with 1,3,5-tri(tretbutyl)benzene was studied.[91] An IR spectroscopic study has shown that this kind of a complex with absorption maximum at 967 cm^{-1} is less stable than a complex with absorption at 973 cm^{-1}. Thus, the polychronic nature of the kinetics of decomposition of a low-temperature (967 cm^{-1}) complex was demonstrated.[92]

In summary, the results obtained in the late 1980s and early 1990s on the reactions of metal particles in low-temperature co-condensates made it possible to generalize and formulate several specific features for reactions involving two or more substances.[93]

These features include:

- the presence of mechanically stressed nonequilibrium states;
- the existence of a molecular organization in low-temperature condensates;
- the presence of nonequivalence in energy, kinetics, and thermodynamics; and
- sufficient mobility of reactants at the instance of co-condensation.

An analysis of the data on systems involving metal particles allowed the following conclusions to be drawn.[94]

(1) Nanosize metal particles are systems with stored energy, which is determined by uncompensated bonds of surface and near-surface atoms, contributions of metastable states, latent heat of phase transitions, and energy of defects.
(2) The concentration of vacancies increases with a decrease in the particle size. These effects manifest themselves in the changes in the temperature of polymorphous transitions, a decrease in the lattice parameters, and an increase in the compressibility and solubility.

Classic thermodynamics does not describe particles smaller than 1 nm, and its application is complicated owing to the fundamental problem of determination of dimensional boundaries between different phases and the demarcation line between homogeneous and heterogeneous states. The energy of a system can be enhanced by its dispersion, while a traditional way consists of increasing the temperature.

The latter statement means that the chemical transformations unrealizable with compact metals are possible for nanosize particles. The analysis of the results obtained allowed a general scheme of studies to be proposed, which is illustrated in Figure 5.11.

In fact, when studying clusters, we deal with a certain size distribution dependent on the conditions of preparation of a system. Hence, of great importance is the kinetic analysis of cluster formation. By a computer simulation of one-dimensional growth of clusters, based on the model of diffusion-controlled aggregation in a system of growing particles, it was shown that the kinetics of cluster formation depends on the initial positions of particles, i.e. uniform or equidistant. At the uniform (random) initial distribution of atoms, in contrast to the equidistant one, clusters in the densest areas can grow in the very beginning of particle motion. In this case, the initial rate of cluster accumulation is higher as compared with the equidistant distribution. The dependence holds for times exceeding the time of diffusion mixing. This effect was interpreted as the system's memory.[95] This model closely resembles the model of diffusion-controlled aggregation,[96] which includes three stages: deposition, diffusion, and irreversible aggregation.

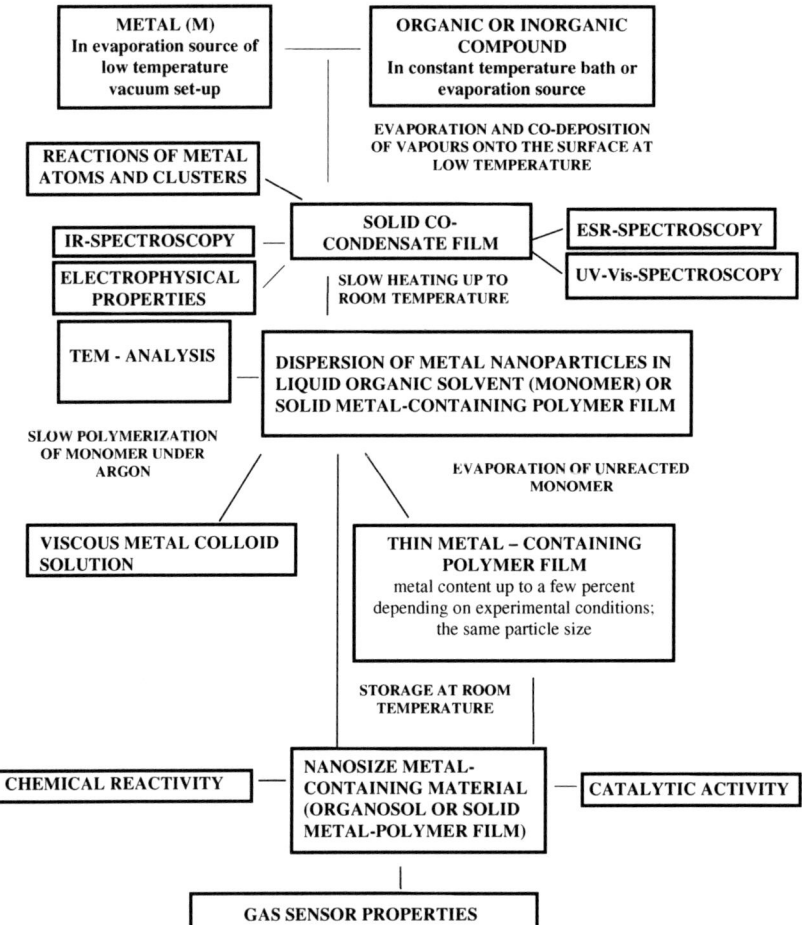

FIGURE 5.11 Cryochemical synthesis of nanosize materials encapsulated into organic and polymeric matrices.

5.4 ACTIVITY, SELECTIVITY, AND SIZE EFFECTS

5.4.1 Reactions at Superlow Temperatures

Inert matrices and superlow temperatures make it possible to gain valuable information on the properties of metal atoms. Metal atoms form clusters and nanoparticles owing to aggregation. The knowledge of initial optical and diffusion properties of atoms in inert matrices lays the basis for understanding the subsequent processes of formation and stabilization of nanoparticles. The reactivity of metal atoms and their trend to form clusters are determined by their interaction with the environment. Toward the beginning of 1990s, spectral and radiospectral techniques made it possible to gain comprehensive knowledge on the behavior of atoms of virtually all elements in inert matrices at low temperatures, except for REE.

The spectra of samarium in argon matrices demonstrate a great number of absorption bands, which could be identified because of the fact that individual transitions were well separated.[97] Heating of samples had practically no effect on the spectrum. The presence of a great number of absorption bands in the samarium spectrum was associated with either the existence of several stable positions of samarium atoms in the argon unit cell or their intense interaction with the inert gas, which lifts prohibitions for certain transitions and band splitting.

Increasing the metal condensation rate and carrying out photoaggregation made it possible to assign the absorption bands to samarium dimers. Light absorption in the wavelength range of most intense bands of atoms resulted in gradual disappearance of the atomic spectrum and the appearance of a new band in the vicinity of 646 nm, which was assigned to coarse aggregates of samarium. An increase in the condensation rate gave rise to absorption bands of the dimer at 352, 541, 599, and 727 nm. The samarium spectrum was sensitive to the substrate temperature. Co-condensation at 15 K resulted in virtually similar intensities for atomic and dimeric bands. For a substrate temperature of 20 K, the spectrum could not be observed owing to the strong increase in the background absorption corresponding to light scattering by coarse particles.

Studies of holmium atoms isolated in an argon matrix showed that this element exhibits a more complicated spectrum as compared with samarium.[98,99] Most bands in the holmium spectrum are grouped in the vicinity of 400 nm. Holmium has a single stable isotope ^{165}Ho, the high magnetic moment of which (I = 7/2) is probably the reason for such a complicated spectrum.

Heating and illumination of a matrix showed that the spectrum of holmium atoms consists of two bands pertaining to atoms isolated in different cells of the argon matrix. One of the cells is thermally instable. The illumination of a matrix induces photoaggregation of atoms and the formation of holmium dimers that absorb at 500 and 570 nm. The energies of the excited states of holmium dimers obtained in quantum-chemical calculations based on the pseudopotential method agreed with those obtained in experiments.[100]

Holmium, like samarium, is sensitive to the substrate temperature. With the increase in condensation temperature in the range 10–20 K, its spectrum revealed a broad absorption band at 450 nm, which was assigned to metal plasmons.

The revealed peculiarities in the behavior of argon-matrix isolated samarium and holmium atoms and small clusters provided grounds for studying their reactions with various ligands.

The electron spectra provide valuable information on the behavior of metal particles but are less informative with regard to the final products, which are conventionally analyzed using IR spectroscopy.

Ab initio quantum-chemical calculations of the geometry, energy, and vibrational frequencies for the magnesium–methane halide systems were carried out using the programs SAMESS and Gaussian-94.[101] The calculations employed the methods of multiconfiguration self-consistent field (MCSCF), the Möller–Plessett perturbation theory of the second order (MP2), and the method of valence bands (VB) used in combination with the electron density functional theory. The reaction of magnesium particles with methane halides was observed experimentally only under the photolysis conditions, although the earlier studies showed that it can also occur during condensation. The main problem was to determine the number of chlorine or bromine atoms that get inserted into the carbon–halogen bonds. The calculations of the energy corresponding to the formation of different channels

$$RMgX \rightarrow RX + Mg$$

$$RMgMgX \rightarrow RMgX + Mg$$

$$RMgMgX \rightarrow RX + Mg_2$$

showed that all compounds are stable and the synthesis of bimagnesium compounds is advantageous for energy reasons. When going from RMgX to RMgMgX, the energy gain was 6–8 Kcal/mol. The nonmonotonic changes in binding energy in the series fluorine–chlorine–bromine were also noted. According to calculations, intense absorption bands in the vibration spectrum should be observed in a range of 600–400 cm^{-1}. For compounds containing one or two magnesium atoms, the number of absorption lines is different. Valence vibrations C–H coincide for all Grignard reagents and are virtually independent of the nature of a halogen atom due to its screening by the metal atom.

Two kinds of equi-intens\e vibrations should appear in the vicinity of 600 cm^{-1}. The first kind of vibrations represents pendular vibrations p(CH$_3$) shifted by magnesium from 1017 cm^{-1} in CH$_3$Cl and 955 cm^{-1} in CH$_3$Br; the second kind is deformation vibrations of C–Mg–X. The incorporation of the second magnesium atom gives rise to low-frequency deformation vibrations in the C–Mg–Mg–X system in the vicinity of 400 cm^{-1}.

Experimental spectra confirmed the presence of only two bands in the vicinity of 550 cm^{-1} and the absence of absorption below 500 cm^{-1}. This agrees with the formation of CH$_3$MgX. When a cut-off filter with a wavelength $\lambda > 300$ nm was used, the light energy proved to be insufficient for excitation of magnesium atoms but sufficient for absorption by small clusters. Thus, the higher activity of clusters was confirmed. The most probable mechanism was associated with the insertion of an excited cluster:

$$CH_3X + Mg_n^* \rightarrow CH_3Mg_nX$$

followed by ejection of a smaller cluster:

$$CH_3Mg_nX \rightarrow CH_3MgX + Mg_{n-1}$$

A more sophisticated experimental and theoretical study of the reactions between methane halides and magnesium made it possible to refine the reaction mechanism. Indeed, as shown by ab initio quantum-chemical calculations, the synthesis of compounds formed at the insertion of a magnesium cluster, e.g. dimer, into a carbon–halogen bond is advantageous for energy reasons. However, IR spectroscopy failed to confirm the formation of the corresponding compounds. Probably, their formation can occur by sequential insertion of two magnesium or calcium atoms. The absence of dimers was demonstrated for magnesium particles obtained by thermal evaporation.[101] No dimers were detected in the course of laser evaporation of magnesium, followed by its reaction with methane halides CH_3X, where X=F, Cl, Br, I.[102] At the same time, magnesium particles obtained by laser and thermal evaporation demonstrated a certain deviation in the reaction products. The interaction of particles synthesized, e.g. by laser evaporation, and the kinetic control over their growth may depend on the gas pressure and temperature. However, as a whole, the problem is not yet solved, and it is still impossible to predict the reactivity.

The study of the relationship between the reactivity and the particle size relies, on the one hand, on physicochemical, particularly spectral, characteristics of atoms, dimers, trimers, and more complex nanoparticles and, on the other hand, on their reactions with molecules specially added to the system. The method of matrix isolation at superlow temperatures provides the best way for combining these two approaches.

As shown by ab initio quantum-chemical calculations, the magnesium–carbon dioxide complex represents a radical–ion pair $Mg^{+\bullet}CO_2^{-\bullet}$, and this pair is metastable, which is ensured by the argon matrix.[33,103,104] In co-condensates of argon and ethylene, absorption bands of ethylene dimers were observed. As demonstrated by calculations, the potential energy surface of the $Mg(C_2H_4)_2$ system has no minimum with C_2 symmetry. A global minimum could be obtained only when the limitations in symmetry were lifted. In the Möller–Plessett (MP2) approximation, the minimum corresponded to a cyclic structure comprising an ethylene dimer and a magnesium atom. The structure of a compound in the magnesium–ethylene–carbon dioxide system was also calculated.

The study of co-condensates of samarium with carbon dioxide showed that samarium forms a complex of an angular structure. The first products are, correspondingly, CO and carbonate anion CO_3^{2-}. A radical anion $CO_2^{-\bullet}$ and a samarium carbonyl of unknown composition were formed. Below, we show a scheme of the reaction between a samarium atom and a carbon dioxide molecule in an argon matrix (straight arrows indicate the processes during condensation, curved arrows designate processes during heating of the condensate):

$$Sm + CO_2 \longrightarrow O\!=\!\!\overset{\overset{\displaystyle Sm}{|}}{C}\!\!=\!\!O \xrightarrow{T=35\,K} Sm^+CO_2^{-\bullet} \xrightarrow{} Sm^+CO_2^{-}$$

$$\longrightarrow CO + CO_3^{2-} \xrightarrow{Sm,\ T=35\,K} Sm(CO)_n$$

Figure 5.12 shows the relative activity of samarium particles in the reaction with carbon dioxide as a function of temperature. The data in the figure point to a higher activity of samarium particles as compared with atoms. A similar dependence was obtained for magnesium particles.

According to IR spectra, in co-condensates of samarium and ethylene, the formation of samarium complexes with ethylene of the sandwich type SmC_2H_4 and $Sm(C_2H_4)_2$ takes place, rather than the formation of cyclic compounds observed for the case of magnesium.

By examining the electron spectra in the system $Sm/C_2H_4 = 1:1500$, it was shown that samarium atoms can be stabilized in ethylene at 14 K.[105] An analysis of temperature-induced changes in the absorption spectra and relevant kinetic studies of samarium atoms demonstrated that an SmC_2H_4 complex is transformed into an $Sm(C_2H_4)_2$ complex, which is stable up to 50 K. The stepwise kinetics was judged from the changes in an absorption band at 386 nm ($f \rightarrow s$) in the temperature range 15–30 K. At such temperatures, ethylene molecules and samarium atoms lack translation mobility; however, this is the temperature interval in which the complete transformation of samarium occurs. High dilution (1:1500) allows one to rule out atomic aggregation and assume that samarium atoms are bound into a complex because of rotational mobility of ethylene molecules. The calculated coefficient of rotational diffusion at 20 K was $0.06 \, s^{-1}$.

Recently, new results on the reactions of magnesium atoms and small clusters with methane polyhalides in the temperature range 12–40 K were obtained.[106–108]

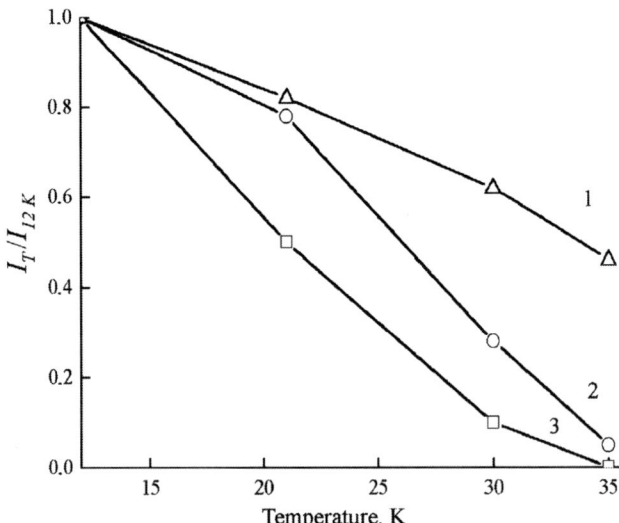

FIGURE 5.12 Normalized integral intensity of samarium particles at different temperatures in the reaction with carbon dioxide: (1) Sm, (2) Sm_x, and (3) Sm_2.

The activities of magnesium atoms, dimers, trimers, and larger particles in their reactions with carbon tetrachloride, trichlorfluoromethane, and trichloromethane were studied. The choice of these compounds was associated with several reasons. Considering the example of CCl_4–Mg, we expected to check the possibility of the reaction at temperatures below the nitrogen boiling point. Moreover, as was already noted for cryoreactions involving magnesium and calcium particles, there is an inconsistency between the activation energy of methane halides and the binding energy of carbon–halogen. The binding energies in CCl_4, $CFCl_3$, and $CHCl_3$ differ significantly.[109] This provided grounds for assuming that the behavior of low-temperature reactions would reveal a substantial difference in binding energies and reflected in the reactivity of compounds under study.

The use of electron spectroscopy allowed us to follow the relative changes in the activity of magnesium particles with different sizes. IR spectroscopy provided information on possible reaction products formed immediately in condensates of reagents at low temperatures. Figure 5.13 shows the spectra of different magnesium particles in an argon matrix in a temperature interval 12–35 K.

The spectra of nontransition metals in a matrix or in the gas phase differ insignificantly, which allows gas-phase spectra to be used for identification of spectra obtained in matrices.

The revealed effect of the temperature on the behavior of different argon-matrix-isolated magnesium particles in the course of their heating allowed studying their activity in reactions with various methane polyhalides.

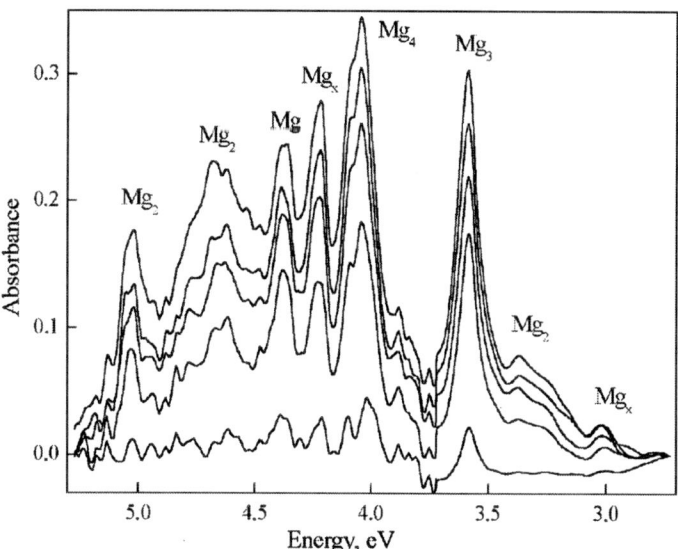

FIGURE 5.13 The effect of temperature in the interval from 12 to 35 K (top–down) on the electron spectra of magnesium particles in argon. Mg/Ar ratio is 5 1:1000.

Figure 5.14a shows electron absorption spectra of co-condensates formed by magnesium atoms and small clusters with carbon tetrachloride in an argon matrix in the temperature range 12–35 K. Figure 5.14b compares the changes in the relative activity of different magnesium particles. The data in figures made it possible to assume that the activity of magnesium

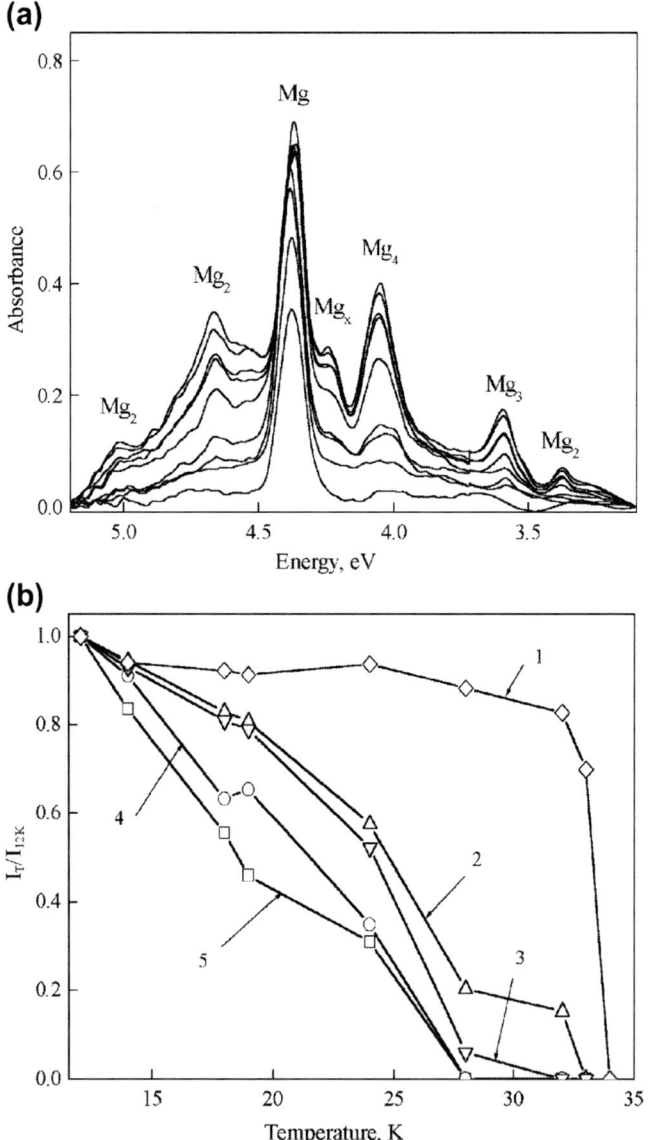

FIGURE 5.14 (a) The changes in UV–visible spectra in the temperature range 12–35 K for Mg/CCl$_4$/Ar system = 1:100:1000 and (b) normalized integral intensity of magnesium particles absorption at different temperatures: (1) Mg, (2) Mg$_4$, (3) Mg$_x$, (4) Mg$_3$, and (5) Mg$_2$.

particles in the reaction with carbon tetrachloride decreases in the series $Mg_2 > Mg_3 > Mg_4 \geq Mg$. The results shown in Figs. 5.14a,b allow the relative reactivity to be plotted as a function of the number of magnesium atoms involved in the reaction. This dependence shown in Fig. 5.14c for the temperature of 18 K clearly demonstrates the presence of the size effect for a solid-phase low-temperature reaction.

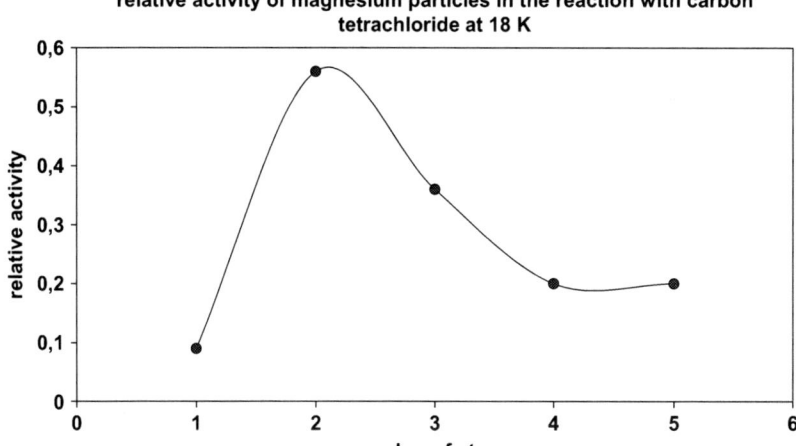

relative activity of magnesium particles in the reaction with carbon tetrachloride at 18 K

According to IR spectra, hexachloroethane and tetrachloroethylene are formed in the system magnesium–carbon-tetrachloride–argon.

The obtained results suggest that the reaction mechanism differs for low and superlow temperatures. At superlow temperatures, no insertion of magnesium particles into the carbon–halogen bond was observed. Theoretical estimates made it possible to assume the detachment (either simultaneous or sequential) of two chlorine atoms and the formation of a complex of dichlorocarbene with magnesium dichloride. Further transformations of the complex yielded the final products by the following scheme:

$$CCl_4 + Mg_n \longrightarrow CCl_3 + MgCl + (n-1) Mg$$

n=1-4
(Concentration of Mg>1%) / CCl_3 \ $MgCl, Mg$

C_2Cl_6 $CCl_2 : + MgCl_2$ $\xrightarrow{CCl_2}$ C_2Cl_4

Evidently, the effect of the particle size on its reactivity is of primary importance for the development of nanochemistry. From our viewpoint, of no less importance is to compare the activity of unisize particles of different substances. Such a comparison was accomplished for atoms and small clusters of samarium and magnesium.[103,110] Their choice was defined by the fact that compact magnesium and samarium at ordinary temperatures react with halogens in the same manner to give OMC.

Table 5.5 shows the products obtained in the reaction of magnesium and samarium particles with different ligands. The interactions of particles of each metal with two different ligands were studied by the matrix-isolation method for the reactions of atoms and clusters of magnesium or samarium with carbon dioxide and ethylene, which were present in the reaction mixture either simultaneously or separately. The experiments were carried out in the temperature range 10–40 K under comparable conditions and at different argon dilutions. The amounts of deposited substances were measured by means of a quartz-crystal microbalance placed inside a vacuum cryostat. The preferential reaction direction in three-component mixtures was studied by considering $Mg-C_2H_4-CO_2$ and $Sm-C_2H_4-CO_2$ systems. Spectroscopic studies of systems involving magnesium particles were combined with quantum-chemical calculations. The interaction of magnesium with a mixture $C_2H_4/CO_2/Ar = 1:1:20$ was studied. After condensation, the IR spectrum of this system represented a superposition of spectra of ethylene and carbon dioxide. Heating induced the appearance of absorption bands at 1592, 1368, and 860 cm^{-1}, which were assigned to a radical anion $CO_2^{-\bullet}$ for the case of $Mg-CO_2$ system. The intensities of these bands were compared to those observed in the absence of ethylene. None of the absorption bands could be associated with the magnesium–ethylene interaction. Heating of an $Mg-CO_2-C_2H_4$ system also gave rise to three new absorption bands: 1786, 1284, and 1256 cm^{-1}, which were assigned to a product of interaction of all three components. Ab initio quantum-chemical calculations suggest

TABLE 5.5 Reaction Products of Magnesium and Samarium with Ligands at 10–40 K

Metal particle	Ligand		
	CO_2	C_2C_4, C_2D_4	$CH_3X, X = Cl, Br$
Mg	$Mg^+CO_2^-$ At matrix annealing	Cycle $Mg(C_2H_4)_2$ at matrix annealing	CH_3MgX at irradiation ($\lambda = 280$ nm)
Mg_{2-4}	$Mg^+CO_2^-$ At co-condensation		CH_3MgX at irradiation ($\lambda = 300$ nm)
Mg_x	$Mg^+CO_2^-$ At matrix annealing		
Sm	$Sm^+CO_2^-$, CO, $SmCO_3$ At matrix annealing	Complexes $Sm(C_2H_4)-(C_2D_4)$ and $Sm(C_2H_4)_2-(C_2D_4)_2$	Methane at co-condensation
Sm_2	$Sm^+CO_2^-$, CO, $SmCO_3$		
Sm_x	At co-condensation		

the formation of a compound $Mg(C_2H_4)CO_2$. This compound, like $Mg(C_2H_4)_2$, was also characterized by the presence of chemical bonds between a ligand and a magnesium atom and between ligands themselves. Figure 5.15 illustrates the structure of compounds and specifies bond lengths (Å) and charges at atoms (shown in parentheses).

In terms of the Möller–Plessett approximation of the second order (MP2), the stabilization energy of $Mg(C_2H_4)CO_2$ is 18 Kcal/mol.

The analysis of IR spectra of $Sm-C_2H_4-CO_2-Ar$ and $Sm-C_2D_4-CO_2-Ar$ systems showed that samarium, like magnesium, reacts preferentially with carbon dioxide. No absorption bands of samarium complexes with ethylene were observed. As in a binary $Sm-CO_2$ system, an absorption band of CO and two absorption bands pertaining to $CO_2^{-\bullet}$ radical anion were observed. Thus, studies of ternary systems metal–carbon dioxide–ethylene have shown that both magnesium and samarium react with CO_2 and do not form any complexes with ethylene. A comparison of results obtained for ternary systems with those found for binary systems points to a sophisticated dependence of the relative reactivity on the size of metal particles and on the nature of compounds involved in the reaction.

Considering the example of matrix-isolated binary systems, it was shown that clusters of magnesium and samarium are more active than their atoms in

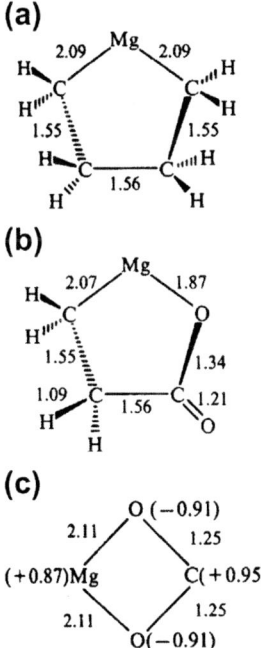

FIGURE 5.15 Bond lengths (Å) and charges on atoms (in parentheses) in magnesium compounds: (a) with ethylene dimer, (b) with ethylene and carbon dioxide, and (c) with carbon dioxide.

the reactions with carbon dioxide and ethylene. In the system magnesium–carbon dioxide, a single radical–ion pair was formed. Under comparable conditions, in the samarium–carbon dioxide system, carbon monoxide and samarium carbonate were also formed in addition to the radical–ion pair. Furthermore, the complexes of magnesium and samarium with ethylene were also different.

Table 5.5 shows that superlow-temperature reactions with methane halides (CH_3Cl, CH_3Br) involve the insertion of magnesium particles into the carbon–halogen bond to give Grignard reagents. Under similar conditions, samarium particles reduced methylchloride and methylbromide to methane. Thus, in contrast to compact metals, the behavior of magnesium and samarium at the level of nanoparticles is different, which points to the high specificity of reactions with participation of atoms and small metal clusters.

However, the results obtained for ternary systems (metal–carbon dioxide–ethylene) revealed both the absence of specificity (magnesium and samarium behave in the same manner, reacting solely with carbon dioxide) and a high selectivity (the reaction only with carbon dioxide and the absence of reactions with ethylene). In our opinion, further studies in multicomponent systems will provide information on the relationship between activity and selectivity of metal nanoparticles of different sizes.

In chemistry and, in particular, in chemistry of free radicals, it is known that in same reactions, more active particles exhibit a lower selectivity and, vice versa. The aforementioned results concerning the reactions of magnesium and samarium particles with ligand mixtures point to the presence of nonunique relationships between the activity and selectivity of nanosize metal particles.

Therefore, in nanochemistry, along with a problem of how the number of atoms in a particle affects the possibility of a reaction (the size effect), a problem of the relationship between activity and selectivity can be formulated. The presence of such "fine" effects is typical of nanochemistry of metals and requires further experimental and theoretical studies.

5.4.2 Reactions of Silver Particles of Various Sizes and Shapes

When synthesizing metal clusters and nanoparticles, high-molecular substances are widely used as stabilizers. Studies in this field have shown that macromolecules not only ensure a high stability of the resulting dispersed system but also immediately participate in its formation by controlling the size and shape of growing nanoparticles. For instance, because of the presence of ionized carboxyl groups, polycarboxylic acids of the acryl series interact with cations of metals, particularly, silver, by (1) binding them into strong complexes, (2) reducing them immediately in this complex under the action of light, and (3) stabilizing small charged clusters and nanoparticles of metals, which are progressively formed in the course of synthesis:

$$R\text{-}COO^- + Ag^+ \to R\text{-}COO^{-\bullet}Ag^+ \qquad (1)$$
$$R\text{-}COO^{-\bullet}Ag^+ \xrightarrow{h\nu} R\text{-}COO^{-\bullet}Ag^0 \qquad (2)$$
$$R\text{-}COO^{-\bullet}Ag^+ + Ag^0 \to R\text{-}COO^{-\bullet}Ag_2^+ \qquad (3)$$
$$\to R\text{-}COO^{-\bullet}Ag_4^{2+} \xrightarrow{h\nu} \cdots \xrightarrow{h\nu} R\text{-}CO^{-\bullet}Ag_n$$

Thus, the overall process of formation of nanoparticles from original cations to a final particle occurs at the immediate contact with the polymeric matrix. One of the major factors determining this process is the presence of ionized carboxy groups in the polymer, the number of which can be varied by changing the molecular mass (M) of polycarboxylic acids, their ionization degree, or by using co-polymers. The effect of these factors on the main stages of formation of silver nanoparticles has been described.[111]

The potentiometric analysis has shown that binding of silver cations by polyacrylate anions (PA) with molecular masses of 450,000 u ($PA_{450,000}$) and 1,250,000 u ($PA_{1,250,000}$) and an ionization degree $\alpha = 1.0$ proceeds by a cooperative mechanism, i.e. with an increase in the silver content in the solution, the concentration of chains saturated with Ag^+ ions increases. When an aqueous solution of $Ag^{+\bullet}$ PA complex obtained under conditions mentioned above was acted on by full light from a high-pressure mercury lamp, the photoreduction of Ag^+ cations occurred. A spectrum measured in the course of this process revealed, first of all, the appearance of an absorption band with the maximum at 700 nm and a shoulder in the vicinity of 300 nm, which was assigned to small charged silver clusters Ag_8^{2+}. In the absence of UV light, Ag_8^{2+} clusters were stable for several weeks. With further irradiation, the absorption bands corresponding to Ag_8^{2+} clusters vanished, and absorption bands at 370 and 460 nm appeared, which were attributed to the formation of coarser clusters Ag_{14}^{2+} and silver nanoparticles, respectively. The resulting colloid solution was stable for at least several weeks. An increase in the total silver content throughout the cooperative binding region resulted in a proportional increase in the concentration of clusters and nanoparticles, while the dynamics of their formation remained unchanged. TEM studies revealed spherical silver nanoparticles of the average size independent of silver concentration throughout the cooperative binding region and, as follows from the data in Table 5.6, equal to 1–2 nm for $PA_{450,000}$ and 4–5 nm for $PA_{1,250,000}$.[112] Figure 5.16 shows a typical microimage of these particles and their size distribution.

It should be mentioned that throughout the cooperative binding region, the particle diameter did not exceed the theoretical value calculated for the particle, which was formed as a result of the reduction of all the silver cations bound with a single macromolecule. Thus, under conditions of highly efficient cooperative binding of silver cations by a polyanion, the reduction of cations and

TABLE 5.6 Average Diameter (nm) of Spherical Silver Nanoparticles Formed as a Result of Photoreduction of Cation in a Complex $Ag^{+\bullet}$ PA (PA – Polyacrylate ion) ([PA] = 2×10^{-3} M)

MM PA	$[Ag^+] = 2 \times 10^{-4}$ M	$[Ag^+] = 6 \times 10^{-4}$ M	$[Ag^+] = 1 \times 10^{-3}$ M	$[Ag^+] = 1{,}5 \times 10^{-3}$ M
1.25×10^6	4 ± 2	5 ± 2	5 ± 2	5 ± 7
4.5×10^5	1.3 ± 0.9	13 ± 0.8	1.3 ± 0.7	2 ± 1
2×10^3	1.5 ± 0.7	4 ± 3	13 ± 11	

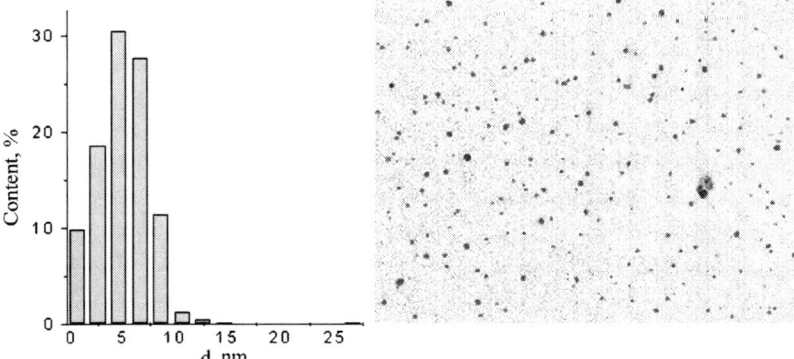

FIGURE 5.16 Microphotograph and size distribution of silver particles synthesized by 15-min irradiation of an aqueous solution of $Ag^{+\bullet}$ $PA_{1{,}250{,}000}$ at [PA] = 2×10^{-3} M, $[Ag^+] = 6 \times 10^{-4}$ M. Scale bar, –1 cm 83 nm.

the growth of particles proceed within a coiled macromolecular globule, which represents a nanoreactor of photochemical synthesis of spherical nanoparticles. One can regulate the size of formed nanoparticles by changing the molecular mass of PA or, in other words, the number of cation-binding centers on the polymeric chain.

Binding of silver cations to PA with a molecular mass 2000 occurs in an uncooperative fashion, i.e. with an increase in Ag content, the uniform filling of macromolecules is accompanied by a substantial increase in the concentration of free silver cations in solution.[113] Photoreduction under these conditions also resulted in the formation of sols containing small charged silver clusters (λ_{max} = 355 nm) and nanoparticles (λ_{max} = 460 nm). As follows from Table 5.6, the size and polydispersion degree of nanoparticles substantially increase with an increase in silver content, apparently owing to the increasing role played by free silver cations in their formation. Thus, under conditions of noncooperative binding of cations, the molecular mass of a polymer does not determine the nanoparticle size.

To decrease the relative content of carboxy groups γ in a polymer, a dilute solution of sodium polyacrylate with pH 9.0 was photolyzed in air. The study of the interaction of silver cations with a decarboxylated polyanion (PAγ) made it possible to find a threshold value γ ~ 0.7 below which the cooperative character of binding of Ag^+ ions is lost, and the effective constant of dissociation of an $Ag^{+} \bullet PA_\gamma$ complex increases. At $1.0 < \gamma < 0.7$, the dynamics of spectral changes that accompany the photoreduction of cations in $Ag^{+} \bullet PA_\gamma$ complexes remained practically unchanged, and the synthesis produced spherical silver nanoparticles; however, their size increased from 1.3 ± 0.9 to 3 ± 2 nm. The process of nanoparticle formation was observed to undergo qualitative changes at a degree of PA decarboxylation below the threshold value. Thus, spectroscopic studies of photolyzed $Ag^{+} \bullet PA_{\gamma=0.5}$ complexes failed to reveal the formation of small silver clusters, while aggregates of coarse (10–30 nm) silver particles with a broad absorption band at 350–550 nm were observed. During further photolysis, an absorption band with the maximum at 375 nm appeared, and the long-wave absorption edge shifted to the near-IR region. Such spectra are typical of systems containing elongated silver particles. Indeed, as seen in a microimage shown in Figure 5.17, nanorods with a thickness of 20–30 nm and a length of several micrometers and their aggregates were preferentially formed under these conditions.

Studying how the degree of PAA ionization affects the formation and properties of silver nanoparticles also revealed the formation of rod-like particles for $\gamma < 0.7$.[333] Thus, the shape of silver nanoparticles is determined by the content of ionized carboxy groups in a polymer regardless of how it was changed (by either photodecarboxylation or protonation of a polyanion). Under experimental conditions, nanorods can be formed as a result of both light-induced aggregation of

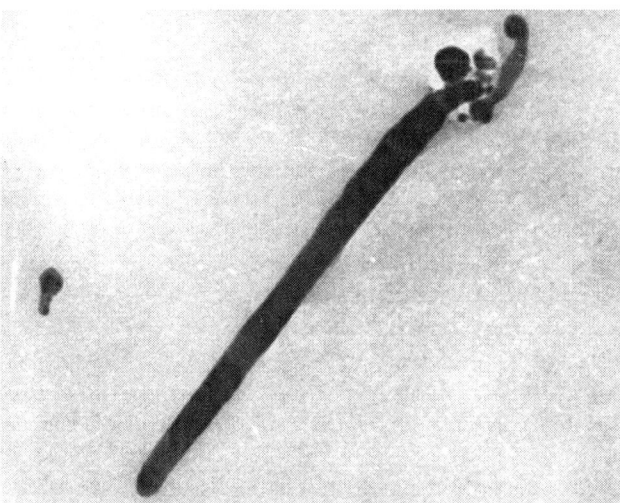

FIGURE 5.17 Microphotograph of silver particles obtained during 15-min photoreduction of 2×10^{-4} $AgNO_3$ solution in the presence of 2×10^{-3} M $PA_{\gamma=0.5}$. Scale bar, –1 cm 200 nm.

primary spherical particles and photoreduction of silver cations on their surfaces. In order to find which of these processes dominates, the effect of light on the size distribution of sols formed by spherical silver nanoparticles and containing no Ag$^+$ ions was studied.[114] The complete reduction of silver cations was achieved by using sodium borohydride as the reducing agent.

Reduction of AgNO$_3$ (6×10^{-4} M) with sodium borohydide (1.2×10^{-3} M) in the presence of photodecarboxylated PAγ=0.5 (2×10^{-3} M) resulted in the formation of a stable silver sol consisting of spherical particles (6 ± 3 nm). Irradiation with full light from a DRSh-250 lamp had no effect on the absorption spectrum and, hence, on the size distribution of the sol. Provided that AgNO$_3$ or Al(NO$_3$)$_3$ (3×10^{-4} M) was preliminarily added to the sol, the photolysis with $\lambda > 363$ nm resulted in the formation of elongated, particularly, rod-like, particles with a length reaching 500 nm. The same result was obtained if the original sol was acted on with light ($\lambda > 455$ nm) passing through an interference filter with a transmission band virtually coinciding with the absorption band of spherical nanoparticles. On the other hand, light with $\lambda > 555$ nm induced no changes in the spectrum. Thus, to transform primary spherical particles stabilized by PA with γ=0.5 into elongated particles, photolysis in a range $363 \leq \lambda \leq 555$ nm, i.e. in the range of their absorption band, was sufficient. Under these conditions, the dipole–dipole interactions between particles strengthen, which apparently is the reason for their photoinduced aggregation. From our viewpoint, the initiating role of cations Ag$^+$ (Al^{3+}) consists triggering the formation of microaggregates of primary spherical nanoparticles—the centers of subsequent photoinduced aggregation.

A qualitatively different result was obtained with a sol synthesized by reducing silver cations with borohydride in the presence of nonmodified PA$_{450,000}$. The incorporation of different amounts of AgNO$_3$ into the sol followed by photolysis in a range $363 \leq \lambda \leq 555$ nm increased the absorption intensity of spherical nanoparticles, which points to the increase in their volume fraction owing to photoreduction of cations. It should be mentioned that light with $\lambda > 363$ nm induced no photoreduction of silver cations in aqueous solutions of Ag$^+ \cdot$ PAA$_{450,000}$ complex, because the absorption band of the latter is at $\lambda < 300$ nm. Thus, the observed photosensitizing effect of silver nanoparticles on the reduction of Ag$^+$ by carboxy groups in PAA consists in shifting the long wavelength cutoff of photoreduction to 555 nm. However, the fact that no rod-like particles were formed under these conditions was explained by the formation of a PA$_{450,000}$ protective coating on the surface of spherical silver nanoparticles, which effectively prevented the aggregation of the latter.

Thus, the control over the size, shape, and size distribution of silver nanoparticles formed as a result of photoreduction of Ag$^+$ cations can be excersized by changing the molecular mass and degrees of ionization and decarboxylation of polycarboxylic acids.

Methods developed for controlling the size of silver particles were used in studying the kinetics of silver-catalyzed reduction of *p*-nitrophenol by sodium borohydride in the temperature range 283–333 K. Silver nanoparticles measuring

4 and 18 nm and stabilized by polyacrylic acid were obtained by chemical and photochemical reduction in the aqueous solutions. It was found that the reaction rate constants per unit catalyst surface, the reaction order with respect to the catalyst, and the activation energy depend on the size of silver particles.

5.5 THEORETICAL METHODS

5.5.1 General Remarks

The whole set of modern theoretical simulation methods are actively used for interpretation of properties and chemical reactivity of molecular systems of interest for nanochemistry. Of paramount importance is the information on the potentials of interactions between particles in a system, according to which the calculation methods are divided into semiempirical and nonempirical (ab initio).

In the former methods, the potentials are written in the analytical form based on the known theoretical expressions, and the parameters of these expressions are fitted to a sampling of experimental data. The most popular expressions are those used in molecular mechanics, classical molecular dynamics, as well as the potentials applied in solid-state physics. Such an approach is employed to simulate the properties of atomic clusters both homogeneous and heterogeneous. In the first stage, the equilibrium geometrical configurations of clusters are calculated as the points in minima on the multidimensional potential-energy surfaces. Here, it should be borne in mind that a nanosystem has a great number of such minima and, correspondingly, a wide range of structural modifications. As a rule, local minima are separated by the potential barriers, which are so low that the transitions between structures can occur even at moderate temperatures. Hence, it is necessary to enumerate quite a number of minima, which is, however, a cumbersome task. Next, from the points determined, the energy characteristics are calculated and the thermodynamic stability of a system is forecasted. A great body of information is provided by the molecular dynamic calculations of the paths of particles within a cluster, which are carried out at a fixed temperature. Analyzing the trajectories and building different functions of distribution and autocorrelation functions allow one to characterize a cluster, observe how far its properties deviate from those of the condensed medium built of the same atoms, and analyze the dependence of the cluster properties on its size. The most significant limitation on the simulations based on the analytical potentials is the fact that they cannot be used for analyzing the chemical reactions involving clusters: the parameters of potentials are not calibrated for describing the changes in the electronic structures of the particles.

The interaction potentials derived by ab initio methods of quantum chemistry are naturally more universal and make it possible to solve in principle all the problems concerned with the structure and chemical reactions in molecular systems. Here, the major limitations are associated with the size of the system. However, modern quantum-chemical methods can provide sufficiently reliable results for particles comprising up to 10 atoms. Moreover, they allow one to find the coordinates of steady-state points on the potential-energy surfaces constructed for the ground electron states of clusters (i.e. the points that correspond to local minima and the potential barriers hindering rearrangements), estimate relative energies in these points, calculate energy profiles of chemical reactions in the system, predict vibrational and electronic spectra, and analyze in detail the electron density distributions. Such calculations are rather expensive, because they require using methods that take into account the electron correlation effects (first of all, the theory of electron density functional), but are technically feasible. For systems comprising 20–30 atoms, ab initio calculations are also technically possible but at the cost of the accuracy of the results obtained. Equilibrium geometrical configurations of low minima in potential-energy surfaces can be determined with sufficient accuracy, and the thermodynamic stability of the clusters can be assessed. However, the prognosis of spectra and the reactivity of the cluster particles are given with lower accuracy, although the qualitative trends are reproduced quite correctly.

The so-called hybrid methods that combine quantum mechanics and molecular mechanics (QM/MM) are actively developed and are very promising for simulating properties of large molecular systems. The main idea of such approaches consists in applying the quantum description only to that part of a subsystem that is assumed to be most important and take into account the structure of the peripheral part of the total system and its effect on the central region by employing empirical and ab initio potentials. In many cases, the size of the central part can be chosen within 20–30 atoms, and ab initio methods of quantum chemistry would ensure a good quantitative description of both the structure and the reactions in the chosen subsystem. Although certain fundamental problems of the QM/MM theory are still far from their final solution, such an approach is finding increasing use in simulating the processes in biosystems and material science.

Below, we show certain concrete examples concerned with simulation of properties, spectra, and reactivity of cluster particles, which illustrate the theoretical studies in this field.

5.5.2 Simulation of the Structure of Mixed Metallic Particles

To estimate equilibrium geometrical configurations and binding energies of clusters $(M_1)_m(M_2)_n$, where M_1, M_2 = Ag, Cd, Cu, Mg, Na, Pb, Sn, Zn ($m+n \leq 4$), ab initio methods of quantum chemistry were used.[51] For each cluster, the electronic-state type (with regard to its spin) with the lowest energy was determined. Then, for a fixed multiplicity, the geometric configuration corresponding to the minimum on the potential energy surface was calculated using the Hartree–Fock–Roothaan method. In all the cases, the effective potentials of the core and the corresponding valence-orbital bases in the approximation of Steven–Basch–Krauss (SBK) were used. By calculating vibrational frequencies, it was checked whether the configuration determined the real minimum. For the coordinates found, the energies of systems were estimated according to the Möller–Plessett perturbation theory of the second order (MP2), which allows taking into account the main contributions of electronic correlation effects. The calculations were carried out using a PC GAMESS program,[115] which represented a version of the well-known quantum-chemical program GAMESS[116] developed for personal computers and stations equipped with Intel processors.

Table 5.7 shows the estimates of binding energies (in kcal/mol) of all possible pair combinations M_1M_2, which also involve homonuclear particles. A dash in a corresponding position indicates that a given diatomic system is not bound. Based on these data, it can be concluded that if we restrict the consideration to atomic pairs only, the most promising compositions are as follows: AgCu, CuPb, CuSn, AgPb, CuNa, SnPb, AgSn, and NaSn.

Studies of triatomic mixed metallic clusters took into account the results obtained for diatomic systems. Naturally, Ag and Cu were of prime interest as partners in mixed clusters such as Ag_nM_m and Cu_nM_m; hence, these combinations were considered first. It should be noted that triatomic systems are characterized by a wide spectrum of isomeric structures in addition to the diversity of spin states. The search for the main isomers of these clusters took into account all these peculiarities. Tables 5.8 and 5.9 show the binding energies of silver- and copper-containing clusters, respectively. The results in these tables consider both the cases of dissociation: to atoms and to a combination of atom + diatomic molecule.

The energy values shown in Tables 5.7 and 5.8 clearly demonstrate that the series of elements M ordered with respect to relative affinity of Ag to M can be presented as Pb > Sn > (Cu, Mg, Na) > Zn > Cd. The data in Tables 5.7 and 5.9

TABLE 5.7 Binding Energies (kcal/mol) of Atomic Pairs

	Ag	Cd	Cu	Mg	Na	Sn	Pb	Zn
Ag	29.4	4.6	39.9	6.5	21.2	18.7	29.6	4.4
Cd		–	6.7	–	–	4.8	8.1	–
Cu			36.8	9.7	23.9	32.2	36.2	6.8
Mg				–	–	5.3	4.5	–
Na					3.2	18.5	15.7	<3
Pb						31.3	21.1	5.5
Sn							29.3	8.3
Zn								–

TABLE 5.8 Binding Energies (kcal/mol) of Silver-containing Clusters

Metal M	Dissociation to atoms			Dissociation to an atom and a diatomic molecule			
	AgM	AgM$_2$	Ag$_2$M	Ag + AgM	Ag$_2$ + M	AgM + M	Ag + M$_2$
Ag	29.4	39.1	39.1	9.7	9.7	9.7	9.7
Cd	4.6	10.0	37.1	32.5	7.7	5.5	–
Cu	39.9	49.0	41.8	8.9	12.4	16.1	12.3
Mg	6.5	11.3	40.5	34.0	11.2	4.8	–
Na	21.2	23.2	38.4	17.2	9.0	2.1	20.1
Sn	18.7	79.8	63.3	36.9	33.9	56.1	48.5
Pb	29.6	86.5	74.7	45.1	45.3	56.9	57.3
Zn	4.4	9.7	37.9	33.6	8.5	5.3	–

TABLE 5.9 Binding Energies (kcal/mol) of Copper-containing Clusters

Metal M	Dissociation to atoms			Dissociation to an atom and a diatomic molecule			
	CuM	CuM$_2$	Cu$_2$M	Cu + CuM	Cu$_2$ + M	CuM + M	Cu + M$_2$
Ag	39.9	41.8	49.0	16.1	12.3	8.9	12.4
Cd	6.7	13.8	46.4	39.7	9.6	7.2	–
Cu	36.8	53.0	53.0	16.2	16.2	16.2	16.2
Mg	9.7	20.0	49.9	40.2	13.1	6.2	–
Na	23.9	25.9	46.3	22.5	9.5	2.0	22.8
Pb	36.2	93.8	87.7	51.5	50.9	57.6	64.6
Sn	32.2	89.6	79.0	46.7	42.2	57.3	58.3
Zn	6.8	14.4	48.8	42.0	12.0	7.6	–

make it possible to find a corresponding series of elements ordered with respect to affinity of Cu to M:Pb > Sn > (Na, Mg, Ag) > Zn > Cd. Thus, the combinations formed by silver or copper with lead or tin proved to be the most promising for the formation of mixed metallic clusters.

TABLE 5.10 Binding Energies (kcal/mol) of Ag_2Pb_2, $AgPb_3$, and Ag_3Pb Clusters

Dissociation	Ag_2Pb_2	$AgPb_3$	Ag_3Pb
To atoms	147.2	132.6	117.2
To diatomic molecules	88.5 ($Ag_2 + Pb_2$) 87.9 (2AgPb)	73.8 ($AgPb + Pb_2$)	58.1 ($AgPb + Ag_2$)
To an atom and a triatomic molecule	72.4 ($Ag_2Pb + Pb$) 60.6 ($Ag + AgPb_2$)	50.3 ($Ag + Pb_3$) 46.1 ($AgPb_2 + Pb$)	42.4 ($Ag + Ag_2Pb$)

Combinations of silver with lead are sufficiently suitable from the viewpoint of employing them in experimental studies. Hence, when studying tetraatomic systems, only compositions Ag_2Pb_2, $AgPb_3$, and Ag_3Pb were considered. For $AgPb_3$, a structure of distorted pyramid was found; Ag_2Pb_2 and Ag_3Pb demonstrated structures of the "butterfly" type. Table 5.10 compares the estimated binding energies with regard to the most important dissociation channels. As seen from these results, all compositions of this kind should be stable; however, a system with equal fractions of silver and lead demonstrates the highest binding energy.

This study involved simulating the structural and electronic properties for mixed silver–lead clusters Ag_nPb_m ($n + m \leq 28$).[117] A study of the reactions of lead atoms with colloidal silver particles in an aqueous solution revealed the formation of mixed heteronuclear systems, as evidenced by the absorption band shifts (in the direction of short wavelengths).[118] The knowledge of spectral characteristics of heteronuclear clusters of the largest sizes possible is necessary for the correct identification of co-condensation products. The problem of forecasting the spectra for heteroclusters $(M_1)_m(M_2)_n$ even with a total size $(n + m)$ of an order of magnitude of 10 is an extremely intricate task from the theoretical standpoint, because it requires considering a vast amount of structures formed when two kinds of atoms are arranged within a cluster of a fixed size. All these isomers correspond to the points of global and local minima in comparatively flat potential energy surfaces, which are difficult to determine by employing the known algorithms. A strategy that does not claim high accuracy for estimates of each individual cluster but makes it possible to formulate the main trends in the properties of bimetallic systems when varying the component ratio was used.[117] The positions of bands in UV spectra of each cluster were determined on the basis of the difference in orbital energies ΔE between the highest occupied and the lowest vacant molecular orbitals by using the extended method of Hückel (EMH) with a program ITEREX-88, which took into account relativistic corrections to the parameters.[119]

The structures of clusters were determined as follows. In the first stage, the geometrical configurations of the global minima and those local minima of individual silver clusters Ag_n ($4 \leq n \leq 28$), which had the lowest energy, were found. To describe the interparticle interaction, the empirical Sutton–Chen potential was employed:

$$U = \varepsilon \sum_{i=1}^{N} \left[\frac{1}{2} \sum_{j \neq 1} \left(\frac{a}{r_{ij}} \right) - c\rho_i \right]$$

where $\rho_i \sum_{i \neq 1} (a/r_{ij})^6$, N is the number of atoms in the system, r_{ij} the distance between atoms i and j, c and a dimensionless parameters, and ε a parameter with the dimensionality of length. The values of the parameters were chosen from experimental data for the compact substance: $\varepsilon = 2.542 \times 10^{-3}$ eV, $a = 4.086$ Å, and $c = 144.41$. The search for steady-state points in the potential-energy surfaces was carried out in terms of the Monte Carlo method using the Metropolis algorithm. In iterations, the temperature was decreased from 60 to 4–10 K according to the relationship $T_i + 1 = \lambda T_i$ with $\lambda = 0.90$–0.98.

In analyzing mixed clusters $Ag_m Pb_{n-m}$, the geometric configurations found for individual silver were assumed to remain unchanged. In every structure Ag_n ($n = 4$–28), one or two silver atoms were changed for lead atoms in a random fashion so that to transform an Ag_n cluster, the structure of which corresponded to the global minimum and several low local minima, into n clusters $Ag_m Pb_{n-m}$ ($m = 0, \ldots, n$). For every $Ag_m Pb_{n-m}$ system, lead percentage was determined.

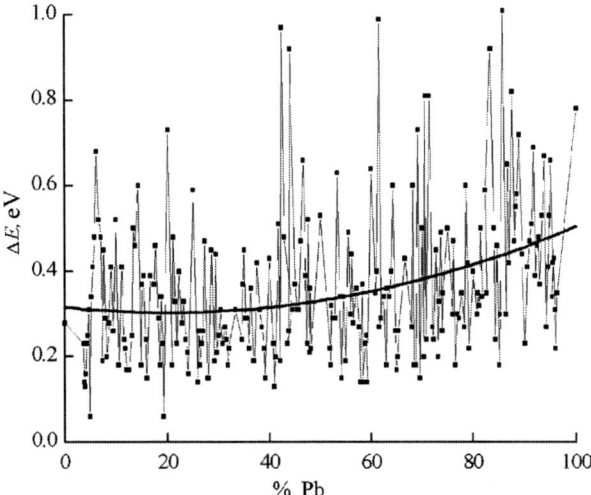

FIGURE 5.18 Dependence of the absolute magnitude of energy difference between the higher occupied and the lower vacant molecular orbitals on the percentage of lead in heteroclusters $Ag_m Pb_{n-m}$.

Then, ΔE values were averaged overall Ag_mPb_{n-m} structures, which contained equal amounts of lead. Figure 5.18 presents the obtained results as the dependence of ΔE on the lead percentage in Ag_mPb_{n-m} clusters. Based on the least-squares technique, a smoothed line was plotted by the background of jump-wise changes in the energy gap for the great series of structures. As seen, with an increase in the lead fraction in mixed clusters, ΔE value increases. Thus, the band in the electron absorption spectra of these systems should shift in the short-wavelength direction. Such a behavior of Ag_mPb_{n-m} clusters agrees with experimental results according to which the chemisorption of lead atoms on the surface of colloidal silver particles shifts the electron absorption band to the short-wavelength range.[118]

5.5.3 Simulation of Properties of Intercalation Compounds

The properties of low-temperature solid matrix systems containing intercalated atoms, molecules, and intermolecular complexes are difficult to describe theoretically. The chief problems are associated with the fact that the interactions between intercalation compounds and the matrix substance are substantially weaker as compared with the interactions within the intercalation system, and, hence, simulating the properties of the system as a whole requires potentials reliable over a very wide scale of interactions. Usually, simulating the matrix systems employs a cluster approximation, which involves a direct consideration of all particles in the matrix and the intercalated molecule. Thereafter, this approximation utilizes either the quantum-chemical methods for simulating the constructed heterocluster or the approaches that combine molecular dynamics and quantum chemistry.[120]

When considering small lithium clusters (Li_3, Li_4, Li_5) isolated in argon matrices at low temperatures, attention was focused on the charge distribution in lithium particles, which is important for predicting the reactivity of metal clusters. In the first stage, using the ab initio methods of quantum chemistry, the following equilibrium geometric configurations of the planar lithium clusters were determined: an isosceles triangle for Li_3, a rhomb for Li_4, and a trapeze for Li_5. This geometry of lithium clusters remained unchanged in the course of subsequent molecular-dynamic calculations. Next, heteroclusters Li_nAr_m ($n = 3, 4, 5$; $m = 18$–62) were constructed in such a way that metal particles were completely surrounded by inert-gas molecules. The potentials of Li–Ar and Ar–Ar interactions were approximated by Lennard–Jones functions with parameters found experimentally or taken from the literature. These potentials were employed in calculations of molecular-dynamic trajectories at time steps of 10^{-15} s, when each trajectory involved up to 10^4 steps measured after thermoisolation of the system. The temperature interval was 3–25 K. For each characteristic heterocluster structure, the contributions made by the surrounding argon atoms to the electron density of lithium particles were calculated along the trajectories. For this purpose, the Fock matrix of a lithium cluster was supplemented with the effective one-electron potentials of each argon atom, the parameters of which were preliminarily found in terms of the density-functional theory. The new Fock matrix thus derived was used for estimating the electron distribution by employing the conventional programs of quantum chemistry. In particular, the effective charges on lithium atoms were calculated. The plotted fluctuations of charge distributions along the molecular-dynamic trajectories[121] look very spectacular—the evolution of the system is accompanied by sufficiently pronounced oscillations of charges on the metal centers when the charge variations in the absolute values can reach 0.2 of an atomic unit. Moreover, a distinct correlation between the partial charges and the position of a metal cluster within an inert-gas sheath is observed; particularly, the charge sign can change when a cluster escapes to the surface.

Such an approach was used in a study[122] devoted to calculations of matrix shifts in the electron spectrum of a Na_2 molecule in low-temperature krypton matrices. According to experimental results,[123] the surrounding krypton induces shifts of bands in the electron spectrum of Na_2 in different directions for different transitions: a blue shift by 12 nm (+523 cm^{-1}) is typical of the B – X($^1\Pi_u - {}^1\Sigma_g^+$) band, while a red shift by 17 nm (406 cm^{-1}) is characteristic of the A – X($^1\Sigma_u - {}^1\Sigma_g^+$) band. In this model,[122] the matrix was simulated as a Na_2Kr_{62} heterocluster, and the band position in the Na_2 electron spectrum was associated with the difference in the orbital energies in a diatomic molecule: $\pi_u - \sigma_g$ for the B–X transition and $\sigma_u - \sigma_g$ for the A–X transition. To calculate the electronic structure of a molecule intercalated into a matrix, ab initio methods of quantum chemistry (Hartree–Fock–Roothaan approximation with 3-21G or 6-31G* bases) were used. The configurations of heteroclusters Na_2Kr_{62} were calculated using molecular dynamic methods. The resulting "instantaneous" structures were employed in calculating the electron distribution in an Na_2 molecule located within a heterocluster, i.e. the Fock matrix elements were supplemented with the effective one-electron potentials that modeled the effect of krypton atoms; then, the orbital energies were calculated and the positions of the spectral bands were assessed. Table 5.11 shows some

TABLE 5.11 Experimental and Theoretical Matrix-induced Shifts (cm^{-1}) of Bands in the Electron Spectrum of an N_2 Molecule in Krypton

N_2 molecule in krypton	A–X	B–X
Experiment (Hoffmann)	−406	+523
Na_2Kr_{62}, T=0 K	−461	+636
Na_2Kr_{62}, T=4 K	−658	+549
Na_2Kr_{62}, T=5 K	−395	+944

results, from which it is evident that the simulation allows estimating the matrix effects on the qualitative level. Thus, under the effect of krypton surrounding, the bands in the Na_2 electron spectrum shift in different directions for A–X and B–X transitions. The estimated shifts were close to the experimental ones and, generally speaking, a better agreement could hardly be achieved.

Theoretical works[121,122] revealed fine effects in nanosystems, namely the effect of solvate sheaths formed by inert gases on the electronic properties of matrix-isolated particles. Being performed in the early 1990s, these works are also interesting from the viewpoint of the calculation techniques, because, in fact, they used the combined (or hybrid) approaches of quantum and molecular mechanics (QM/MM). The central part (metal cluster) was considered on the quantum level, the effects of the periphery (solution atoms) on the central structure were taken into account by one-electron potentials, and the configurations of the whole system were determined using the methods of classical molecular dynamics. As pointed out in the beginning of this section, today, a decade later, such QM/MM approaches are among the most popular methods for simulating large molecular systems.[124]

A total ab initio analysis of a certain electronic problem, which followed the molecular-dynamic simulation, was applied in simulating the properties of an NBr molecule in argon clusters.[125] Within the framework of this model with a number of argon atoms approaching 170, it was possible to quantitatively reproduce a matrix-induced shift (about $19\,cm^{-1}$) for the vibrational band of an NBr molecule.

5.5.4 Simulation of Structural Elements of Organometallic Co-condensates

Reactions of transition metals with organic mesogenic molecules are of special interest because of their possible use in synthesizing new materials based on liquid crystals. Such liquid-crystal materials exhibit important electrophysical characteristics and can find applications in optoelectronics. Quantum-chemical calculations of electron and vibrational spectra of model silver complexes with a central cyanophenyl fragment that enters into the compositions of mesogenic CBs and their certain derivatives made it possible to directly compare the results with the experimental data on low-temperature co-condensation of 5CB with silver atoms in inert matrices.[62,64,68,103]

$$C_5H_{11} - \langle\bigcirc\rangle - \langle\bigcirc\rangle - C \equiv N$$

A CB molecule contains two active centers: aromatic rings and a polar CN group conjugated with the aromatic system. A transition metal can form complexes with CBs by the π-donor and π-acceptor mechanisms. Reactions between transition metal atoms and monomers or dimers of CBs can yield linear and sandwich complexes of different compositions. According to the experimental IR spectrum, the formation of a π-complex during the co-condensation of 5CB and silver on a cooled support is evidenced by a substantial decrease in the vibrational frequency Δn (C≡N) as compared with individual 5CB. In the condensed phase, CBs form dimeric structures of the "head-tail" type. It can be assumed that the complex is formed by incorporation of a silver atom between two 5CB molecules in the dimer. Inasmuch as the complex formation was experimentally proved by the shifts of vibrational frequencies of CN groups, it is worthwhile to consider a model system that contains the most important central cyanophenyl fragment of the 5CB–silver complex with the proposed structure.

The complete optimization of the geometry and the analysis of vibrations were performed in terms of ab initio Hartree–Fock approximations for a 5CB molecule and its most important fragment, cyanobenzene PhCN. The estimates of the strong IR band corresponding to the CN-group vibrations, viz. 2497 cm^{-1} (5CB) and 2501 cm^{-1} (PhCN), were required for comparing them with the CN-vibration band in silver complexes. The complete optimization of the geometry and the analysis of vibrations performed for fragments (PhCN)Ag, (PhCN)Ag$^+$, and (PhCN)Ag$_2$ led to a conclusion that the global minimum in these systems corresponds to σ-complexes with a linear arrangement of the fragment C–N... Ag with C–N-vibration frequencies shifted as compared with PhCN by +9, 7, and +20 cm^{-1}, respectively. For this reason, such a configuration could hardly be realized in the Ag–5CB films.

Later, the structures corresponding to the π complexes were considered. According to the X-ray diffraction analysis, the molecules of CB derivatives in crystals are packed in pairs according to the "head-to-tail" principle.[126] That is the reason we considered the model of a silver complex with two cyanobenzene molecules, which is shown below:

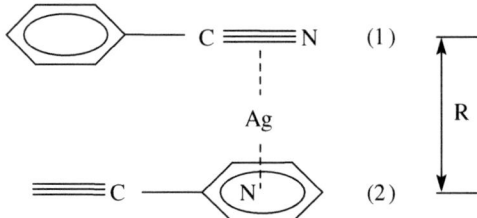

The system has a planar symmetry with a silver atom placed at equal distances from ligands, namely, from the center of a CN fragment in ligand (1) and from the benzene ring center in ligand (2). In calculations, all geometrical parameters were fixed and were equal to the values found for a free PhCN molecule; only the distance R between the planes of ligands (1) and (2) was varied.

A calculation using the MCSCF method with two electronic configurations arrived at two competitive solutions in the ground state: the first corresponded to $(PhCN)_2Ag$, the second corresponded to the charge–transfer configuration $(PhCN)_2^- Ag^+$. For great R ($R < 4.7 Å$), a neutral silver atom was inserted into the space between the ligands. In the vicinity of $R < 4.7 Å$, a radical reconstruction in the electron density distribution took place: an electron was transferred from the silver atom to ligands, and at $R < 4.7 Å$, the charge–transfer configuration predominated. For $R < 4.7 Å$, the almost total electron density was transferred from silver to the π^* system of the benzene ring in ligand PhCN (2) (the lower one in the scheme), while only a small part of electron charge passed to the π^* orbitals of CN group in ligand PhCN (1) (the upper one in the scheme).

The red shift of vibrational frequencies of the CN group in silver–CBs complexes is qualitatively understandable. Upon the formation of the film, silver atoms enter the void between the pairs of organic ligands in such a way as to get an asymmetrical surrounding with respect to CN groups (a simplified version is illustrated by the scheme). The donation of a small part of electron density from silver to the antibinding CN orbitals of one ligand results in an increase in the corresponding internuclear C–N distance and a decrease in the curvature of potential-energy surface along this coordinate, i.e. in a decrease in the vibrational frequency. The donation of the greater part of charge from Ag to the π^* system of the other ligand leads to the electron density redistribution in the resulting negative ion with a decrease in the corresponding frequency of CN vibrations.

Moreover, the matrix-induced shift of vibrational frequency was estimated at the quantitative level.[64] For this purpose, a model illustrated by the above scheme was used. Three characteristic interligand distances R were taken, i.e. $R = 4.4 Å$ (charge–transfer configuration), $R = 4.8 Å$ (neutral configuration), and $R = 4.7 Å$ (the critical point of electron transfer), and potential-energy surface sections corresponding to the C–N coordinate were analyzed. For distances 4.4 and 4.8 Å, a limited Hartree–Fock method for open shells on corresponding configurations was applied. For the distance 4.7 Å, the MCSCF method was employed, which involved averaging over two configurations. First, a partial nongradient optimization of the geometry with respect to C–N and C–C distances (between the neighboring carbon atoms in the benzene ring and in the CN group) was carried out with all other parameters fixed. Then, the points in the potential energy surface in the vicinity of the found equilibrium coordinate C–N were estimated, and the numerical estimates of the potential surface curvature and the vibrational frequencies of CN groups were obtained. Table 5.12 shows the results of this simulation, which implies that the qualitative conclusions drawn above are confirmed by numerical results. For the CTC ($R = 4.4 Å$), the red shifts of CN vibrational frequencies (-150 and $-175 cm^{-1}$) correlated well with those experimentally observed for Ag5CB films (-150 and $-200 cm^{-1}$).

Furthermore, a calculation of an electron spectrum for a model structure of CB shown in the above scheme was carried out using the configuration interactions (CI) method of the first order. Of interest were the differences in energies of ground and excited electron states of the $Ag(C_6H_5CN)_2$ complex, the corresponding strengths of oscillators, and the charges on the fragments of the complex. In our case, the CI method of the first order considered the single excitations of electrons with respect to the initial wave function that consisted of 44 configurations and was constructed in terms of the MCSCF scheme. As a result, the wave function in the CI method contained 96,844 configuration state functions (CSF).

Figure 5.19 compared the calculated electron spectrum of the model complex with the experimental UV absorption spectrum for the co-condensate

TABLE 5.12 Calculated Frequencies of Harmonic CN Vibrations (w) in $(PhCN)_2Ag$ Complex as a Function of Interligand Distance R and the Corresponding Frequency Shift (Dw) as Related to 2501 cm^{-1} in a Free-PhCN Molecule Frequencies (w) and Shifts R (Å)

$(\Delta \omega)$ cm^{-1}	4.4	4.7	4.8
PhCN (1)	2351 (−150)	2392 (−109)	2416 (−85)
PhCN (2)	2326 (−175)	2412 (−89)	2471 (−30)

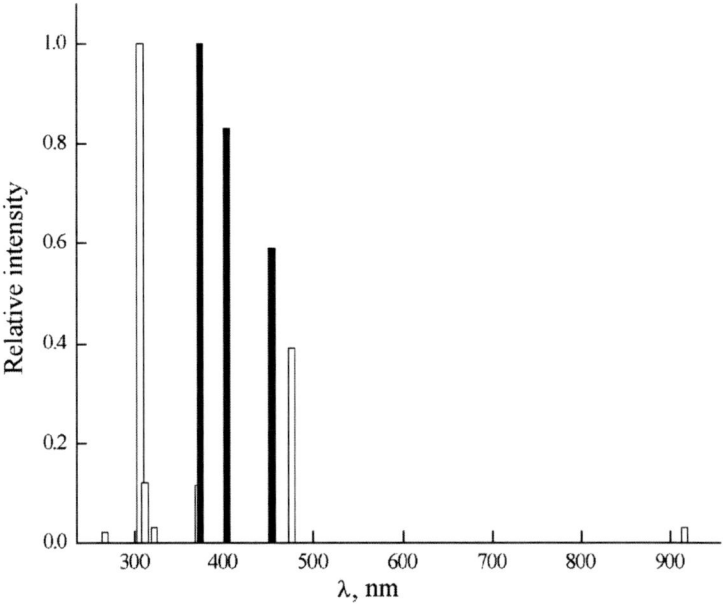

FIGURE 5.19 Electron spectrum of $Ag(C_6H_5CN)_2$ complex. Dark bands correspond to experimental UV spectrum of co-condensate of 4-pentyl-49-CB (5CB) with silver, which are absent in the spectra of individual 5CB and silver. Light columns correspond to fragments of calculated $Ag(C_6H_5CN)_2$ spectrum. The line intensity in the experimental spectrum and the oscillator strength in the calculated spectrum are normalized to unity.

Ag–5CB–decane (1:1:10). The comparison makes it possible to single out the bands comparable with those observed experimentally in the range 300–450 nm (2.5–3.5 eV), which correspond to the product of the 5CB–silver interaction. According to the charges on atoms in the excited states, which were estimated via the scheme of natural binding orbitals, the lines in the mentioned region correspond to the charge–transfer transitions of either ligand–metal or ligand–ligand kinds.

The DFT-calculation based on hybrid B3LYP potential was carried out to study the equilibrium structures of Eu–CB complexes. The model structures including one and two metal atoms with antiparallel disposition of ligand molecules are considered.

Thus, in aggregate, the results of ab initio quantum-chemical calculations allow one to characterize both vibrational and electronic spectra of silver co-condensates with CBs, interpret the observed spectral bands, and, eventually, substantiate the conclusions drawn based on experimental studies.

The theoretical simulation of physicochemical properties and chemical reactivity of metal nanoparticles is actively developed. In the immediate future, attention will probably be focused on modernization of the combined use of quantum-chemistry and molecular-dynamics methods for analyzing the properties of ligand-free clusters. Such clusters are as a rule unstable and cannot be synthesized in considerable amounts, which complicates both their studies and the comparison of experimental results with theoretical models.

Different versions of the cryochemical synthesis of nanoparticles measuring <1 nm should be developed further. By combining experimental data with theoretical simulations and developing new methods for the synthesis and stabilization of cryoparticles and also for exercising control over their sizes, it is possible to pose and solve the challenging problems of nanochemistry—namely, how the number of atoms in a metal nanoparticle, the temperature, the nature of stabilizing ligands, and self-assembling processes affect the chemical activity and selectivity of nanosystems.

REFERENCES

1. Semenov, N. N.; Shal'nikov, A. I. *Russ. Chem. J. (Russ.)* **1928,** *LX,* 303–308.
2. Kargin, V. A.; Kabanov, V. A. *Mend. Chem. J. (Russ.)* **1964,** *9,* 602–619.
3. Sergeev, G. B. *Russ. Chem. Rev.* **2001,** *70,* 809–826.
4. Sergeev, G. B. *Chemical Physics in Front of XXI Century (Russ.)*; Nauka: Moscow, 1996; 149–166.
5. Sergeev, G. B.; Batyuk, V. A. *Cryochemistry,* Mir **1986,** 326.
6. Sergeev, G. B. Cryonanochemistry—the new scientific field. In *Science and Humanity 95–97*; Izd. Znanie: Moscow, 1997; pp 58–63.
7. Sergeev, G. B. Physical chemistry of ultradispersed particles, parts I and II. In *Conference Proceeding*; UrO RAN: Ekaterinburg, 2001; pp 12–23.
8. Sergeev, G. B. *J. Nanopart. Res.* **2003,** *5,* 529–537.
9. Sergeev, G. B. *Mol. Cryst. Liq. Cryst.* **2005,** *440,* 85–92.

10. Kabanov, V. A.; Sergeev, G. B.; Zubov, V. P.; Kargin, V. A. *Polymer Science A (Russ.)* **1959**, *1*, 1859–1861.
11. Sergeev, G. B.; Smirnov, V. V.; Shilina, M. I.; Rostovshikova, T. N. *Mol. Cryst. Liq. Cryst. Inc. Nonlin. Opt.* **1988**, *161*, 101–108.
12. Sergeev, G. B.; Kimmel'feld, Ya. M.; Smirnov, V. V.; Shilina, M. I. *Chem. Phys. (Russ.)* **1983**, *5*, 703–705.
13. Sergeev, G. B.; Smirnov, V. V.; Zagorskii, V. V.; Kosolapov, A. M. *Doklady AN SSSR* **1981**, *256*, 1169–1172.
14. Sergeev, G. B.; Smirnov, V. V.; Zagorskii, V. V. *J. Organomet. Chem.* **1980**, *201*, 9–20.
15. Sergeev, G. B. *Low Temperature Chemistry and Cryochemical Technology*; Izd. MGU: Moscow, 1987, 107–125.
16. Sergeev, G. B.; Zagorskii, V. V.; Badaev, F. Z. *Chem. Phys. (Russ.)* **1984**, *3*, 169–175.
17. Sergeev, G. B.; Smirnov, V. V.; Badaev, F. Z. *J. Organomet. Chem.* **1982**, *224*, 29–30.
18. Ashby, E. C.; Nackashi, J. *Ibid.* **1974**, *72*, 11–20.
19. Sergeev, G. B.; Zagorskii, V. V.; Badaev, F. Z. *J. Organomet. Chem.* **1983**, *243*, 123–129.
20. Ivashko, S. V.; Zagorskii, V. V.; Sergeev, G. B. *Mol. Cryst. Liq. Cryst.* **2001**, *356*, 443–448.
21. Sergeev, G. B.; Smirnov, V. V.; Zagorskaya, O. V.; Zagorskii, V. V.; Popov, A. V. *Vestn. Mosk. Un-ta, ser. 2, Him.* **1982**, *23*, 232–236.
22. Sergeev, G. B.; Zagorskaya, O. V.; Zagorskii, V. V.; Leenson, I. A. *Chem. Phys. (Russ.)* **1986**, *5*, 1384–1932.
23. Vedeneev, V. I.; Gurvich, L. V.; Kondrat'ev, V. N.; Medvedev, V. A.; Frankevich, E. L. *Energies of Chemical Bonds. Ionization Potentials and Electron Affinities. (Russ.)*; Izd. AN SSSR: Moscow, 1962, 215.
24. Sergeev, G. B.; Zagorskaya, O. V. *Vestn. Mosk. Un-ta, ser. 2, Him.* **1987**, *28*, 362–364.
25. Solov'ev, V. N.; Policarpov, E. V.; Nemukhin, A. V.; Sergeev, G. B. *J. Phys. Chem. A* **1999**, *103*, 6721–6725.
26. Sergeev, G. B.; Zagorskaya, O. V. *Vestn. Mosk. Un-ta, ser. 2, Him.* **1985**, *26*, 575–579.
27. Boronina, T.; Klabunde, K. J.; Sergeev, G. B. *Environ. Sci. Technol.* **1995**, *24*, 1511–1517.
28. Boronina, T. N.; Lagadic, I.; Sergeev, G. B.; Klabunde, K. J. *Environ. Sci. Technol.* **1998**, *32*, 2614–2622.
29. Sergeev, G. B.; Zagorskii, V. V.; Kosolapov, A. M. *Chem. Phys. (Russ.)* **1982**, *1*, 1719–1721.
30. Efremov, M. Yu.; Komarov, V. S.; Sergeev, G. B. *Low Temperature Chemistry and Cryochemical Technology*; Izd. MGU: Moscow, 1990, 114–120.
31. Efremov, M. Yu. Crytical Phenomena Occuring During the Formation of Low-temperature Reactive Co-condensates. Abstract of PhD thesis, Moscow: MSU. (Russ.), 1995; p 19.
32. Sergeev, G. B. *Vestn. Mosk. Un-ta, ser. 2, Him.* **1999**, *40*, 312–322.
33. Efremov, M. Yu.; Batsulin, A. F.; Sergeev, G. B. *Vestn. Mosk. Un-ta, ser. 2, Him.* **1999**, *40*, 194–197.
34. Efremov, M. Yu.; Batsulin, A. F.; Sergeev, G. B. *Inorg. Mater. (Russ.)* **1999**, *35*, 1007–1009.
35. Efremov, M. U.; Batsulin, A. F.; Sergeev, G. B. *Mendeleev Commun.* **1999**, *9*, 7–9.
36. Kabanov, V. A.; Sergeev, V. G.; Lukovkin, G. M.; Baranovskii, V. Yu. *Doklady AN SSSR* **1982**, *266*, 1410–1414.
37. Sergeev, V. G.; Baranovskii, V. Yu.; Lukovkin, G. M.; Kabanov, V. A. *Polymer Sci. B (Russ.)* **1984**, *24*, 65–66.
38. Sergeev, G. B. *Mend. Chem. J. (Russ.)* **1990**, *35*, 566–575.
39. Sergeev, G. B.; Komarov, V. S.; Bekhterev, F. Z.; Mash'yanov, M. N. *Theor. i Eksp. Khimiya (Russ)* **1986**, *6*, 673–679.
40. Sergeev, G. B.; Komarov, V. S.; Zvonov, A. V. *Doklady AN SSSR* **1988**, *299*, 665–668.

41. Komarov, V. S.; Sergeev, G. B. *Polymer Sci. A (Russ.)* **1988**, *30*, 2147–2152.
42. Zagorskii, V. V., Petrukhina, M. A., Sergeev, G. B., Rozenberg, V. I., Kharitonov, V. G. A Method of Producing the Film Materials, Containing metal Clusters. (Patent RF 2017547); 15-8-1994.
43. Sergeev, G.; Zagorsky, V.; Petrukhina, M. *J. Mater. Chem.* **1995**, *5*, 31–34.
44. Sergeev, G. B.; Gromchenko, I. A.; Petrukhina, M. A.; Prusov, A. N.; Sergeev, B. M.; Zagorsky, V. V. *Macromol. Symp.* **1996**, *106*, 311–316.
45. Sergeev, G. B.; Nemukhin, A. V.; Sergeev, B. M.; Shabatina, T. I.; Zagorskii, V. V. *Nanostr. Mater.* **1999**, *12*, 1113–1116.
46. Gromchenko, I. A., Petrukhina, M. A., Sergeev, B. M., Sergeev, G. B. (Patent 2040321); 24-12-1992.
47. Sergeev, G.; Zagorsky, V.; Petrukhina, M.; Zav'yalov, S.; Grigor'ev, E.; Trakhtenberg, L. *Anal. Comm.* **1997**, *34*, 113–114.
48. Sergeev, G. B., Zagorskii, V. V., Petrukhina, M. A., Zav'yalov, S. A., Grigor'ev, E. I., Trakhtenberg, L. I. Ammonia-sensitive layer (Russ.). (Patent 2097751); 17-4-1995.
49. Bochenkov, V. E.; Stephan, N.; Brehmer, L.; Zagorskii, V. V.; Sergeev, G. B. *Colloids Surf. A* **2002**, *198–200*, 911–915.
50. Sergeev, B. M.; Sergeev, G. B.; Prusov, A. N. *Mendeleev Commun.* **1998**, 1–2.
51. Ermilov, A. Yu.; Nemukhin, A. V.; Sergeev, G. B. *Russ. Chem. Bull. (Russ.)* **1998**, *62*, 1169–1173.
52. Zagorskii, V. V.; Ivashko, S. V.; Bochenkov, V. E.; Sergeev, G. B. *Nanostr. Mater.* **1999**, *12*, 863–866.
53. Zagorskii, V. V.; Bochenkov, V. E.; Ivashko, S. V.; Sergeev, G. B. *Mater. Sci. Eng. C* **1999**, *8–9*, 329–334.
54. Zagorskii, V. V.; Ivashko, S. V.; Sergeev, G. B. *Vestn. Mosk. Un-ta, ser. 2, Him.* **1998**, *39*, 349–352.
55. Sergeev, B. M.; Sergeev, G. B.; Kasaikin, V. A.; Litmanovich, E. A.; Prusov, A. N. *Mol. Cryst. Liq. Cryst.* **2001**, *356*, 121–129.
56. Badaev, F. Z.; Batyuk, V. A.; Golubev, A. M.; Sergeev, G. B.; Stepanov, M. B.; Federov, V. V. *Russ. J. Phys. Chem.* **1995**, *69*, 1119–1123.
57. Batyuk, V. A.; Shabatina, T. I.; Boronina, T. N.; Sergeev, G. B. *Itogi nauki i tehn. VINITI. Ser. Kinet. i katal. (Russ.)* **1990**, *21*, 1–120.
58. Batyuk, V. A. In *Himiya nizkih temperatur i kriohimicheskaya tehnologiya*; Sergeev, G. B., Ed.; Izd. MGU: Moskva, 1987; pp 163–185.
59. Batyuk, V. A.; Shabatina, T. I.; Boronina, T. N. In *Himiya nizkih temperatur i kriohimicheskaya tehnologiya*; Sergeev, G. B., Ed.; Izd. MGU: Moskva, 1990; pp 69–79.
60. Batyuk, V. A.; Shabatina, T. I.; Morozov, Yu. N.; Sergeev, G. B. *Doklady AN SSSR* **1988**, *300*, 136–139.
61. Vovk, E. V.; Shabatina, T. I.; Vlasov, A. V.; Sergeev, G. B. *Supramolec. Sci.* **1997**, *4*, 509–514.
62. Shabatina, T. I.; Vovk, E. V.; Ozhegova, N. V.; Morosov, Y. N.; Nemukhin, A. V.; Sergeev, G. B. *Mat. Sci. Eng. C* **1999**, *8–9*, 53–56.
63. Alexanyan, V. T.; Lokshin, B. V. *Kolebatelnye spektry p-komplexov perehodnyh elementov*; Stroenie molecul i himicheskaya svyaz: Moskva, 1976, 177.
64. Ozhegova, N. V.; Shabatina, T. I.; Nemukhin, A. V.; Sergeev, G. B. *Mendeleev Commun.* **1998**, *6*, 218–220.
65. Michlik, J.; Yamada, D. R.; Brown, D. R.; Kevan, L. *J. Phys. Chem.* **1996**, *100*, 4213–4218.
66. Shabatina, T. I.; Timoshenko, V. A.; Belyaev, A. A.; Morozov, Yu. N.; Sergeev, G. B. *Doklady Phys. Chem.* **2002**, *387*, 267–279.

67. Timoshenko, V. A.; Belyaev, A. A.; Morosov, Yu. N.; Shabatina, T. I.; Sergeev, G. B. *Mol. Cryst. Liq. Cryst.* **2005**, *440*, 79–83.
68. Polikarpov, A. V.; Shabatina, T. I.; Sergeev, G. B.; Nemukhin, A. V. *Vestn. Mosk.Un-ta, ser. 2, Him.* **2000**, *41*, 283–285.
69. Shabatina, T. I.; Vovk, E. V.; Morosov, Y. N.; Timoshenko, V. A.; Sergeev, G. B. *Mol. Cryst. Liq. Cryst.* **2001**, *356*, 143–148.
70. Shabatina, T. I.; Vovk, E. V.; Timoshenko, V. A.; Morosov, Y. N.; Sergeev, G. B. *Colloids Surfaces A-Physi. Eng. Aspects* **2002**, *198–200*, 255–259.
71. Shabatina, T. I. *Liquid Cryst. Appl. (Russ.)* **2002**, *1*, 58–64.
72. Sergeev, G. B.; Smirnov, V. V.; Zagorskii, V. V.; Badaev, F. Z.; Zagorskaya, O. V. *Book of Abstracts of XII Mendeleev Symposium on General and Applied Chemistry. #3 (Russ.)*; Nauka: Moscow, 1981, 112–113.
73. Evans, W. J.; Coleson, K. M.; Engerer, S. C. *Inorg. Chem.* **1981**, *20*, 4320–4325.
74. Evans, M. D.; Engerer, S. C.; Coleson, K. M. *J. Am. Chem. Soc.* **1981**, *105*, 6672–6677.
75. Cloke, F. G. N. *Chem. Soc. Rev.* **1993**, *22*, 17.
76. Brennan, J.; Cloke, F. G. N.; Sameh, A. A.; Zalkin, A. *J. Chem. Soc. Chem, Commun.* **1987**, *24*, 1668–1669.
77. Luebcke, M.; Sonntag, B.; Niemann, W.; Rabe, P. *Phys. Rev. B* **1986**, *34*, 5184–5190.
78. Kosolapov, A. M.; Kondakov, S. E.; Sergeev, G. B.; Zagorskii, V. V. In *Low Temperature Chemistry and Cryochemical Technology (Russ.)*; Izd. MGU: Moscow, 1990, 127–138.
79. Zagorskii, V. V.; Kondakov, S. E.; Kosolapov, A. M.; Sergeev, G. B.; Solov'ev, V. N. *Metallorg. Khim.* **1992**, *5*, 533–536.
80. Sergeev, G. B.; Zagorskii, V. V.; Grishechkina, M. V. *Vestn. Mosk. Un-ta, ser. 2, Him.* **1988**, *29*, 38–42.
81. Sergeev, G. B.; Zagorskii, V. V.; Grishechkina, M. V. *Metallorg. Khim.* **1988**, *1*, 1187–1189.
82. Sergeev, G. B.; Zagorskii, V. V.; Grishechkina, M. V. *Metallorg. Khim.* **1988**, *1*, 820–825.
83. Sergeev, G. B.; Komarov, V. S.; Tarkhanova, I. G. *Kinet. Catal.* **1990**, *31*, 209–213.
84. Komarov, V. S.; Tarkhanova, I. G.; Sergeev, G. B. *Low Temperature Chemistry and Cryochemical Technology*; Izd. MGU: Moscow, 1990, 121–126.
85. Vlasov, A. V.; Shabatina, T. I.; Konyukhov, S. V.; Ermilov, A. Yu.; Nemukhin, A. V.; Sergeev, G. B. *J. Struct. Chem.* **2004**, *45*, 382–387.
86. Shabatina, T. I.; Vlasov, A. V.; Konyukhov, S. V.; Ermilov, A. Yu.; Nemukhin, A. V. *Mol. Cryst. Liq. Cryst.* **2005**, *440*, 309–316.
87. Vlasov, A. V.; Shabatina, T. I.; Ivanov, A. Yu.; Sheina, G. G.; Nemukhin, A. V.; Sergeev, G. B. *Mendeleev Commun.* **2005**, *15*, 10–11.
88. Shabatina, T. I.; Vlasov, A. V.; Vovk, E. V.; Stufkens, D. J.; Sergeev, G. B. *Spectrochim. Acta A-Mol. Biomol. Spectr.* **2000**, *56*, 2539–2543.
89. Shabatina, T. I.; Vlasov, A. V.; Sergeev, G. B. *Mol. Cryst. Liq. Cryst.* **2001**, *356*, 149–154.
90. Vlasov, A. V.; Shabatina, T. I.; Sergeev, G. B. *Russ. J. Phys. Chem.* **2002**, *76*, 1784–1788.
91. Bochenkov, V. E.; Zagorskii, V. V.; Sergeev, G. B. *Vestn. Mosk. Un-ta, ser. 2, Him.* **2000**, *41*, 327–330.
92. Bochenkov, V. E.; Zagorskii, V. V.; Sergeev, G. B. *Mol. Cryst. Liq. Cryst.* **2001**, *356*, 299–309.
93. Sergeev, G. B. *Mol. Cryst. Liq. Cryst.* **1992**, *211*, 439–443.
94. Sergeev, G. B.; Solov'ev, V. N. *Modern Trends in Low Temperature Chemistry, Conference Proceeding*; Moscow Univ. Publ. House: Moscow, 1994, 148–155.
95. Shestakov, A. F.; Solov'ev, V. N.; Zagorskii, V. V.; Sergeev, G. B. *J. Phys. Chem. (Russ.)* **1994**, *68*, 155–158.
96. Jensen, P.; Barabase, A. L.; Larralde, H. *Nature* **1994**, *368*, 22–24.

97. Klotzbuecher, W. E.; Petrukhina, M. A.; Sergeev, G. B. *Mendeleev Commun.* **1994**, *4*, 5–7.
98. Petrukhina, M. A.; Klotzbuecher, V. E.; Sergeev, G. B. *Vestn. Mosk. Un-ta, ser. 2, Him.* **1995**, *36*, 360–364.
99. Klotzbuecher, W. E.; Petrukhina, M. A.; Sergeev, G. B. *J. Phys. Chem. A* **1997**, *101*, 4548–4554.
100. Nemukhin, A. V.; Ermilov, A. Yu.; Petrukhina, M. A.; Klotzbuecher, W. E.; Smets, J. *Spectrochim. Acta A-Mol. Biomol. Spectr.* **1997**, *53*, 1803–1812.
101. Solov'ev, V. N.; Sergeev, G. B.; Nemukhin, A. V.; Burt, S. K.; Topol, I. A. *J. Phys. Chem. A* **1997**, *101*, 8625–8630.
102. Bare, W. D.; Andrews, L. *J. Am. Chem. Soc.* **1998**, *120*, 7293–7301.
103. Sergeev, G. B.; Shabatina, T. I.; Solov'ev, V. N.; Nemukhin, A. V. *Spectrochim. Acta A-Mol. Biomol. Spectr.* **2000**, *56*, 2527–2537.
104. Polikarpov, E. V.; Granovsky, A. A.; Nemukhin, A. V. *Mendeleev Commun.* **2001**, *2001*, 150–151.
105. Tulegenov, A. N.; Solov'ev, V. N.; Sergeev, G. B. *Book of Abstracts of MIFI Scientific Session (Russ.).* 2002, Vol. 9, pp 203–204.
106. Mikhalev, S. P., Soloviev, V. N., Sergeev, G. B. International Conference on Low-Temperature Chemistry, Finland, 2002; pp 100–101.
107. Mikhalev, S. P.; Solov'ev, V. N.; Sergeev, G. B. *Mendeleev Commun.* **2004**, *14*, 48–50.
108. Rogov, A. V.; Mikhalev, S. P.; Granovsky, A. A.; Soloviev, V. N.; Nemukhin, A. V.; Sergeev, G. B. *Vestn. Mosk. Un-ta, ser. 2, Him.* **2004**, *45*, 219–224.
109. Chase, M. W. NIST-JANAF. Thermochemical tables. *J. Phys. Chem. Ref. Data* **1998**, *27*, 1–1954.
110. Solov'ev, V. N. Abstract of PhD thesis, M: MSU, 1998; p 24.
111. Kiryukhin, M. V. Synthesis of Silver Nanoparticles in Aqueous Solutions of Polycarbonic Acids, PhD thesis, Moscow, 2002; p 137.
112. Sergeev, B. M.; Kiryukhin, M. V.; Bakhov, F. N.; Sergeev, V. G. *Vestn. Mosk. Un-ta, ser. 2, Him* **2001**, *42*, 308–314.
113. Kiryukhin, M. V.; Sergeev, B. M.; Prusov, A. N.; Sergeev, V. G. *Polymer Sci. B* **2000**, *42*, 324–338.
114. Sergeev, B. M.; Kiryukhin, M. V.; Prusov, A. N. *Mendeleev Commun.* **2001**, *11*, 68–70.
115. Granovsky, A. A. http://lcc.chem.msu.ru/gran/gamess.html, 2002.
116. Schmidt, M. W.; Baldridge, K. K.; Boatz, J. A.; Elbert, S. T.; Gordon, M. S.; Jensen, J. H., et al. *J. Comp. Chem.* **1993**, *14*, 1347–1363.
117. Kokyukhov, S. S.; Polykarpov, E. V.; Nemukhin, A. V. *Chem. Phys. (Russ.)* **1999**, *18*, 67–70.
118. Henglein, A.; Holzwarth, A.; Janata, J. *Ber. Bunsenges. Phys. Chem.* **1993**, *97*, 1429–1434.
119. Larsson, S.; Pyykkoe, P. *Chem. Phys.* **1986**, *101*, 355–369.
120. Nemukhin, A. V.; Grigorenko, B. L.; Sergeev, G. B. *Vestn. Mosk. Un-ta, ser. 2, Him.* **1995**, *36*, 379–382.
121. Grigorenko, B. L.; Nemukhin, A. V.; Sergeev, G. B.; Stepanyuk, V. S.; Szasz, A. *Phys. Rev. B* **1994**, *50*, 18666–18669.
122. Nemukhin, A. V.; Grigorenko, B. L.; Sergeev, G. B. *Can. J. Phys.* **1994**, *72*, 909–912.
123. Hofmann, M.; Leutwyler, S.; Schulze, W. *Chem. Phys.* **1979**, *40*, 145–152.
124. Gordon, M. S.; Freitag, M. A.; Bandyopadhyay, P.; Jensen, J. H.; Kairay, V.; Stevens, W. J. *J. Phys. Chem. A* **2001**, *105*, 293–307.
125. Nemukhin, A. V.; Grigorenko, B. L. *Chem. Phys. Lett.* **1995**, *233*, 627–631.
126. Zeynalov, V. A.; Blinov, L. M.; Grebennkin, M. F.; Ostrovskii, B. I. *Crystallography (Russ.)* **1988**, *33*, 185.

Chapter 6

Chemical Nanoreactors

Chapter Outline
6.1 General Remarks 155
6.2 Alkali and Alkaline-Earth Elements 160
6.3 Transition Metals of Groups III–VII in the Periodic Table 169
6.4 Elements of the Group VIII of the Periodic System 179
6.5 Subgroups of Copper and Zinc 191
6.6 Subgroup of Boron and Arsenic 198

6.1 GENERAL REMARKS

The high activity of metal clusters and particles is associated, first of all, with uncompensated surface bonds, which pose many problems associated with protection against aggregation. In fact, atoms, clusters, and particles of metals are divided into free or unligated and isolated or solvated.

Naturally, such particles have different stabilities and activities. As a rule, transformation of original particles into the reaction products involves overcoming a potential barrier, which is called the activation energy (E) of a chemical reaction. The presence of a potential barrier means that every chemical species, viz., a molecule, a radical, an atom, or an ion, represents a more or less stable formation as regards its energy.

The reconstruction of reacting species requires breaking or weakening of certain chemical bonds, which results in an expenditure of energy.

Let us return to a scheme M+L shown in Section 2.4. This is a multifactor process of competitive reactions, consecutive and parallel, which proceed with an activation energy $E \approx 0$.

Evidently, each formation in the scheme can be identified with an initial state of a peculiar kind and considered as a nanoreactor. Gas-phase reactions proceed in a similar way, where the formation of a metal nucleus as a carrier of compact metal properties proceeds with an increase in the number of atoms. The ratio of surface atoms that determine the interactions with the medium, i.e. chemical properties, is also great. The systems obtained by the M+L scheme are nonequilibrium ones. According to the Arrhenius law, the reaction rate should decrease with a decrease in the temperature. This determines the use of low temperatures for stabilizing active particles and the products of their interaction with ligands. Moreover, it is evident that for more active particles, lower temperatures should be used for stabilization. The advantages of low-temperature co-condensation in studying nanoparticles are determined by

- unlimited choice of metals;
- the absence of impurities such as ions and the products of redox reactions; and
- the possibility of studying atoms and small clusters in inert matrices and their reactions in active matrices.

Thus, at low and superlow temperatures, we deal with real nanochemistry.

Atoms of most metals can be stabilized at a temperature of 4–10 K in inert matrices under, e.g. 1000-fold dilution. This is achieved by the method of matrix isolation,[1,2] which is widely used for stabilizing active particles that involve not only atoms and small clusters but also free radicals such as OH, CH_3, and NH_2.

We anticipate the discussion of applications of the matrix-isolation method with the general remarks concerning the effects of temperature and particle size on the reactivity. The exponential dependence of the reaction rate on the temperature manifests itself in different ways over different temperature intervals. A comparison of two parallel bimolecular reactions with activation energies differing by 1 Kcal shows that they cannot be distinguished at room temperature, while they proceed at sufficiently different rates at low temperatures. In the chemistry of low temperatures, this phenomenon is named energy selection and lays the basis for the development of waste-free technologies.[3] For example, chlorination of ethylene in the vicinity of liquid nitrogen temperatures proceeds with the formation of chromatographically pure dichloroethane. Radiation-induced hydroboration of ethylene at low temperatures yields medicinal ethyl bromide, which is used for anesthesia. In both examples, the expensive rectification stage is eliminated. The use of low temperatures for finding the reaction mechanism and the optimal temperatures for its performance was considered by the example of reactions of halogenation and hydrohalogenation of olefins.[4]

High reactivity of such species is closely connected with the size effects for nanometer systems. Such effects appear in general when particle size correlates with the characteristic size for a definite system property, for example, magnetic domain or the electron-free length. The structural and phase nonhomogeneity is characteristic of nanosize systems, and the new coordinate—'dis-persity'—should be added to classical physical and chemical analysis that would result in diagrams "composition–structure property." Thus, the particle size (the number of atoms in the particle's structure) becomes the active thermodynamic property, determining the system state and its chemical activity. The dependence of the relative chemical activity of metal species upon their size is presented schematically in Figure 6.1.

According to this scheme, the chemical activity decreases in the following direction: free metal atoms→oligomeric clusters→nanoparticles→aggregates→bulk metal. The atoms and small metal clusters possess the highest reactivity. Their properties are usually studied using a special technique for their stabilization in inert matrices at superlow temperatures. An increase of the size of the reactive species leads to decrease of the number of active surface atoms. It is followed by the proper decrease of the system's chemical activity per atom going from dimers and trimers to nanosized particles and then to rougher dispersions and

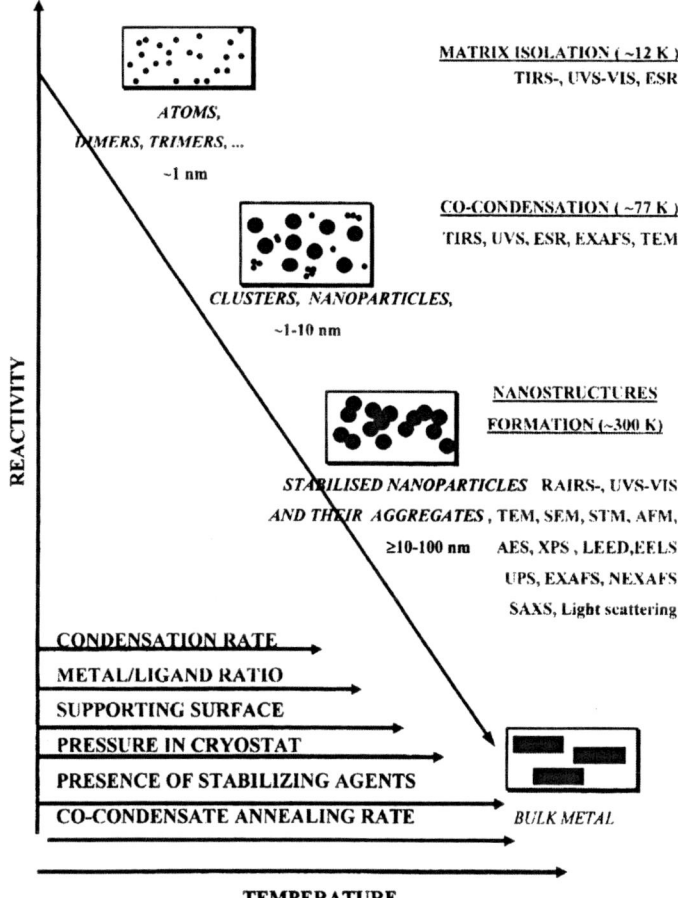

FIGURE 6.1 General scheme describing the relative chemical activity of metal species vs. their size. According to this scheme, the chemical activity decreases in the following direction: free metal atoms→oligomeric clusters→nanoparticles aggregates→bulk metal. The main factors determining the size of metal species formed in the chemically active systems and some experimental techniques used for their characterization are also indicated: TIRS; UVS–VIS; RAIRS; ESR; EXAFS; NEXAFS; TEM; SEM; AFM; AES; XPS; UPS; LEED; EELS; SAXS; and various sorts of light-scattering techniques.

bulk metals. It should be mentioned that the plot of chemical activity against reactive species size is not a monotonic one. There should be some extremes due to the higher stability of metal clusters with definite (magic) atom numbers, which are specific for individual metals and their combinations. The loss in chemical activity is more significant for growing small particles and less sharp for the final transition from rough dispersions to bulk.

The size of metal particles forming such systems and their reactivity are determined by the combination of different experimental conditions (Figure 6.1). The main experimentally controllable factors are the support temperature, the metal/ligand ratio, reagent condensation rate, and the rate of sample annealing. It was shown that the lower the temperature of the support surface, the lesser the diffusion-controlled interactions between reactive particles and the more possibility for the formation of high-energetic and high-reactive species at low temperatures. The metal/ligand ratio greatly affects the size of metal particles obtained via low temperature co-condensation. Raising this ratio and annealing of the sample usually lead to an increase of the part of the clusters and higher aggregated metal particles. The component condensation rate has a complex effect on the properties of low temperature co-condensate film. The lifetime of highly active species (metal atoms, their dimers, or trimers) during co-condensation at the cooled surface is inversely proportional to the condensation rate. It depends on the nature of the relaxation processes in the co-condensate system. The intensity of the particle beam determines the number of collisions of the molecules with the surface and with each other. The temperature of the surface determines the relaxation processes. Together with the chemical nature of the reagents, these factors determine the pathway of processes leading to or not leading to the reaction. The processes, which occur during the real condensation, are more complicated than the given scheme. The experimental techniques usually applied for the study of structure, texture, and particle size in co-condensate films are also shown in Figure 6.1.

The most spectacular results of nanochemistry were obtained in the gas phase by using the matrix-isolation method. This method consists in accumulating a substance under conditions that hinder reactions. For example, in a solid inert substance at low temperatures, the matrix prevents diffusion so that active particles are virtually immobile (stable), frozen in a medium that cannot react with them.

In the matrix-isolation method, IR and UV spectroscopies are the main studying techniques. Hence, there are certain requirements to matrices:

(1) Rigidity, i.e. the absence of noticeable diffusion of stabilized particles, which should rule out their recombination and aggregation.
(2) Inertness, i.e. the absence of chemical reactions with the matrix. These requirements reflect the inconsistencies between the matrix-isolation method and those used in preparative cryochemistry, as mentioned above.
(3) Optical transparency, which makes it possible to study condensates by spectral and radiospectral techniques.

As matrices, inert gases were widely used. The melting points T_m (K) of inert gases at different pressure are as follows:

	Ne	Ar	Kr	Xe
Atmospheric pressure	25	83	116	161
$P = 10^{-3}$ mmHg	11	39	54	74

For $P = 10^{-3}$ mmHg, temperatures of the higher limit are shown; however, it is advisable to work at a temperature equal to $\frac{1}{3}T_m$. Optimal temperatures T_{opt} for the listed gases are 8, 28, 40, and 50 K, respectively.

As the surfaces (supports) for measuring IR spectra, NaCl, KCl, CaF_2, and CsI were used. Calcium difluoride was used in the IR and UV regions. Matrix-isolation units use vacuum of 10^{-5}–10^{-7} mmHg.

The formation of metal atoms is an endothermic process. The heat of formation varies from 60 to 800 KJ/mol (e.g. 62 for Hg and 791 KJ/mol for Re). The values shown determine the difference in energies between the solid and atomic states. The energy of metal atoms in low-temperature inert matrices changes insignificantly as compared with the gas phase.[12] This means that the kinetic energy liberated during condensation is only a small part of the total energy. The change in the energy in the course of reaction can be estimated from the binding energies. For example, the heat of a reaction $M_i + L_i \rightarrow ML_i$ is approximately equal to the energy of the M–L bond ("i" is the index of stabilized structures similar to those in the gas phase). The processes are thermodynamically permitted.

A compound ML can be extracted in the individual form as the product, if its decomposition is hindered for kinetic reasons. Thus, the number of compounds, which can be obtained under matrix-isolation conditions, exceeds the number of compounds obtained in preparative synthesis. For instance, a compound $Ni(N_2)_4$ exists in an argon matrix and was never observed at 77 K.

Thus, we can formulate the methodological significance of low temperatures in nanochemistry. This consists in synthesizing new, earlier unknown compounds, improving the methods of preparation of already known compounds, and refining the mechanisms of reactions. The examples are shown in the following sections of this chapter.

The interaction between a ligand and a nanocrystalline inorganic (metal) core can be considered as the formation of a core–ligand complex. The stability of such complexes is determined by the strength of the interaction between ligands and surface atoms of the core. Moreover, the stability of complexes determines the possibility of their practical application. In connection with this, a constant quest for new types of stabilizing ligands was carried out. For this purpose, a number of hydrophilic organic dendrons and nanosize voids were proposed.[5,6] Moreover, new approaches to studying stabilization based on the formation of labeled ligand coatings were developed.[7–10] Typical ligands represent amines, thiols, and phosphoric acid derivatives. For these ligands, a situation where a decrease in pH can have an effect on the core–ligand interaction is possible. For example, a ligand can be protonated and passed into solution, which destabilizes the particle. The appearance of such a situation is critical for systems with biological functions. The average pH of human digestive juices is close to 2. In this case, separation of ligands can result in the appearance of toxic nanocrystals.[8,11]

Carrying out reactions in inert and active matrices poses the problem of revealing the relationship between a reaction and the number of atoms in a particle. The subsequent presentation of the chemistry of nanoreactors is carried out in accordance with the groups of elements in the Periodic Table; moreover, this involves only those elements for which most clear and interesting results were obtained.

6.2 ALKALI AND ALKALINE-EARTH ELEMENTS

Stabilization processes and spectral properties of alkali and alkaline-earth atoms have been surveyed in detail.[1] However, studies that prove that several atoms, i.e. a metal particle of a definite size, can enter into a reaction are few. Alkali metals were used as reagents in the reduction of halide compounds of other metals. At low temperatures in the liquid phase, the following abstraction reaction was carried out:[12]

$$MoCl_5 + 5K_{nap} + 2\,[naphthalene] \rightarrow [\,[naphthalene]\,]_2 \; Mo + 5KCl$$

Such reactions are remarkable as a new method for synthesizing *bis*-arene compounds of metals. In this case, a substitution of easily evaporated potassium for molybdenum that is difficult to evaporate takes place. It was shown that this reaction does not occur with compact potassium, which is probably due to the fact that the presence of locally excessive potassium can favor decomposition of the resulting product.

Particles Li_3, Li_7, Na_7, and K_7 isolated in low-temperature matrices were studied by the EPR technique.[13–15] Triatomic alkali-metal particles formed the most stable planar structures, which were shaped as triangles with D_{3h} symmetry, and generated EPR signals.

Five atoms formed a planar structure with C_2 symmetry but revealed no signals in the EPR spectra. This was strange because particles with odd numbers of atoms having spin $\frac{1}{2}$ should exhibit paramagnetic properties and hence demonstrate EPR spectra. The possible absence of spectrum was associated with their complex anisotropy and complicated superfine structure.[16] Particles comprising more than five atoms formed three-dimensional structures. As shown for the deposition of silver cations Ag_n^+ ($n = 50 - 200$) with energy of 250–2500 eV on graphite surfaces, these particles could be stabilized even at room temperature.[17]

Among alkali and alkaline-earth metals, magnesium particles attract the keenest attention. This is explained by several reasons:

- Magnesium is easily evaporated without any impurities, because the latter have higher melting points.
- Its gas-phase spectra are well studied.
- Of great importance are the organometallic compounds of magnesium, which served for synthesizing the first Grignard reagent.
- Magnesium atoms and small clusters can be simulated by ab initio quantum-chemical methods.
- By the involvement of magnesium, Grignard reagents with new properties were synthesized at low temperatures. In contrast to solvated (normal) reagents that yield *tert*-alcohols in their reaction with acetone, the reaction of low-temperature nonsolvated reagents with acetone in its enol form-produced hydrocarbons.[1]
- Magnesium is a component of valuable alloys and materials.

The information on atomic aggregation is of prime importance for understanding reactions that involve magnesium particles. Studies in inert-gas matrices make it possible to gain insight into the properties of metal atoms and the possibility of their aggregation during either condensation or heating of the matrix. Absorption spectra of magnesium atoms isolated in matrices of different inert gases were repeatedly studied and compared with their gas-phase spectra. Based on the data of the interatomic distances for van der Waals complexes magnesium–inert gas and the diameters of inert-gas unit cells, it was found that a magnesium atom can occupy a unique position in the argon matrix.[18] Absorption of magnesium atoms at 285 nm was assigned to the $3p^1P_1 \leftarrow 3s^1S_0$ transition. By varying the metal–inert gas ratio, considerable amounts of magnesium dimers and coarser particles, which absorb at 200–240 nm, can be obtained at low temperatures.

The proved presence of magnesium particles in argon matrices provided grounds for an assumption that they are formed at the instant of condensation when atoms are still mobile. However, attempts to stabilize magnesium atoms even by additional dilution with inert gas failed: their adsorption spectra always contain bands at 345 nm pertaining to Mg_x particles and also bands corresponding to dimers. This circumstance and also the fact that absorption of magnesium atoms and coarser clusters dramatically decreases with an increase in the temperature allow us to assume that it is the magnesium particles that react at 77 K, rather than magnesium atoms and, hence, a nanochemical reaction is observed.

Matrix-isolation method allowed other alkali-earth metals to be studied. Atoms of all these metals were found to form dimers; moreover, mixed dimers such as CaMg and BaMg were observed.[19] The absorption regions of the latter compounds lie between the absorption regions of their components. Table 6.1

TABLE 6.1 Absorption Maxima of Alkaline-Earth Metal Dimers in Inert Gases

Dimer	Ca_2	Sr_2	CaMg	SrMg	BaMg	SrCa
Argon, λ_{max} (nm)	648	710	545	596	635	685
Krypton, λ_{max} (nm)	666	730	557	604	638	–

shows the positions of absorption maxima for dimers isolated in argon and krypton matrices.

Dimers of alkali-earth elements are van der Waals particles, and their atoms, as a rule, do not absorb light. Aggregation of alkali-earth atoms was studied in inert-gas matrices; however, calcium atoms were observed in benzene matrices at the dilution of 1:1000 over a temperature range of 55–110 K.[20] In diphenyl, triphenyl, decane, and certain other hydrocarbons, calcium atoms remained stable up to 120 K. The effect of diffusion on the formation of calcium dimers was demonstrated.[21] In the course of heating, no dimers were observed in a krypton matrix up to 45 K and a xenon matrix up to 65 K. The dimer ratio in a matrix increased with an increase in the condensation temperature. The effectiveness of atomic stabilization as a function of inert-gas nature decreased in the order Xe > Kr > Ar > Ne.[165] Such a dependence was explained by the formation of complexes between the metal and inert-gas atoms due to van der Waals forces. Most polarizable inert-gas atoms were shown to form most stable complexes. The diffusion of a metal bound in such a complex either ceases or is dramatically hindered.[1]

The high activity of magnesium atoms was observed for the first time when co-condensed vapors of magnesium and diethyl ether were heated.[22] In place of the expected colloidal solution of magnesium, an unidentifiable organomagnesium compound was obtained. Successful experiments on reactions of low-temperature co-condensates containing magnesium particles were accomplished.[1,23] The latter reference shows many examples of reactions involving alkali and alkali-earth metals, which were published before the 1980s. Among the reactions involving magnesium, the synthesis of Grignard reagents attracts the keenest attention. The unabated interest in synthetic problems is evident from the materials cited in a recently published monograph.[24] As follows from this work, of greatest importance is the possibility of synthesizing Grignard reagents that involve several magnesium atoms. After heating and removal of the excess solvent, low-temperature co-condensates of benzyl halides and magnesium formed colorless crystals. Hydrolysis, elemental analysis, and time-of-flight and mass spectrometry techniques made it possible to assume that a Grignard reagent that contains four magnesium atoms was formed.[25] An XRD study of the obtained crystals of Grignard reagents would additionally prove the structure of such unusual compounds.

Among other interesting reactions in which magnesium particles participate, the interaction with methane and acetylene should be mentioned. Magnesium atoms in low-temperature co-condensates react with methane only in their excited state.

Upon illumination of a co-condensate with light of $\lambda = 270\text{–}290$ nm, a magnesium atom is inserted into a C–H bond of methane.[26] The reaction proceeds according to the following scheme:

$$\begin{array}{c} CH_3\text{-}H \\ + \\ Mg \end{array} \longrightarrow \begin{array}{c} H_3C\text{-}H \\ \diagdown \diagup \\ Mg \end{array} \longrightarrow CH_3MgH$$

The formation of methylmagnesium hydride was confirmed by IR spectra. Further studies have shown that the insertion of a magnesium atom competes with its emission from the excited triplet state.[27] In solid perdeuteromethane, an almost five-fold increase in emission intensity was accompanied by a corresponding decrease in the insertion product yield. This suggests that the insertion process involves magnesium atoms in the triplet state and that photophysical and photochemical processes are closely interrelated. The observed isotope effect was associated with the presence of a reaction barrier and the difference in the zero-point energies of CH_4 and CD_4. Methylmagnesium hydride was also obtained in the co-condensation of methane and in laser-evaporated magnesium.[28] Excited magnesium was shown to react with acetylene to produce MgCCH, CCH, MgH, and $(C_2H_2)_2$. Two possible reaction mechanisms were proposed:[29]

$Mg^* + HCCH \rightarrow (HMgCCH)^* \rightarrow MgH + CCH$
($\Delta E = +106$ kcal/mol)

$Mg^* + HCCH \rightarrow (HMgCCH)^* \rightarrow H + MgCCH$
($\Delta E = +65$ kcal/mol)

The energy estimates were obtained by the density-functional method. From the viewpoint of the authors, the second mechanism is preferential, because an excited magnesium atom has an energy of 63 Kcal/mol.

Reactions of magnesium atoms involving the formation of clusters of several atoms or their interactions with other chemicals depend on the combination of several conditions. The reactivity of particles increases with an increase in the degree of dispersion of magnesium. However, such particles require special stabilization procedures, including the use of superlow temperatures, which, together with a stable valence shell and a high ionization potential of magnesium, have an inhibiting effect on the activity of magnesium particles. As a result, reactions with their participation depend on many factors such as the method of metal evaporation, the condensation rate, and the temperature of the condensation surface.

For example, magnesium particles obtained by thermal evaporation did not form oxidation products in the course of their co-condensation with oxygen–argon mixtures, whereas under a laser beam, quite a number of different magnesium oxides were observed.[30] The products contained minimum amounts of MgO. The main products were a linear isomer of magnesium oxide MgOMgO and a linear compound OMgO. The use of isotope-substituted oxygen and magnesium and also the estimates of vibrational frequencies made it possible to assume that the reaction $Mg + O_2 \rightarrow OMgO$ is the main stage. According to the estimates, this reaction is endothermic with a barrier of 12 Kcal/mol. For the ground-state magnesium atoms produced by resistive evaporation, such a barrier was high and the reaction did not occur. Laser-evaporated atoms are in the excited state and easily react with oxygen during co-condensation. It was assumed that magnesium oxide is formed in a reaction $Mg + O_2 \rightarrow O + MgO$, while dimers are formed by a reaction $Mg + OMgO \rightarrow MgOMgO$.

Electron spectroscopy allows identification of atoms, dimers, trimers, and more complex particles of alkali-earth metals in argon matrices at low temperatures. This technique was used for a comparative study of activities of magnesium and calcium particles in their reactions with different methyl halides of the CH_3X type, where $X = F$, Cl, Br, and I.[31,32] The behavior of the systems argon–calcium and argon–magnesium, containing small amounts of various methyl halides and conditioned for a certain time at 9 K, was compared. Figure 6.2 shows the results for calcium particles in argon and a mixture of argon with methyl chloride. As seen, the addition of CH_3Cl has a different effect on the reactivity of different calcium particles. Dimers and Ca_x particles react with methyl chloride, whereas the amount of calcium atoms in the argon–methyl chloride system further increases.

Similar experiments with different methyl halides and magnesium helped to reveal unusual trends in the reactivity of particles of different sizes. These

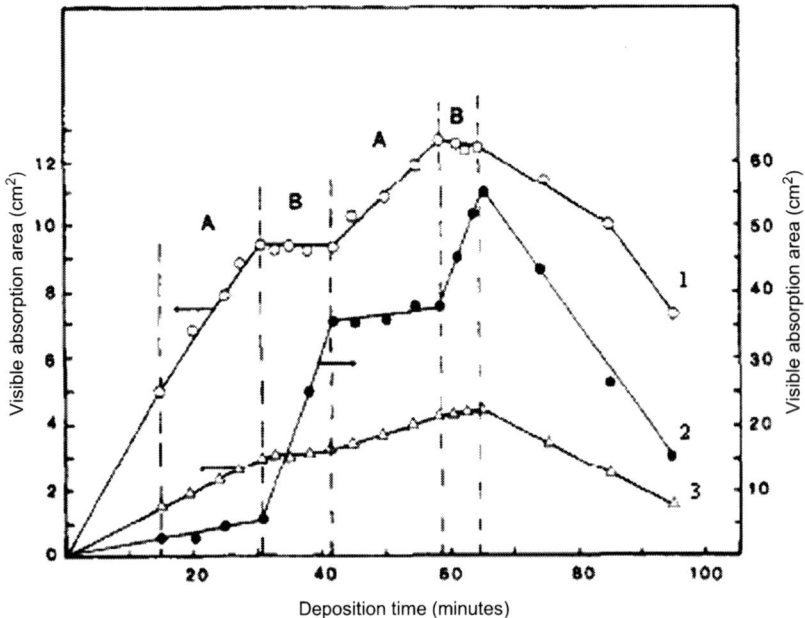

FIGURE 6.2 Reactivity of calcium particles in argon and argon-chloromethane mixture[32] (1) $Ca_x, \lambda = 472$ nm; (2) $Ca, \lambda + 415$ nm; (3) $Ca_2 + Ca_x, \lambda = 505$ nm. Domain A—argon, domain B—argon and chloromethane mixture.

trends are reflected in Table 6.2, which allows us to arrange calcium and magnesium particles in the following order: $Ca_x \approx Ca_2 > Mg_x > Mg_4 > Mg_3 \approx Mg_2 > Ca > Mg$.

A magnesium atom turns out to be the least active particle. The activities of methyl halides also change in an unusual way—out of any correlation with the strength of carbon–halogen bonds known from gas-phase data. Activity of methane halides in argon matrix is arranged in the following order: $CH_3I > CH_3F > CH_3Br > CH_3Cl$.

None of the magnesium and calcium particles under study reacted with methane.

To interpret the results obtained, it has been speculated that it is the magnesium clusters that are inserted into the carbon–halogen bond, rather than its atoms. The first ab initio quantum-chemical calculations have shown that reactions with clusters are more advantageous from the viewpoint of thermodynamics.[33,34] It is also possible that the lower energy of cluster ionization can favor the initial electron transfer, which precedes the bond rupture. For example, the process takes the following path:

$$Mg_2 + CH_3Br \rightarrow [Mg^+CH_3Br^-] \rightarrow CH_3Mg_2Br$$

A more detailed scheme of possible reactions is shown below:

$$Ca + Ar \xrightarrow{k_1} \text{(isolated atom)}$$
$$Ca + Ca \xrightarrow{k_2} Ca_2$$
$$Ca + Ca_2 \xrightarrow{k_3} Ca_x \quad (x \approx 3)$$
$$Ca + CH_3Br \rightarrow \text{no reaction}$$
$$Ca_2 + CH_3Br \rightarrow CH_3Ca_2Br$$
$$Ca_x + CH_3Br \rightarrow CH_3Ca_xBr$$

Here, k_1, k_2 and k_3 are the rate constants of the corresponding reactions.

TABLE 6.2 Chemical Reactivity of Calcium and Magnesium Particles with Methane Halides in Argon Matrix

Alkyl halide	Ca	Ca$_2$	Ca$_x$	Mg	Mg$_2$	Mg$_3$	Mg$_4$	Mg$_x$
CH$_3$I	+	+	+	−	+	+	+	+
CH$_3$Br	−	+	+	−	+	+		+
CH$_3$Cl	−	+	+	−	−	−	−	−
CH$_3$F	+	+	+	−	−	−	+	+
CH$_4$	−	−	−	−	−	−	−	−

The formation of a calcium atom is possible, e.g. by the reaction

$$CH_3Ca_3Br \rightarrow CH_3Ca_2Br + Ca$$

Reactions of atoms are kinetically limited. Of great importance for a reaction are the ionization potentials of particles of different sizes, the binding energies of metal–metal and metal–halogen, and the affinity of methyl halide to electron. The appearance of a charge on the particles due to the electron transfer profoundly changes the binding energies. For example, in an uncharged magnesium dimer, the energy of Mg–Mg bonding is approximately 1.2 Kcal, and the particle represents a van der Waals' complex. In a charged magnesium dimer, the binding energy increases to 23.4 Kcal. The reaction is also contributed by the energies of metal–halogen bonds formed.

Reactions of solvated metal ions (M^+Sn), where $M^+ = Mg$, Ca, Sr, and Ba, were studied in the gas phase by using a time-of-flight mass spectrometer.[35] The following reactions were observed:

$$Mg^+ + (CH_3OH)_m \rightarrow [Mg^+(CH_3OH)]^* \rightarrow Mg^+(CH_3OH)_n + (m-n)CH_3OH$$

An $Mg^+(CH_3OH)_n$ particle is stabilized due to either evaporation of a CH_3OH molecule or a collision with a gas carrier. For $n \geq 5$, an $Mg^+(CH_3OH)_n$ particle undergoes the transformation

$$Mg^+(CH_3OH)_n \rightarrow MgOCH_3^+(CH_3OH)_{n-1} + H$$

When n reaches 15, the following reaction occurs:

$$MgOCH_3^+(CH_3OH)_{n-1} \rightarrow Mg^+(CH_3OH)_n$$

The interpretation of synthesized particles was based on ab initio quantum-chemical calculations. Similar reactions were observed earlier for interactions with water.[36]

Recently, magnesium oxide nanoparticles were applied in a number of catalytic reactions. Crystals of MgO measuring 4 nm were treated with different amounts of potassium vapors. On the oxide surface, K^+ and e^- sites were formed. Sites of electron-surface binding generated the appearance of superbasic zones.[37] Interactions with alkenes resulted in the proton abstraction and the formation of allyl anions, which took part in the alkylation of ethylene by the following reaction:

$$H_3CCH = CH_2 + CH_2=CH_2 \xrightarrow{K-MgO} CH_3CH_2CH_2CH=CH_2 + \text{heptenes}$$

A comparison of the degree of conversion for nano- and microcrystals of magnesium oxide doped with potassium vapors has shown that at 210 °C, the yield increases from 15% for microcrystals to 60% for nanocrystals. The effect observed was related with the increase in the surface area for nanocrystals and correspondingly, the increase in the number of basic surface sites. This example shows that the specific catalytic sites are easily attainable for nanocrystals as compared with microcrystals.

Catalytic chlorination of alkanes was accomplished on magnesium oxide nanocrystals.[38] When substantial amounts of chlorine gas are adsorbed on the surface of magnesium oxide, the properties of a chlorine molecule approach those of a chlorine atom. It is known that in dark, chlorine molecules do not chlorinate alkanes, whereas the system $MgO-Cl_2$ formed by the method described above can chlorinate hydrocarbons. It should also be mentioned that the selectivity of the $MgO-Cl_2$ adduct is sometimes higher as compared with chlorine atoms. In this case, an analogy with the well-known activity and selectivity of radical chlorination in aromatic solvents can be traced.[39] By considering the example of photochemical chlorination of hydrocarbons with primary, secondary, and ternary bonds in the benzene medium, the high selectivity of the reaction was demonstrated. Chlorination occurred largely for the weakest ternary C–H bond. This phenomenon was associated with the formation of a complex of a chlorine atom with a benzene molecule, which had a lower activity than a free chlorine atom and, hence, could selectively interact with the weakest bond.

The peculiarities of the $MgO-Cl_2$ system mentioned above allow us to suggest that dissociative adsorption of a chlorine molecule takes place on the MgO nanocrystals. This results in stabilization of an atom-like particle, the electron density of which is shared with the oxygen anions (O^{2-}) on the surface. The chlorination of methane and other hydrocarbons probably proceeds under specific catalytic conditions when secondary reactions $MgO + 2HCl \rightarrow MgCl_2 + H_2O$ proceed more slowly as compared with chlorination and oxide regeneration. Below, we show a possible scheme of successive reactions for alkane chlorination on magnesium oxide nanocrystals:

$$\begin{array}{c}
\overline{MgO\ MgO} \xrightarrow{Cl_2} \overline{\begin{array}{cc} Cl & Cl \\ | & | \\ MgO & MgO \end{array}} \\
\swarrow_{-RCl} \qquad \searrow^{RH} \searrow_{-HCl} \quad 2HCl \\
\overline{\begin{array}{cc} R & Cl \\ | & | \\ MgO & MgO \end{array}} \qquad \overline{MgO\ MgO} \xrightarrow[-H_2O]{} \overline{\begin{array}{cc} MgCl & MgO \\ & | \\ & Cl \end{array}}
\end{array}$$

To fully understand the chemical mechanism of this reaction, it should be elucidated whether the interaction of alkyl radicals R with chlorine atoms occurs on the surface or in the gas phase and whether R is bound with Mg^{2+} or with O_2^{2-}.

As particles with well-developed surfaces, nanocrystals should exhibit an enhanced ability to adsorb various molecules. For instance, MgO nanocrystals at room temperature and a pressure of 20 mmHg strongly chemisorbed 6 SO_2 molecules/nm^2 of their surface, whereas a microcrystal absorbed only 1.8 SO_2 molecules/nm^2 (see Ref. 40). Moreover, for nanocrystals, the formation of single bonds prevailed, while double bonds were preferentially formed on microcrystals. This is explained by the morphological peculiarities of the two kinds of crystals and is schematically illustrated by the structures shown below:

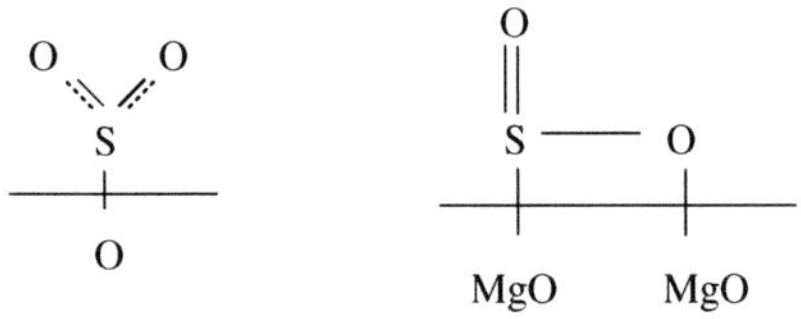

| a nanocrystal that adsorbs a monolayer at a pressure of 20 mmHg | a microcrystal that adsorbs a ⅓ of a monolayer at a pressure of 20 mmHg |

With an increase in pressure, molecules adsorbed on a microcrystal are bound more weakly, which results in the formation of ordered and condensed multilayers because of the predominance of a certain direction of interaction. Thus, for adsorption too, the size and shape of a crystal play a significant role. Similar peculiarities were observed for adsorption on MgO nanocrystals of different gases with acidic properties, namely, CO_2, HCl, HBr, and SO_3.[41] However, when multilayer adsorption occurs at elevated pressures, the ordering of a microcrystal surface becomes one of greater importance. It was assumed that both polarization and morphology change during the adsorption. In such a case, magnesium oxide can be considered as a participant of a stoichiometric process. In 4-nm magnesium oxide crystals, up to 30% of the total number of ions pertain to their surfaces. Precisely, this accessible 30% of magnesium oxide is used, i.e. in fact, a common gas–solid reaction occurs.

The reactions of magnesium oxide nanocrystals with aldehydes, ketones, and alcohols were observed to involve destructive adsorption.[42] The reaction of magnesium nanooxide with aldehydes is endothermic, and nearly a mole of acetaldehyde per mole of oxide undergoes destructive adsorption. Presumably, the reaction proceeds as follows:

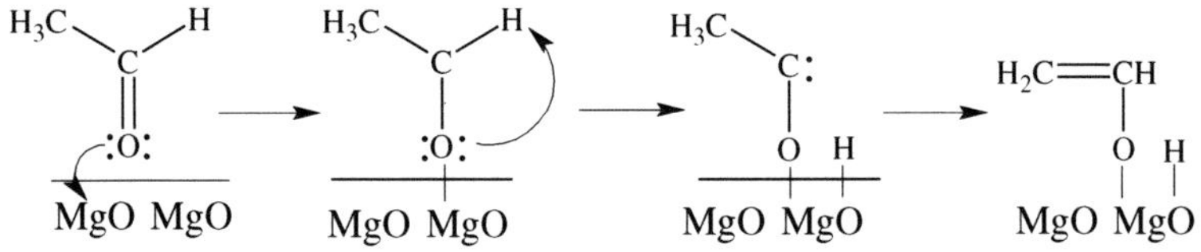

The surface reactions under discussion involve the loss of a proton by an aldehyde molecule owing to coordination of carbonyl oxygen with Mg^{2+} and the subsequent transfer of a proton to the surface. An analysis of the IR spectra showed that the absorption intensity rapidly decreases for the C–H bond in aldehyde and gradually increases for C=O and C=C bonds. It was assumed that the reaction with new aldehyde molecules results in the formation of polymers, which suggests that the amount of destructively adsorbed aldehyde exceeds a monolayer. No destruction was observed on carbonaceous adsorbents with highly developed surfaces. Hence, nanocrystalline metal oxides with highly active polar surfaces exhibit new valuable properties.

Nanocrystalline oxides, particularly MgO, were also used for the destruction of different bacteria, viruses, and toxins in the processes, employing chlorinated adducts.[43] The mechanism of the interaction of biologically active particles with nanosize oxide–chlorine systems, which results in their destruction, requires further studies to be conducted. These processes are of interest for exercising control over biological weapons and various biotoxicological diseases.

Recently, the studies of magnesium particles once again attracted the attention of theorists. One of the main reasons for this is that in contrast to most substances, small magnesium particles (dimers, trimers, and tetramers) are bound by weak van der Waals forces. This stems from the quasi-closed nature of the ground state of magnesium atoms.

Density-functional calculations have shown that in an Mg_3 particle, which forms an isosceles triangle, the binding energy is 0.14 eV per atom, whereas in a tetrahedral Mg_4 particle, this energy is 0.3 eV per atom, i.e. twice higher. The bond length in Mg_3 is 329 pm, which is smaller than 319 pm in a compact metal.[44]

The density-functional method was used for analyzing how the degree of solvation by diethyl ether affects both the structure of Grignard reagents RMgX, where R = CH_3, C_2H_5, C_6H_5, and X = Cl, Br, and the strength of the carbon–magnesium bond.[45] It was noted that the strength of the Mg–C bond depends on the organic substituent and the solvation process, rather than on the halogen nature. For example, the C_6H_5–Mg bond is stronger than the C_2H_5–Mg bond by 70 KJ/mol and increases further by 40 KJ/mol in the course of solvation to yield 322 KJ/mol for $C_6H_5MgBr \cdot 2(C_2H_5)_2O$.

Small magnesium particles present a challenge for quantum chemistry, and different calculation methods invoked for finding the binding energies in such particles provide only qualitative agreement with the experiments.[44] Theoretical studies of small magnesium clusters have also been accomplished.[45,46] Extensive theoretical studies have made it possible to make the following conclusions:

- Mg_4, Mg_{10}, and Mg_{20} particles are highly stable;
- the changes in metal properties observed with an increase in the size are non-monotonic.

Thus, additional studies of the structure of particles of different sizes and the processes of their transformation into compact metals are required.

6.3 TRANSITION METALS OF GROUPS III–VII IN THE PERIODIC TABLE

These groups include such important elements as titanium, vanadium, chromium, molybdenum, manganese, etc. Among their reactions, the interactions with carbon dioxide were studied most comprehensively. Co-condensation of Ti and certain other elements of groups III–VII involves the electron transfer and insertion of a metal into the C–O bond. The intermediate of the O–M–CO type thus formed reacts with carbon dioxide.[47] Reactions with titanium are illustrated by the following scheme:

$$Ti + CO_2(15\ K) \rightarrow O=Ti-C\equiv O + Ti=O + Ti-C\equiv O + C\equiv O,$$

$$O=Ti-C\equiv O + CO_2 \rightarrow \underset{\underset{O}{\overset{O}{\diagdown}}}{\overset{O}{\diagup}}Ti\underset{O}{\overset{C\equiv O}{\diagdown}}$$

Only few studies revealed a distinct dependence between the chemical activity and particle size, which compels us to use indirect results. For instance, when no chemical changes are observed during the co-condensation of metal and ligand vapors on a surface cooled to 77 K, then stabilization of metal atoms and clusters can be claimed with high probability. If further heating gives rise to chemical reactions, then the formation of nanoparticles of different sizes and their reactions with the ligand, which, as a rule, is present in excess, can be asserted.

The process of low-temperature co-condensation of metal vapors and vapors of various organic substances is actively used by chemists for synthesizing new organometallic compounds with unusual properties. Many scientists from different countries are successfully working in this direction.[48,49]

Here, the details of such nanochemical experiments should be considered once again. Japanese scientists thoroughly examined different reactions between various silicon derivatives and Ca and Ge, using the procedure described above. They succeeded in synthesizing products of insertion into Si–Cl and Ge–Cl bonds, namely, compounds of the Si–M–Cl type, and in carrying out their reactions with different ligands. At the same time, they observed the cases of magnesium insertion into Si–Cl bonds in its reaction with, e.g. $(CH_3)_3SiCl$. However, we have shown that under molecular-beam conditions, joint condensation of Mg and $(CH_3)_3SiCl$ yields a silicon analog of Grignard reagent, in contrast to the classical organic synthesis. The following scheme of competitive reactions was put forward:

$$(CH_3)_3SiCl + Mg \xrightarrow{77\ K} (CH_3)_3Si - Si(CH_3)_3 + MgCl_2$$

$$R_3SiCl + Mg \rightarrow R_3SiMgCl \xrightarrow{H_2O,\ HCl} R_3SiH$$

The synthesis of an organosilicon analog of Grignard reagent was evidenced by the reactions of $R_3SiMgCl$ with water vapors, hydrogen chloride, and *tert*-butanol. The resulting R_3SiH was identified by a characteristic frequency $\nu = 2120\ cm^{-1}$ of the Si–H bond.

Below, a possible reaction mechanism is shown:

$$R_3SiCl + Mg \rightarrow [R_3SiCl^{-\cdot}\ Mg^{+\cdot}] \rightarrow R_3SiMgCl$$
$$R_3SiMgCl + R_3SiCl \rightarrow R_3Si-SiR_3 + MgCl_2$$

A new stage is the formation of a radical–ion pair and silylmagnesium chloride, which interacts with the initial reagent to give disilane.

Using a setup shown in Figure 2.6, the following reaction was carried out:

$$\text{Re} + \text{BrCH}_2\text{CH}_2\text{Br} \xrightarrow[77\text{ K}]{\text{THF}} \cdots \xrightarrow{300\,°C} \text{Re}_3\text{Br}_9(\text{THF})_3$$

A solvated halide of a transition metal was obtained in the absence of water.[50]

Particles containing a large number of atoms, which are of special interest, are still scantily studied. Mention should be made of the synthesis of Cr_4 and Cr_5 particles and their EPR studies in argon matrices.[51] It was assumed that a weakly bound vertex chromium atom in Cr_4 and Cr_5 is a habitat of virtually all unpaired spins responsible for the appearance of 16 lines in the spectrum. In the authors' opinion,[51] this result was quite unexpected.

At present, most detailed gas-phase studies were carried out with Nb_x particles. The use of beams intermittent through nozzles made it possible to obtain Nb_x particles, where $x = 5–20$. Several chemical reactions were accomplished with these particles. The interaction of Nb_x particles with benzene followed the scheme:

$$Nb_x + C_6H_6 \rightarrow Nb_x - C_6H_6 + Nb_xC_6$$

Nb_x particles were obtained by combining laser evaporation with supersonic expansion. A pulse method at a helium pressure of 3–5 atm was used. Collisions with helium cooled hot niobium atoms, thus slowing down their rate. Clusters were formed in a flow, which passed through a reactor, where it was mixed with different reagents (in this case, benzene) introduced together with helium. Then, the mixture was expanded to prevent collisions and was subjected to the selection procedure in a time-of-flight mass spectrometer.[52]

The additive-induced loss in the intensity of a peak corresponding to a particle of a certain size was measured. It was shown that the reaction with benzene actively proceeds starting from clusters with $x = 4$.

The minima at $x = 8$ and $x = 10$ correspond to the highly stabile Nb_8 and Nb_{10} particles. The observed sharp increase in reactivity for $x = 4$ and 5 can be explained by both thermodynamic and catalytic reasons. For the reaction to start, the formation of a certain minimum number of Nb–C bonds was vital. This is the thermodynamic factor. The catalytic effect was associated with the necessity of a simple anchor-type bonding of benzene rings in order to promote the dehydrogenation process.

Based on the studies with benzene and other aromatic molecules, the following conclusions can be drawn:

- the initial molecule should have at least one double bond;
- the loss of an even number of hydrogen atoms was observed.

The first statement implies a mechanism that involves the π-electron system into the reaction. The second conclusion states the formation of evaporated hydrogen molecules.

By the example of a reaction between Nbx and BrCN, it was shown that the cluster size affects the selectivity of the process. A niobium particle abstracts either a bromine atom or a CN radical from a BrCN molecule:

$$Nb_x + BrCN \rightarrow Nb_xBr + Nb_xCN$$

Small clusters promote the abstraction of CN groups. For $x \geq 7$, both pathways cease to depend on the particle size. The results obtained were explained based on different kinds of collisions between niobium clusters and BrCN molecules. For small clusters, brief collisions prevail; for large particles, the formation of complexes, the decomposition of which yields both products, was assumed.

The steric effect in this reaction also changes with the particle size. Figure 6.3 shows the results obtained for two alkyl bromides, namely, CH_3CH_2Br and $CH_3CHBrCH_3$, in the reactions with Nbx clusters. As seen from the figure, for small niobium clusters ($x \leq 4$), the Nb_xBr yield is independent of the bromide nature. However, when the reaction involves particles with $x \leq 5$, the yield of isopropyl bromide substantially decreases (by ca. 20%), which is explained by steric effects. In this case, the approach of a cluster to bromine is hindered, which is aggravated with an increase in the cluster size for Nb_5, Nb_6, Nb_7, etc. In other words, for large particles, the number of effective collisions decreases.

By the example of niobium particles, the effect of particle size on the reaction pathway was demonstrated. Such an effect was observed when the ligand–cluster interaction resulted in the formation of a complex, which could react to give two different products. Moreover, if one of the pathways is preferred for energy or steric reasons, we can expect certain changes in the product distribution depending on the cluster size. This kind of a phenomenon does indeed take place, e.g. during the reaction of niobium particles with halogen-containing olefins.

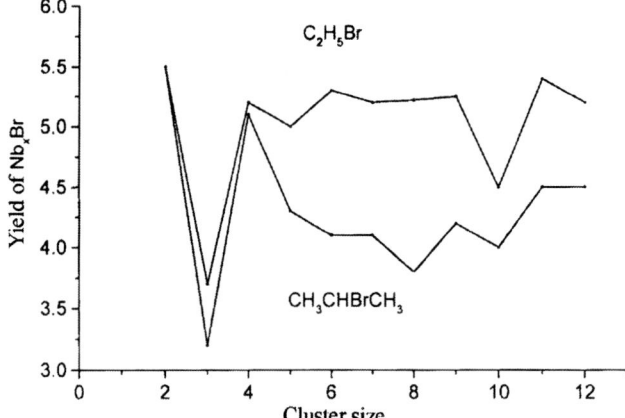

FIGURE 6.3 Effect of the size of niobium particles and the steric properties of ligands on chemical reactivity.[52]

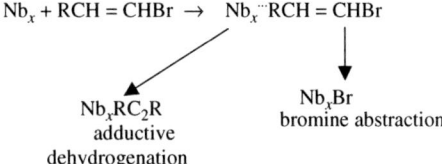

It was shown that coarse clusters favor the formation of Nb_xRC_2R. Apparently, this means that small clusters readily attack bromine. Probably, coarse clusters, in turn, more easily form π-complexes with double bonds, which result in dehydrogenation.

Yet another interesting example of the particle-size effect on the reaction pathway was observed for the reaction of niobium clusters with CO_2.

The reaction is illustrated by a scheme:

$$Nb_x + CO_2 \rightarrow [OCNb_xO] \begin{array}{c} \xrightarrow{x>7} OCNb_xO \quad I \\ \xrightarrow{x<7} Nb_xO + CO \quad II \end{array}$$

Apparently, a common intermediate product is formed in this reaction, because the yields of products I and II are inversely related. This was demonstrated by the example of isotope-labeled carbon dioxide. Thus, small clusters with $x = 3$–7 favor the formation of Nb_xO, whereas large clusters promote the formation of $OCNb_xO$. Probably, this can be explained by the fact that large clusters stabilize "hot" intermediate $OCNb_xO$ due to energy dissipation, whereas adducts of CO_2 with small clusters undergo decomposition.

Examples shown above, on the one hand, clarify certain questions concerning the reactivity of nanoparticles and, on the other hand, pose new problems. The procedure mentioned for synthesizing niobium nanoparticles involves several steps that can influence the corresponding reactions. Thus, a serious drawback of this procedure is associated with the fact that the temperature of a cluster is unknown; moreover, during the reaction, large clusters can be fragmented by a scheme $Nb_x + R_x \rightarrow Nb_yR + Nb_z$. Such a process should affect collisions and reactions. Some uncertainties are associated with the laser-detection step, especially for the production of neutral particles when fragmentation and side processes are possible.

Several general remarks should be made. The interaction of Nb_x clusters with hydrocarbons such as cyclohexane, cyclohexene, and cyclohexadiene points to their sensitivity to unsaturated bonds. Small clusters do not react with saturated hydrocarbons but actively dehydrogenate alkenes and dienes.

For Nb_x clusters with $x = 1$–3, complexes Nb_x–C_6H_6 were observed. For large clusters with $x = 4$–9, ions prevailed, especially for $x = 9$. This is why large clusters more actively dehydrogenate benzene and form Nb–C bonds to produce carbide-like structures. For benzene, it was found that Nb_5, Nb_6, Nb_7, Nb_9, and Nb_{11} were more active in converting C_6H_6 to C_6. This intriguing result was confirmed by the fact that Nb_8 and Nb_{10} do not interact with hydrogen and nitrogen. Dissociative chemisorption of hydrogen on Nb_7^+, Nb_8^+, and Nb_9^+ was studied. Similar to neutral clusters, Nb_7^+ was more active than Nb_8^+. The equal activities of neutral and charged clusters do not agree with the simple electrostatic model of dissociative chemisorption.

Studies of the activity of niobium clusters made it possible to outline several general problems for nanochemistry. It is most likely that large clusters can be used for controlling the high-energy process of dehydrogenation via concurrent reactions. Particles Nb_8 and Nb_{10} with closed shells are apparently structurally inactive. To exercise control over the process, a certain number of Nb–C bonds should be formed.

$$Nb_x + C_6H_6 \rightarrow (C_6H_6)Nb_x + C_6Nb_x + H_2$$

Generally, we have more questions than answers, and the major question is: why are some particles active while others are not.

By the example of niobium particles ($n = 2$–20) under commensurable conditions at different temperatures, a kinetic comparison of their reactivities toward hydrogen, nitrogen, and deuterium was carried out.[53] The experimental results were compared with the estimates made within the framework of the electron-density functional model. Figure 6.4 shows the results on the interaction of nitrogen and deuterium with niobium particles at different temperatures. Their reactivity was observed to strongly depend on the particle size; moreover, an unusual dependence on the temperature was observed. The reaction rate decreased with temperature over a range of 280–370 K. Such dependence was explained by the formation of an intermediate weakly bound complex according to the scheme:

$$Nb_n + N_2 \underset{k_b}{\overset{k_a}{\leftrightarrow}} Nb_n(N_2) \xrightarrow{k_c} Nb_nN_2$$

where k_a, k_d, and k_c are the rate constants of association, dissociation, and chemisorption, respectively.

A similar scheme was proposed for nitrogen interactions with molybdenum particles,[54] and the quantity $\Delta E = E_d - E_c$, where E_d is activation energy and E_c is chemisorption energy, was determined for clusters of different sizes. It was found that ΔE is 8.0 Kcal/mol for Mo_{15}, 5.0 Kcal/mol for Mo_{16}, and 6.5 Kcal/mol for Mo_{24}.

An analysis of the temperature dependence for the above scheme of reactions with Nb particles made it possible to conclude that N_2 and D_2 retain their molecular bonding within their complexes, whereas in the reaction products, their bonding is dissociative.[54] The possibility of direct conversion of reactant without the formation of an intermediate complex was considered. Summing up, the calculated and experimental results make it possible to conclude that the scheme that involves the complex is necessary but insufficient for describing the intricate temperature dependences of reactions of nitrogen and deuterium with niobium particles of different sizes. Density-functional calculations of the reactions of niobium atoms, dimers, and trimers with N_2 and H_2 made it possible to determine the conformation, symmetry, binding energy, and charge transfer for particles Nb_2N_2, Nb_3N_2, and Nb_3H_2.

The ionization potential usually correlates with the reactivity of a cluster. In the reaction of niobium particles ($n = 8$, 10, and 16) with nitrogen, an anticorrelation dependence was observed. This concrete case was explained by the presence of a barrier at the intersection of the neutral potential of repulsion and the ionic potential of attraction.[53] For the reactivity of a cluster, the location of

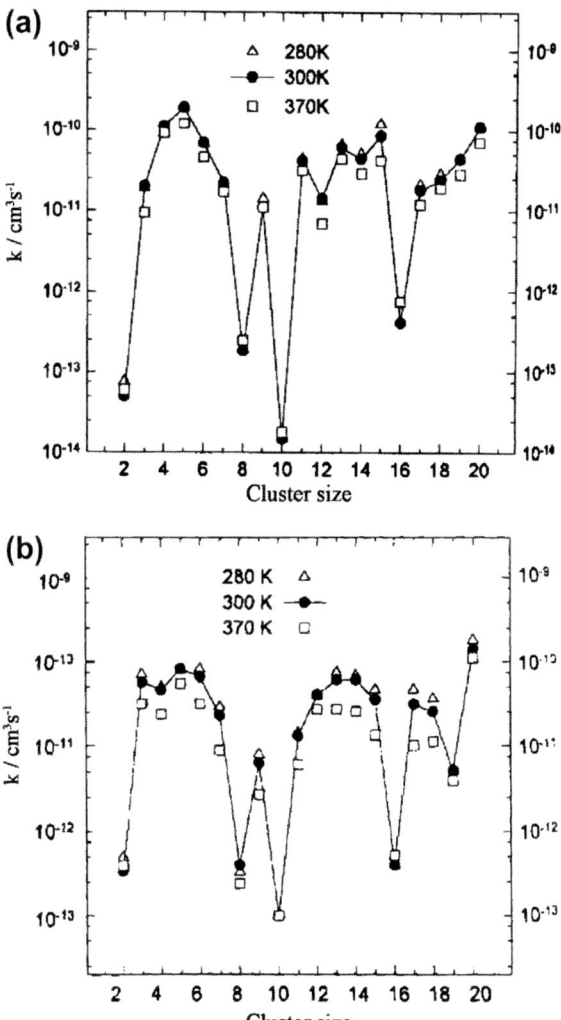

FIGURE 6.4 Dependence of the rate constant of second-order reactions of (a) deuterium and (b) nitrogen with Nb_n particles in a cluster at 280, 300, and 370 K.[53]

the charge on its surface is also important. The charge is usually located near the center of mass but can be removed from the reaction site on the surface. This means that the surface of coarser particles has a lower stabilizing ionic potential, even though the ionization potential decreases with an increase in the particle size. Hence, large clusters should be less reactive, which is not the case for niobium particles. This inconsistency was removed by the introduction of an effective ionization potential, which takes into account the radius of a cluster and the energy of its polarization. Based on the aforementioned analysis, it was noted that the contribution of electrostatic energy to ionization potential has no effect on the reactivity of a particle. The process is controlled by other factors determined by the cluster structure. The correct interpretation of the reactivity of niobium particles requires considering the potential of uncharged particles. This is in line with the concepts of the temperature effect and the participation of intermediate complexes in the reaction.

By the example of a reaction of niobium particles with deuterium, it was shown that anions, cations, and neutral clusters have similar reactivities.[55,56] Such a feature has been mentioned earlier for other metals.[57–59] The similar reactivity for neutral and charged particles makes it possible to conclude that the electron-transfer models used as the basis of explanation of reactivity require refining; particularly, the geometry of nanoparticles can define the adsorption processes and the activity of metal clusters.[60]

Particles with magic numbers of atoms are stable because of the presence of closed shells, either electronic or geometric. Such particles tend to have high binding energies per atom, high ionization potentials, and widely deviating highest occupied molecular orbital (HOMO) and lowest unoccupied molecular orbital (LUMO) energies. Thus, it was shown that niobium particles Nb_8, Nb_{10}, and Nb_{16} with closed electron shells are active toward hydrogen adsorption.[61] The same authors stressed that the no dependence of niobium activity on the presence of a charge substantiates the importance of the geometric structure. Apparently, the charge-transfer model and a model considering the geometric structure of a particle can supplement one another. Unfortunately, the geometric structure of small clusters cannot yet be studied by direct experimental methods and is determined only by calculations.

A wide difference in ionization potentials of neutral and charged particles indicates that the ionization potential is not the main factor that controls the reactions of clusters. In real cases, more intricate phenomena were observed. In particular, the presence of charges on clusters perturbs their potentials of interaction with molecules; moreover, the existence of cluster isomers with different ionization potentials and different activities is possible. The formation of structural isomers and their different activities were evidenced by the existence of biexponential kinetics.[58,62]

The high reactivity of niobium particles of different sizes toward nitrogen and hydrogen is determined by the presence of a relatively small number of valence electrons. In the corresponding reactions of molybdenum clusters, which have a higher number of valence electrons, the interaction of orbitals, which is associated with repulsion processes, plays the key role. In this case, steric effects defined by the cluster geometry become the decisive factor.

Under conditions of single collisions, the molecular and dissociative adsorption of nitrogen on tungsten particles in a range W_{10}–W_{60} was studied.[63] The probability of the reaction with the first and second nitrogen molecules was measured for tungsten particles at room and liquid-nitrogen temperatures. The results in Figure 6.5 reflect the higher probability of the interaction of tungsten particles W_{10}–W_{60} with the first nitrogen molecule at a temperature of ca. 80 K as compared with 300 K. A pronounced nonmonotonic behavior of the reactivity was also observed. At room temperature, the maxima for clusters W_{15}, W_{22}, and W_{23} were observed. The W_{10}–W_{14} interval corresponded to the low reactivity at

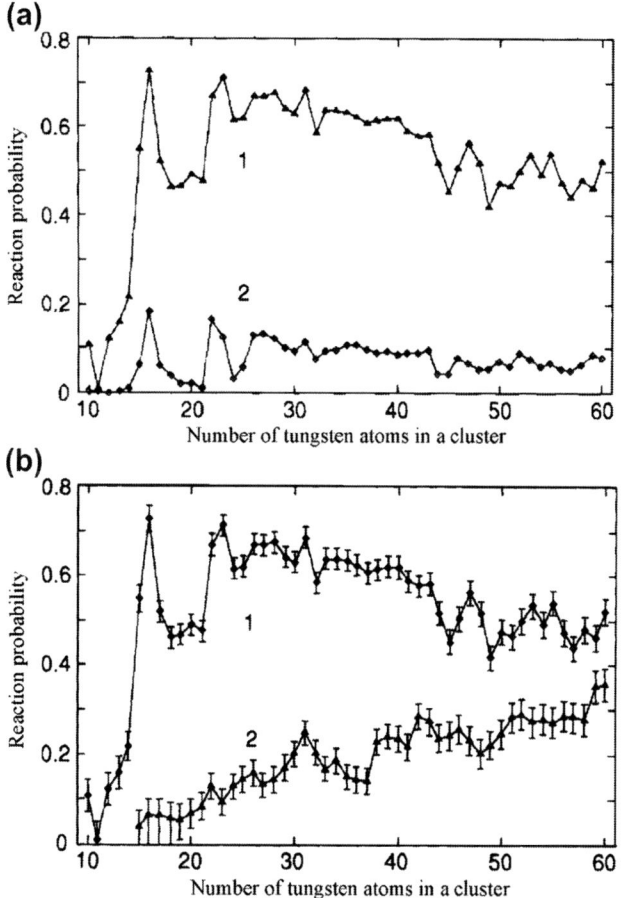

FIGURE 6.5 Dependence of the probability S of a reaction of (a) one and (b) two nitrogen molecules with W_n particles at (1) 80 and (2) 300 K on the number of tungsten atoms in a cluster.[63]

room temperature. On the other hand, at low temperatures, only W_{11} particles exhibited a low activity, while the highest activity corresponded to the W_{15} clusters. The activity of all tungsten particles studied was higher at low temperatures as compared with room temperature. At room temperature, virtually no interaction was observed between tungsten particles and the second nitrogen molecule, while at low temperatures this interaction was weaker as compared with the interaction with the first nitrogen molecule. As seen from Figure 6.5, for tungsten particles W_{20}–W_{60}, the probability of the reaction slowly increases with an increase in n; however, this does not correlate with the activity vs. n dependence observed for the first nitrogen molecule.

Tungsten particles with nitrogen molecules were heated using a pulsed excimer laser. Samples obtained at room temperature and 77 K behaved in different manners. No nitrogen desorption was observed at room temperature, whereas at low temperatures, substantial desorption of nitrogen molecules was observed for all the clusters studied. The number of nitrogen molecules left on the surface after the desorption approached their number observed at room temperature. The fraction of nonadsorbed nitrogen molecules was higher for the more active particles. Thus, we can infer the existence of the sites on tungsten clusters that differ in the energy of nitrogen bonding.

At low temperatures, the first nitrogen molecule reacts with tungsten particles, while the second nitrogen molecule is bound by W_nN_2 particles. The fact that the number of adsorbed secondary molecules is smaller than that of primary molecules may be a consequence of the negative dependence of the reactivity on temperature. One of the explanations was based on the higher temperature of W_nN_2 particles as compared with W_n due to the possible dissociation of a part of the nitrogen molecules at low temperatures. Dissociative bonding of nitrogen at low temperatures can also be caused by their more effective absorption on the molecular precursors. If nitrogen molecules diffuse over the cluster surface, they can be localized in sites with the high binding energy. In this case, the probability of dissociation increases. On the other hand, the dissociation process has an activation barrier. Hence, it cannot be ruled out that dissociation or desorption can occur during the laser beam-induced heating.

For vanadium particles V_{10}–V_{60}, the dependences of their reactivity on the number of atoms and temperature of clusters were revealed in reactions with CO, NO, O_2, D_2, and N_2.[64] It was shown that low-temperature reactions with both the first and second CO, NO, and O_2 molecules are virtually independent of the number of vanadium atoms. For deuterium and nitrogen molecules, temperature dependences were observed. For these molecules, a strong effect of the particle size on the reactivity at room temperature was observed; moreover, sawtooth dependences were revealed. For particles containing less than 20 vanadium atoms, adsorption of CO, NO, and O_2 resulted in fragmentation of clusters, which probably proceeded via evaporation of a metal atom. It was shown that V_{13} and V_{15} particles are more stable than V_{14}.

In the gas phase, reactions of oxide ions separated using a mass spectrometer ($V_xO_y^\pm$, $Nb_xO_y^\pm$, $Tl_xO_y^\pm$) with ethylene and ethane were studied.[65] It was demonstrated that the highest activity corresponds to vanadium oxide cations $(V_2O_5)_n^+$ that have a specific stoichiometry and gave up their oxygen to hydrocarbons. Anions of oxides of vanadium, niobium, and thallium did not enter into this reaction.

Chromium particles Cr_n ($n = 1$–25) were studied by X-ray adsorption spectroscopy.[66] Mass-spectrally separated chromium particles were subjected to mild deposition (without fragmentation) on the (001) surface of Ru oxide. Strong interaction of particles with the substrate inhibited the formation of islets.

Germanium nanowires of a diameter of ~4 nm doped with boron (p-type) or phosphorus (n-type) were synthesized by CVD from germanium tetrahydride (GeH_4) at 275 °C. The process was catalyzed by gold particles of 2–20 nm diameters. The size of germanium nanowires was determined using TEM; evaporation was performed on gold-plated grids.[67] Germanium wires thus synthesized were used for creating the field effect in field transistors.

Gas-phase studies of the physicochemical peculiarities of isolated clusters of definite sizes are important for understanding the properties of particles incorporated into a matrix or stabilized in it. Moreover, the properties of particles containing several atoms, which determine the chemical peculiarities of the system, are of prime interest.

Hybrid nanomaterials based on organic and inorganic components find increasing application. This is explained by the fact that organic chemistry allows synthesizing a vast diversity of compounds with a wide spectrum of physical properties. Moreover, the properties of hybrid structures depend not only on organic or inorganic components but also on acquiring interface properties, which can be tuned with more sophistication using organic materials. Lamellar structures containing highly ordered crystalline yttria layers separated by organic layers were synthesized by the reaction between yttrium alkoxides and benzyl alcohol.[68] The yttria layer thickness was 0.6 nm, while the thickness of the organic layer depended on the nature of the organic substance and was 1.74 nm for benzyl alcohol and 2.21 nm for 4-*tert*-butylbenzyl alcohol. Doping of the structure with Eu^{31} ions resulted in the appearance of strong luminescence in the spectral red range, which is typical of oxide matrices.

6.4 ELEMENTS OF THE GROUP VIII OF THE PERIODIC SYSTEM

Group VIII of late transition elements is represented by iron, cobalt, nickel, ruthenium, rhodium, palladium, osmium, iridium, and platinum.

By the example of palladium, an important cycle of studies devoted to synthesizing nanoparticles of definite stoichiometric compositions was accomplished.[69] A particle of the $Pd_{561}L_{60}(OAc)_{180}$ composition, where L is

1,10-phenanthroline and OAc groups form a ligand shell, was identified. The formation of a palladium cluster proceeded in two steps:

$$Pd(OAc)_2 + L + H_2 \rightarrow (1/n)[Pd_4H_4(OAc)L]_n + AcOH$$
$$[Pd_4H_4(OAc)L]_n + O_2 + AcOH \rightarrow Pd_{561}L_{60}(OAc)_{180} + Pd(OAc)_2 + L + H_2O$$

The synthesized palladium particles are "magic" particles, i.e. they contain strictly definite numbers of metal atoms, namely, 13, 55, 147, 309, 561, etc. Such numbers correspond to closed shells of cubic octahedral clusters.

The successful synthesis of particles containing 561 palladium atoms initiated the studies on synthesizing palladium clusters with different numbers of filled shells. Magnetic properties of palladium particles with different numbers of atoms were considered.[70] The mechanism of synthesis of particles with a fixed number of atoms is still unclear. For a cluster with a Pd_{561} core, it was assumed that the metal particle forms a crystalline lattice in the last stage of thermal treatment when it acquires the icosahedron shape. The principles of building icosahedral metal clusters based on a model generalization were described.[71]

Below, we show the scheme of a reaction of coordination-unsaturated cobalt compounds with ethylene.[72] This process is of interest as a method of elongating carbon chains, and it pertains to chemistry of one- and two-carbon molecules C_1 and C_2.

$$C_2H_4 + Co \xrightarrow[77\ K]{co\text{-}condensation} (C_2H_4)_m Co_n \xrightarrow{CH_3CHO} (C_2H_4)_m \underset{H}{Co_n}\!-\!\overset{O}{\overset{\|}{C}}CH_3$$

with branches via H_2 and CO leading to:

$(C_2H_4)_m \underset{H}{Co_n} CO$

$CH_3CH_2CHO + (C_2H_4)_{m-1} Co_n$

$(C_2H_4)_{m-1} Co_n\!-\!\overset{O}{\overset{\|}{C}}CH_3$
$\qquad\qquad\quad |\ CH_2CH_3$

$CH_3\overset{O}{\overset{\|}{C}}CH_2CH_3 + (C_2H_4)_{m-1}Co_n$

Here, the product of low-temperature co-condensation of Co and C_2H_4 is designated as $(C_2H_4)_mCo_n$. It readily reacts with acetaldehyde CH_3CHO, formaldehyde HCOH, and, more importantly, with CO–H_2. The use of deuterium-labeled compounds made it possible to refine this scheme. Reactions with the addition of other substrates proceed successfully only if the melting point T_m of the latter is sufficiently low to provide the mobility in the Co–C_2H_2 matrix up to the decomposition of the $(C_2H_4)_mCo_n$ complex. Unfortunately, these reactions are nonselective and a part of the products are formed spontaneously.

It is interesting that under similar conditions (77 K), a reaction with iron produced more mobile complexes $(C_2H_4)_2Fe_2$.[73] The synthesis and reactions of the complexes are illustrated by the following scheme:

$$(C_2H_4)_2Fe_2 \rightleftarrows \underset{\underset{C_2H_4}{|}}{\underset{Fe \text{——} Fe}{CH_2=CH \quad H}} \rightleftarrows \underset{\underset{CH_2CH_3}{|}}{\underset{Fe \text{——} Fe}{CH_2=CH}}$$

Complexes of $(C_2H_4)_m Fe_n$ type are stable only at a temperature below 18 K. Nickel complexes $Ni(C_2H_4)_3$ were synthesized at 77 K; however, they were not extracted.[74]

The method of low-temperature co-condensation of vapors of metals and various ligands gave rise to the appearance and successful development of new directions in the chemistry of organic compounds of metals and other elements. Among a wide diversity of directions in this field, we briefly consider only dispersions of solvated metal atoms. They are usually synthesized at 77 K. The heating of such co-condensates is accompanied by migration of atoms and the formation of clusters and nanoparticles. The cluster growth competes with the interaction of growing clusters with the medium material.

A certain control over the size of the obtained particles can be provided by the appropriate choice of the matrix material (xenon, hydrocarbons, and aromatic compounds) and the heating conditions. At the same time, the unlimited growth is a common problem. Nonetheless, the method proved to be very fruitful for obtaining new catalytic and bimetallic systems. At present, a great attention is paid to the physical properties of M_x particles. Thus, a nonmonotonic variation in the ionization energy with size was observed. The ionization energy was 5.9 eV for an iron dimer and ca. 6.4 eV for its trimer and tetramer. However, for Fe_{9-12} particles, this energy was lower than 5.6 eV, whereas for Fe_{13-18}, it exceeded 5.6 eV.[75]

Nonmonotonic changes were also observed for Ni_x clusters with $x=3$–90.[76] The results are rather in poor agreement with theoretical predictions.[77] It should be stressed that magnetic properties of Fe, Co, and Ni nanoparticles also depend on their sizes.[78–80]

The dependence of the reaction of Fe_n with H_2 on the particle size and the annealing temperature was studied. An interesting phenomenon was observed: an increase in the temperature of the cluster–helium flow resulted in the reduction of particle activity. The possible explanation assumes that the initial growth of clusters is kinetically controlled and, probably, leads to the formation of more defective and hence more active clusters. Under annealing, a cluster can acquire another form, e.g. collapse to give a structure more stable with respect to thermodynamics, which can be less active. The activity of the interaction of Fe_x with H_2 changes with the size of the particles. Furthermore, the particle size also determines the sharp changes in the energy of the bonding of adsorbed NH_3 and H_2O. Apparently, all the phenomena described above are of the same origin, being related to the changes in the ligand-free cluster structure. For Fe_n ($n=2$–165), the samples with chemisorbed NH_3 indicated the existence of metastable structures. Reactions of Fen particles with O_2, H_2O, and CH_4 were studied[81]

$$Fe_n + O_2 \rightarrow Fe_n - O_2 (n = 2 - 15)$$
$$Fe_n + H_2O \rightarrow FeO + H_2$$
$$Fe_n + CH_4 \rightarrow \text{no reaction}$$

It was also noted that iron atoms react with none of these reagents and the activity of particles increases with their size; however, for particles that involve more than six atoms, the activity ceases to depend on the size.

Under commensurable conditions, the chemisorption of deuterium on neutral and positively charged iron clusters was studied.[82] Particles Fe_n and Fe_n^+ demonstrated a nonmonotonic dependence of the reaction rate on the number of atoms, n. As seen in Figure 6.6, the activity changes by four orders of magnitude in a range

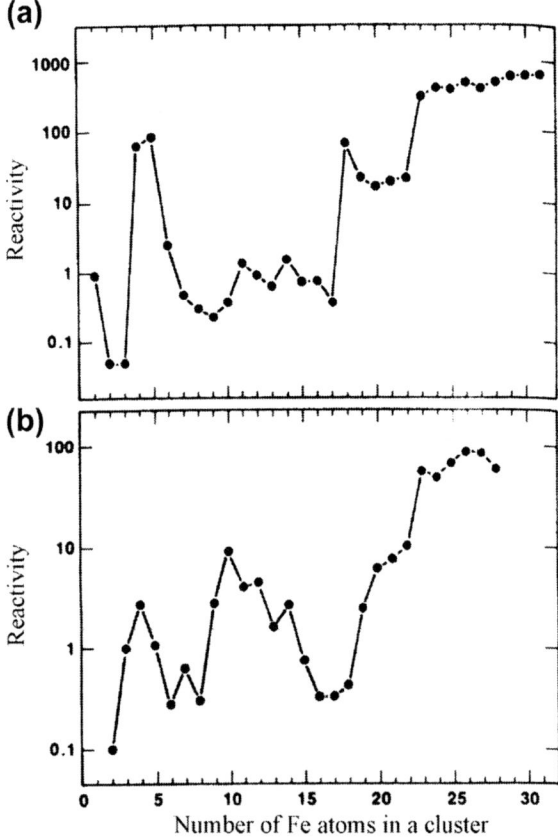

FIGURE 6.6 Effect of size on the activity of (a) charged and (b) neutral iron particles.[82]

from $n=1$ to 31. The presence of a charge also affects the rate. These results were qualitatively explained by the changes in the valence electron structure of clusters. The geometric structure of a cluster, which determines the number, energy, and spatial orientation of the valence orbitals capable of effectively reacting with hydrogen, is also important.

Figure 6.6 shows that the activity of samples is similar for Fe_n and Fe_n^+ particles for $n \geq 20$. Strong deviations are observed for $n < 20$. An Fe_n^+ particle demonstrates a sharp peak at $n=4-6$. At the same time, the activities of Fe_4^- and Fe_5^- particles are approximately 1000 and 1200 times higher, respectively, as compared with Fe_3^+. It is comparable with the activity of Fe_8^+ and approaches the activity of Fe_{23-31}^+. The activity of Fe_{9-14}^+ clusters is suppressed as compared with that of the nearby clusters, which, as seen in the figure, is in contrast to the uncharged particles that have a wide maximum in the same range with a peak at Fe_{10}.

The chemical synthesis of nanoparticles and nanomaterials with controlled shapes has only achieved limited success. Only a few authors reported on the preparation of heterostructures smaller than 20 nm.[83,84] A chemical method for synthesizing heterodimers was also proposed.[85] To produce Fe_3O_4–Ag heterodimers, first, Fe_3O_4 nanoparticles were prepared. These particles were then dissolved in , as for example, hexane or dichlorobenzene, the resulting solution was mixed with aqueous silver nitrate solution, and the mixture was subjected to ultrasound to form a microemulsion. Presumably, ultrasound also induced self-organization processes at the liquid–liquid interface. In the process, few Fe(II) sites on the interface acted as catalytic centers for the reduction of the Ag^+ ions and nucleation of the Ag nanoparticles. After a 30 min-reaction, the organic phase, separated by centrifugation, contained finely divided Fe_3O_4–Ag heterodimers measuring 13.5 nm. According to TEM studies, the diameters of Fe_3O_4 and Ag particles in a dimer were 8 and 5.5 nm, respectively. Using the same procedure, FePt–Ag and Au–Ag dimers were obtained. The advantage of this method is that it allows attaching different biomolecules to silver.

The geometry and the binding energy in Ni_n particles ($n \leq 23$) were determined by employing ab initio multiparticle potential and molecular-dynamics simulations.[86] The average interatomic distance, which was found to be a function of the particle size, was shown to monotonically change up to $n=9$, after which small oscillations were observed. The average distance in Ni_{23} was 2.25 Å, which is 10% longer than the average distance in the compact metal. In a nickel dimer, the bond length is 2.01 Å. It is interesting to note that the equilibrium geometries of Ni_{12}–Ni_{16} particles are similar to those observed in inert-gas clusters and clusters with closed electron shells, such as in Mgn. In contrast to alkali metal particles, for nickel particles, the binding energies per atom demonstrate no magic numbers. As compared with other particles, Ni_2 and Ni_{13} are more stable.

The electronic structure of transition metals such as Ni_n, Co_n, and Fe_n is complicated by strong correlation effects. In particular, these particles do not form closed electron shells.[87]

Interesting results were obtained for a reaction of carbon monoxide with a cluster of nickel ions in the gas phase.[88] Separation was performed on a quadrupole mass spectrometer. Nickel–carbonyl complexes $Ni_n(CO)_k^+$ and $Ni_nC(CO)_l^+$, where $n = 1$–13 and k and l change as a function of the cluster size, i.e. n, were observed. For clusters of definite sizes, the stability was calculated as a function of the valence electron number, which allowed the number of bounded carbon monoxide molecules to be predicted and fit the XRD data. As a whole, such correlations are obeyed.

In the gas phase, chemisorption of CO by clusters of different metals was studied for V, Co, Ni, Nb, Mo, Ru, Rh, Pd, W, Ir, Pt, and their particles M_n, where $n = 1$–14.[89] Clusters were obtained by laser evaporation and analyzed on a time-of-flight mass spectrometer. For the majority of transition metals, the following dependence was observed: for $n \leq 5$, the clusters readily chemisorbed CO, whereas the activity of coarser clusters was 2–3 times weaker. It was noted that metal atoms and particles M_2, M_3, and M_4 are comparatively weakly active with respect to carbon monoxide. This phenomenon was explained as a possible competition between monomolecular destruction and stabilization upon collisions.

The deviations in the reactivity of clusters of transition metals such as Pd and Pt in the gas phase were observed for their reactions with H_2, D_2, N_2, CH_4, CD_4, C_2H_4, and C_2H_6; the following general trend in the activity variation was found: D_2, $H_2 > N_2 > C_2H_4 > CD_4$, CH_4, C_2H_6.[90,91]

For platinum clusters, their reactions with hydrocarbons heavier than ethane were studied.[92,93] The degree of dehydrogenation was shown to increase with the cluster size. Thus, for cyclohexane, a benzene diadduct $(C_6H_6)_2Pt_n$ was formed, which was followed by the liberation of six hydrogen molecules. At the same time, benzene itself could be dehydrated on particles containing three or more platinum atoms. The temperature at which platinum particles reacted with a hydrocarbon was estimated as ca. 300–600 K, and the time was estimated as 100 μs. It is interesting to study such reactions at lower temperatures and at greater times. It was noted that studies of this kind cannot be carried out with pulsed beam techniques—a solid-jet method is more suitable.

Now, we consider briefly the chemistry in matrices and cold liquids. As was noted above, low temperatures provide natural conditions for stabilizing metal atoms, clusters, and nanoparticles and in studying their chemical activity. However, here we face various problems associated with the fact that only extremely low temperatures (4–10 K) in such inert media as argon allow us to infer the existence of free unligated clusters. In contrast to the gas phase, in the liquid phase, we deal with solvated particles, which can naturally affect their chemical activities. Nonetheless, real materials represent solid and liquid systems—hence, the attention paid by scientists to physicochemical properties of metal particles in solid and liquid phases.

Below, we show the general scheme of the synthesis of solvated metal particles in organic media.[94]

Chapter | 6 Chemical Nanoreactors

```
M   +   S      77K        M–S–solvate complex   heating      M–S solvated
metal solvent  co-condensation  (colored)                     metal atom

                                                              further
                                                              heating ↓

(M)ₙ–S–black nanocrystalline or  ← evaporation of    (M)ₙ–S–suspended
amorphous solvated metal particles  excessive solvent at  solvated particles
                                    300K
```

The first stage is co-condensation at 77 K, which usually produces a weakly solvated complex. The latter is, as a rule, colored due to charge transfer. This is either a CTC or a donor–acceptor complex. Such complexes play an important role in cryochemistry. They transform a two-component system into a single-component one and lift diffusion limitations. In many low-temperature cases, no chemical processes can proceed without the formation of complexes. The formation of complexes is the necessary condition for chemical reactions.[95,96] The second stage is heating, which gives rise to an M-solution system. The third stage is the further heating, resulting in the formation of M_n clusters located in a viscous sticky medium resembling liquid clay. And, finally, the fourth stage is the removal of the solvent excess to produce an M_n-solvent system, which can be either amorphous or crystalline. Stages 1–3 are often accompanied by reactions of metal clusters with the solvent.

The resulting final size of a particle and its state, e.g. crystalline, depend on the metal concentration, the solvent activity, and the heating conditions. In fact, the growth of clusters can be accompanied by unusual and unexpected reactions with solvent at low temperatures. Particularly, systems formed with nickel contained small particles lacking ferromagnetic properties. The growing crystals actively interacted with alkanes (solvents) and activated C–H and C–C bonds when heated to 150 K.[97]

At this point, it is worth pointing that the reaction occurs at low temperatures and the growing clusters are more active on the clean metal surfaces. The latter factor infers the formation of defects during the growth of clusters. The unusual conditions of cluster growth in organic media give rise to new peculiarities. The kinetic control results in the appearance of defects and voids. The formation of such sites on a growing particle can be beneficial for the better combination of orbital energies and favors the cleavage of C–H bonds. Reactions with solvents are also possible. The formation of particles depends on the metal concentration, the solvent activity, the heating conditions, and the presence of surface-active additives (surfactants).

Particles comprising two different metals were also obtained by using various solvents. The scheme below illustrates such a synthesis for manganese and cobalt that applies the solvated metal atoms dispersion (SMAD) method.[98]

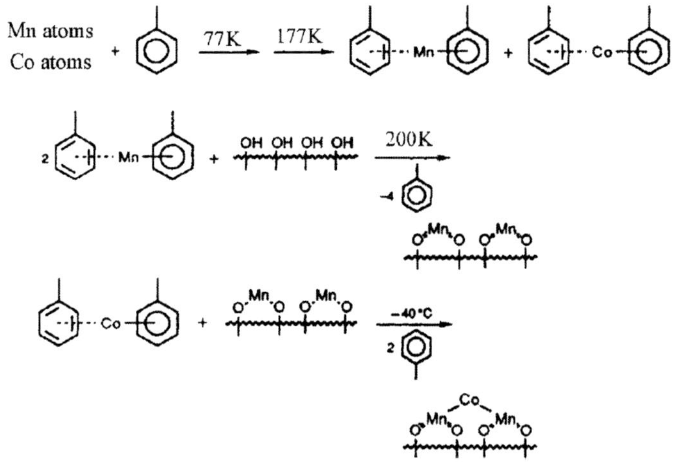

The obtained system was used for catalyzing the hydrogenation of butene-1 at 213 K.[99] The hydrogenation rate turned out to be higher with a bimetallic catalyst as compared with individual cobalt and manganese. It this case, we observe a synergistic phenomenon. Such catalytic systems are either introduced into zeolite pores or synthesized immediately in the pores.[100]

For metals of group VIII (M = Ru, Os, Rh), the reactions of oxidative addition were observed. Thus, the reactions such as $M+CH_4 \rightarrow [M-H-CH_3] \rightarrow CH_3MH$ proceed easily without activation energy at a temperature of 10 K in the argon matrix. The reaction is favored by a high (50–60 Kcal/mol) energy of the metal–hydrogen bond. The reactivity of particles correlate with the evaporation rate of metals and, hence, the kinetic energy of atoms acquires great importance.

Cobalt, nickel, and iron react with methane only in the presence of light of a wavelength $\lambda \approx 300$ nm, whereas light with $\lambda \approx 430$ nm initiates the reverse process.[101,102] For iron, this process occurs via the following scheme:

$$Fe + CH_4 \underset{420 \text{ nm}}{\overset{300 \text{ nm}}{\rightleftharpoons}} CH_3FeH$$

A metal particle attacks a C–H bond. The possible reaction mechanism is associated with the formation of a σ-complex. Other hydrocarbons such as C_2H_6 and C_3H_8 react in a way similar to methane. Iron reacts with cyclopropane by attacking the carbon–carbon bond.[103] According to IR spectra, the following compounds were detected in the products of the reactions with benzene: $Fe(C_6H_6)$, $Fe(C_6H_6)_2$, and $Fe_2(C_6H_6)$.[104] In argon matrices at a temperature of 12 K, the photolysis initiates the following reaction of iron with 1,4-cyclohexadiene:

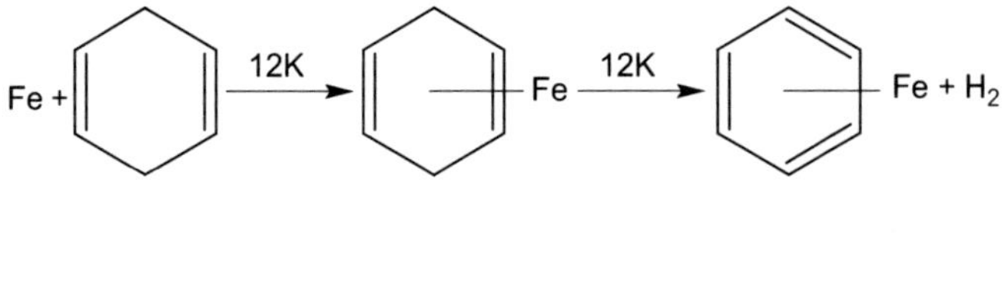

In argon matrices, iron and nickel enter into a reaction with diazomethane CH_2N_2.[105,106] Particles of $M{=}CH_2$ or N_2MCH_2 were formed as the products. Illumination of the matrix with light of wavelength $\lambda = 400\text{--}500$ nm increased the yield of N_2NiCH_2. Presumably, photoinduced diffusion plays an important role.

Interesting reactions of particles of group-VIII metals with triple bonds were observed.[107,108] In argon at 15 K, iron and nickel react with acetylene to give MC_2H_2 compounds. Moreover, it is most likely that iron forms a σ-complex rather than a π-complex, because IR spectra revealed no changes corresponding to the triple bond, while a change in the C–H bond was observed at $n = 3270\,\text{cm}^{-1}$. Presumably, this process proceeds according to the following scheme:

$$\text{Fe} \cdots \overset{\text{-H}}{\underset{\text{C}{\equiv}\text{CH}}{}} \longrightarrow \text{H}{-}\text{Fe}{-}\text{C}{\equiv}\text{CH}$$

In the case of Ni, the formation of a π-complex is preferential.

The above examples demonstrate the sensitivity of similar reactions to regrouping and light action.

Interaction of iron atoms with alkenes produced substituted ferrocenes.[109] Presumably, this reaction follows the scheme:

$\text{Fe} + \text{RC}{\equiv}\text{CR} \rightarrow \text{(co-condensation, heating)}\ (C_5R_5)\text{Fe} + \text{trimers and tetramers}$
of alkines.

Such an unusual reaction requires the cleavage of at least one triple bond. The reaction mechanism is unclear.

An analysis of the reaction of nickel with CH_4 and H_2O was carried out. An interesting method for synthesizing new trinuclear compounds based on the reactions of metal atoms with organometallic compounds was proposed.[110] Particularly, the following co-condensation reaction was described:

$$\text{Co} + C_6H_3(CH_3)_3 + \text{Fe(CO)}_5 \xrightarrow{77\,\text{K}} [C_6H_5(CH_3)_3]_2\text{FeCo}_2(CO)_5$$

The latter example confirms once again the complexity of this problem. It is still difficult to write a stoichiometric reaction equation and determine the number of atoms in a metal particle that enters into a chemical reaction. It is quite unlikely that such active particles as cobalt atoms form no coarser clusters immediately during the co-condensation process at 77 K. However, the uncertainty in the metal particle size should not restrain chemists from studying new chemical reactions and synthesizing new compounds.

The geometry, the electronic structure, and magnetic properties of Co_n particles ($n = 2\text{--}8$) were considered in terms of the nonlocal density functional.[111] It was shown that small cobalt clusters can be described by a set of various geometric structures that do not differ too widely in energy. In this study, the following equation was proposed:

$$\lg \nu = kn(E_b - E_0)$$

It describes the relative rate ν of the reaction of cobalt particles with deuterohydrogen as a function of the number of cobalt atoms n and the binding energy E_b (E_0 and k are constants). This equation was used for processing the data on the reaction of Co_n with D_2.[57] Figure 6.7 shows the results obtained.

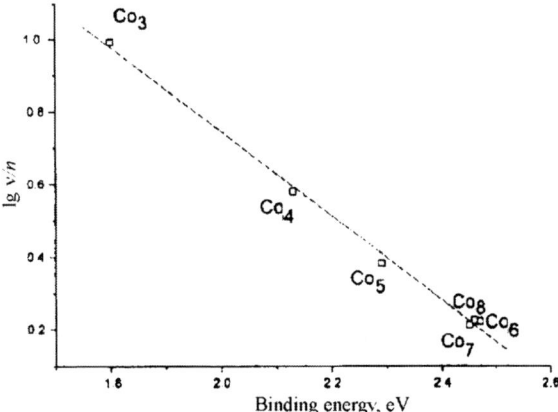

FIGURE 6.7 Dependence of the relative rate v of a reaction between cobalt particles and deuterohydrogen on the number of cobalt atoms, n, and the binding energy.[57]

The low activity of the Co_6 cluster can be explained by its higher binding energy and enhanced stability. In turn, the particles Co_7 and Co_8 have lower binding energies per atom as compared with Co_6, which may result in their higher activity.

Yet another type of chemical reactors is represented by various kinds of micelles. Thus, cobalt particles measuring 3 nm were obtained by using toluene and surfactants.[112] The following scheme illustrates this synthesis. In fact, the size of the particles obtained in such synthesis is determined by the size of the void in a micelle. As shown in the scheme, cobalt particles tend to form rod-like particles 11-nm long and 3–4 nm in diameter.

Under the effect of ultrasound, in alcohol solutions of inverse micelles based on cobalt porphyrins, nanorods with a distance of 0.25 nm between cobalt atoms and a length of 200 nm were prepared. The aggregated particle included nearly 800 cobalt porphyrin molecules.[113]

Cobalt-doped titanium dioxide nanocrystals measuring 4.4±0.5 nm were synthesized in inverse micelles. Thin films formed from these nanocrystals exhibited strong ferromagnetic properties at room temperature.[114] The synthetic method involved the use of an ionic precursor of cobalt and an oxidation reaction, which resulted in metal cobalt.

Unique particles were obtained using organic solvents and two different metals. In this case, it is remarkable that the authors managed to mix the metals immiscible under usual conditions. The scheme, presented below, exemplifies such an experiment.[115] Atoms of evaporated iron and lithium were trapped in cold pentane at 77 K. Further heating resulted in aggregation of atoms. Clusters of Fe–Li were formed. The kinetic control over their growth in cold liquid pentane led to the formation of 3-nm α-Fe crystals surrounded by a noncrystalline lithium medium. The total size of particles was 20 nm. The resulting powder exhibited pyrophoric properties. Its surface area was 140 m²/g. Controlled oxidation (to prevent ignition) and heating of such a cluster yielded an onion form of a hybrid core–shell structure. The core consisted of α-Fe crystals, whereas the shell consisted of metal lithium or its oxide.

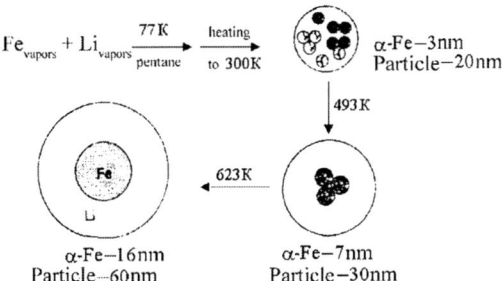

By choosing the conditions of heating and oxidation, we can control the size of α-Fe crystals in the range 3–32 nm. It is important that the particles were covered by protective coatings of Li_2O or Li_2CO_3 and, thus, were stable in air for several months. Particles Fe–Mg were synthesized in a similar way.

Interesting results were obtained for iron nanoparticles with the average size of ~7 nm. The particles were stabilized by oleic acid or AOT. The reactivity of Fe^0 particles was assessed by their reaction with oxygen to give Fe_xO_y. The oxygen concentration was controlled according to the lifetime of pyrene fluorescence.[116]

For such core–shell particles, the iron core oxidation at room temperature was inhibited. A possibility of this reaction was opened in a relatively narrow temperature interval of about 110 °C. The specifics and mechanism of the activation process require further investigation.

Preparation of various core–shell nanosystems has attracted keen attention. Of prime interest was the problem regarding how the properties of individual components change in such systems. It was shown that deposition of noble metals on magnetic cores and "vice versa" changes the magnetic, optical, and chemical properties as compared with individual components.[117,118] For example, iron oxide nanoparticles overcoated with dye-impregnated silica shells were shown to retain the magnetic properties of the core, while exhibiting the luminescent optical properties of the organic dye.[119] Recently, a new method for synthesizing bifunctional nanocrystals, which combined the properties of magnetic nanoparticles and semiconductor quantum dots, was proposed.[118] The synthesized core–shell Co–CdSe nanocrystals had 11 nm cores and 2–3 nm shells.

Platinum is the major catalyst in nitric acid production, reduction of automobile exhaust gases, oil cracking, and proton-membrane-exchange fuel cells.[120] In all the applications listed, platinum was usually employed in the finely divided state. In particular, nanotubes and nanowires that possess well-developed and active surfaces as compared with nanoparticles are most promising for catalysis. The synthesis of platinum nanowires with a diameter of 3 nm and a length up to 5 nm, which employed organic–inorganic mesoporous silicon compounds as templates, was proposed.[121] A chemical synthesis of nanotubes as separate crystals was carried out.[122]

In the first stage, H_2PtCl_6 or K_2PtCl_6 were reduced with ethylene glycol at 110 °C in the presence of polyvinylpyrrolidone to yield Pt(II) species. Further reduction proceeded more slowly in air at 110 °C and resulted in the formation of platinum particles with an average diameter of ~5 nm. In this process, ethylene glycol served as both a reducing agent and a solvent. The rate of the process was regulated by small amounts of $FeCl_3$ or $FeCl_2$, which induced the assembling of platinum particles into spherical agglomerates. Moreover, platinum atoms, which continued to form very slowly, started the growth of uniform nanowires on the surfaces of agglomerates so that the latter resembled sea urchins at the end of the process.

Bimetallic FePt nanoparticles attract high interest in view of their applications in memory storage units and as high-performance permanent magnets. These particles were also used in the preparation of water-soluble systems that involve biomolecules.[123,124] Moreover, the structural and magnetic properties of FePt nanoparticles were shown to depend on the size and composition of particles.[125,126] A one-step synthesis of FePt nanoparticles of controlled size and composition was proposed.[127] This can be illustrated as follows:

$$\text{Fe/Pt} \xrightarrow{(a)} \text{Fe/Pt Fe} \xrightarrow{(b)} \text{Fe/Pt} \xrightarrow{(c)} \text{Fe/Pt Fe}_3\text{O}_4$$
$$(1) \qquad\qquad (2) \qquad\qquad (3) \qquad\qquad (4)$$

In the first stage, spherical FePt cores enriched with Pt were obtained. They were formed by the reduction either of platinum acetylacetonate at a temperature exceeding 200 °C or with Fe atoms from decomposed $Fe(CO)_5$, or by both processes. In stage (a), more Fe atoms were coated over the Pt-rich nuclei to form particles (2) measuring ~7 nm. In stage (b), the interphase diffusion occurred. The exposure of these particles to air resulted in the formation of system (4), which represented a core–shell $FePt/Fe_3O_4$ structure measuring 7 nm/1.2 nm.

The synthesis of metal particles containing two metals, and their deposition on solid carriers play an important role in catalysis. This method allows different systems to be obtained. We can deposit either one kind of metal or both, in layers or together. Insofar as the deposition occurs at low temperatures (below 200 K), the formation of metastable compounds of two metals even those that cannot form stable alloys for thermodynamic reasons and the so-called immiscible metals, becomes possible.

Conductive palladium wires of 100-nm diameters (according to SEM data) were synthesized via the electrolysis of palladium acetate solutions between chromium and gold electrodes spaced at 10 µm and 25 µm under the effect of an AC electric signal of $10V_{rms}$ and 300 Hz.

The wires grew spontaneously along the direction of the electric field. The proposed procedure can be used for creating nanosize fuses, device interconnection, and chemical sensors.

In as much as the morphology of metal pairs can strongly differ from one another, it is virtually impossible to predict the properties of the catalytic systems obtained. In bimetallic clusters, one element affects the properties of another. Detailed studies of properties of such systems will allow synthesizing new catalysts with high activity. The synthesis of cryodispersions and their deposition on adsorbents also makes it possible to obtain new catalytic systems.

6.5 SUBGROUPS OF COPPER AND ZINC

These subgroups of the periodic table include several important and interesting metals, namely, Cu, Ag, Au, Zn, Cd, and Hg.

These metals are remarkable owing to their ability to form clusters sufficiently and easily in their zero-valence state, which allows scientists to operate with them in different media. These metals also form classical colloids.

The idea of cryophotochemistry was put forward and realized by considering the example of silver atoms. Silver atoms isolated in an inert-gas matrix are readily crystallized under the action of light. Thus, Ag_2, Ag_3, Ag_4, Ag_5, etc. were obtained. As a result of photoexcitation, the energy is transferred to the matrix. The latter is heated during the relaxation of silver atoms to the ground state. The softening of the matrix promotes atomic motion and the formation of clusters. This is a complex process, which proceeds in consecutive stages. Thus, under the action of light, Ag_3 takes a silver atom and transforms into Ag_4; however, it can also transform into Ag_2 and Ag. An interesting method for synthesizing particles with different sizes was developed.[128] It combines the mass spectral selection with the subsequent condensation of particles in inert-gas matrices. Using this method, dimers and trimers were obtained for many metals. EPR studies of these clusters and also those with higher numbers of metal atoms (M_5 and M_7) provided information on the electron density distribution over individual atoms. The distribution over atoms was shown to be nonuniform, which probably determined the different reactivities of clusters of different sizes.

A linear chain is the simplest structure built by metal atoms. For such structures, the interval between the higher occupied and lower vacant molecular orbitals usually oscillates depending on even or odd numbers of atoms in a cluster.[129]

Transition metals are of interest for studying the formation of a structure with an increase in the particle size. Transition metal atoms include both localized $3d$ electrons and delocalized $4s$ electrons. It is assumed that these states depend in different ways on the particle size. For example, the atomic structure of copper is $3d^{10}4s^1$ and, similar to alkali metals, copper has one unpaired electron in the outer shell. However, the $3d$ sublevel is lower with respect to energy and affects the cluster properties. In compact copper, the $3d$ sublevel makes a substantial contribution to the Fermi level, which results in the higher conductivity of copper. In copper clusters, the 4_s sublevel is delocalized, which gives rise to discrete and size-dependent changes. At the same time, the $3d$ sublevel monotonically changes with the particle size.

In studying the structural and electronic peculiarities of Cu_n ($n \leq 18$) particles deposited on an MgO (100) film, the theory of density functional was used.[130] Calculations considered the complete relaxation of the surface of overlapping levels corresponding to the $3d$–$4s$ orbitals of copper and $2p$ orbitals of oxygen. The competition of interactions of the copper atoms with one another and with the oxide surface was observed. The calculated average adsorption energy per copper atom decreased with the increase in the particle size, whereas the average energy of Cu–Cu interaction increased. This feature explains the preferential formation of three-dimensional structures for $n \geq 5$. In this case, the system gains more energy from the binding of copper atoms than from their interaction with the surface.

The behavior of atoms at the metal–thiolate junction was studied by density-functional method for the interaction of small copper and gold clusters with alkanethiols. The fragmentation energy estimates have shown that for Cu_n–thiolate ($n = 1, 3, 5, 7,$ and 9), the energy required for breaking the S–C bond in the thiolate progressively decreases from 2.9 eV for $n = 1$ to 1.4 eV for $n = 9$. The reduction in the binding energy was attributed to the polarization of the electronic density in the S–C bond.[131] No effects of this kind were observed for gold clusters with methyl and ethylthiolates.

The studying of ligand-free metal clusters evidenced the existence of closed electron shells.[132] They exist for spherical M_n particles, where $n = 2, 8, 18, 20, 34, 40$, etc. It is these clusters that are thermodynamically more stable and comparatively more abundant. These particles also have a higher ionization energy, a weak electron affinity, and a low reactivity as compared with particles with open shells.

A possibility of attaching chemical reagents to clusters with closed shells was studied for copper particles. Clusters Cu_6, Cu_7^+, Cu_{71}^+, etc., were examined. Metal atoms were modeled as one-electron systems by using the effective potential of the nucleus, its polarizability, and chemisorption of carbon monoxide on different sites of a cluster.[468] Table 6.3 shows the results of simulations.

If one considers carbon monoxide as a ligand with two electrons and each copper atom as a donor of one electron, then the closed electron shells should correspond to Cu_6CO and Cu_7^+CO. The results of theoretical simulations and experiments confirm this assumption. Indeed, as seen from table, Cu_6CO and Cu_7^+CO particles have the highest chemisorption energies. A Cu_{17}^+CO particle also has a closed shell and is stable. For small clusters, the effect of the shell type is less pronounced. Probably, this stems from the fact that a single electron pair can add to an unfilled $1p$ shell of a cluster.

Nanowires with a diameter from 40 to 150 nm and a length of several tens of micrometers were obtained on copper and silver foils of large-areas (8×3.6 and $14.3 \times 3.4\,cm^2$) via solid-phase reactions of vapors of AgTCNQ or CuTCNQ (TCNQ = 7,7,8,8-tetracyanoquinodimethane).[133] Metal nanowires of a diameter from 30 to 50 nm, based on copper, silver, and gold, were synthesized.[134] Organic semiconductor nanowires were also synthesized.[135]

The method of low-temperature co-condensation is also suitable for synthesizing colloid particles. Thus, gold atoms were condensed together with acetone, tetrahydrofuran, trimethylamine, dimethylformaline, and dimethylsulfate oxide. The condensates were heated to obtain stable gold particles measuring ca. 6 nm.[136] In all the solvents listed above, no reactions were observed. Only stabilization of gold particles by solvents took place. "Pure" colloid solutions formed could be sputtered to produce film coatings of different thicknesses.

TABLE 6.3 Energy of Chemisorption (eV) of Carbon Monoxide on Neutral and Charged Copper Particles

Cluster	Neutral particle	Cation
Cu_1	0, 10	1, 18
Cu_2	0, 60	1, 06
Cu_3	0, 93	1, 00
Cu_4	1, 03	1, 04
Cu_5	0, 46	1, 19
Cu_6	1, 06	1, 23
Cu_7	0, 59	1, 42
Cu_8	0, 44	0, 93
Cu_9	0, 72	0, 88
Cu_{10}	0, 31	0, 66

Some solvents could be partially incorporated into the films and removed from them by heating.

It was found that gold particles measuring 3 nm exhibit catalytic activity with respect to hydrogenation of double bonds. Thus, for hydrogenation of butene to butane, it was shown that the activity of 3-nm gold particles is highly comparable with that of palladium black, a conventional catalyst of such reactions.

EPR spectroscopy was employed in studying the host–guest interaction of the derivatives of a *para*-substituted benzylhydroalkylnitroxyl radical and gold particles measuring 3.4 ± 0.7 nm stabilized by a monolayer formed by a water-soluble thiol and triethylene glycol monomethyl ether. From the changes in the spectra of nitrogen and the two β-hydrogen atoms, the equilibrium constant of the exchange between free radicals and radicals incorporated in the particle was determined. The synthesis of water-soluble gold nanoparticles was described.[137]

By the example of gold, yet another problem of nanochemistry, which can be conditionally formulated as resolation, i.e. a transition of particles from one solvent to another without changing their size, was solved. As an example, the agglomeration of gold atoms in acetone and perfluor-*n*-tributylamine (PFTA) is depicted in the scheme.[94] Here R_H is an acetone fragment and R_F is a fragment of PFTA.

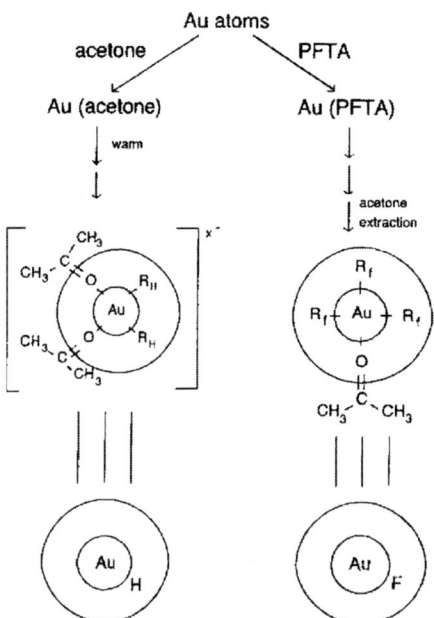

The heating resulted in the formation of a dark-brown colloidal solution of Au in PFTA, which lost its color when extracted with acetone, while the upper acetone layer became purple.

Among other reactions of elements from this subgroup, mention should be made of the reaction of copper with diazomethane at 12 K.[138] In the argon matrix, two products were formed, according to the scheme:

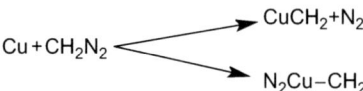

Under the action of light, N_2CuCH_2 can dissociate in argon matrix. A different situation was observed in the nitrogen matrix: the N_2CuCH_2 formed did not undergo light-induced dissociation. The composition of products was revealed by analyzing the IR spectra.

A unique study on the reaction of cadmium, zinc, and mercury with ethyl halides at low temperatures in krypton matrices was performed.[139,140] The choice of metals was dictated by the fact that the above-mentioned metals do not react with ethyl halides in gas and liquid phases and, also, that these systems are amenable to simulations. The metal alkyl halide matrix ratio was 1:100:1000. The differences in the reactions with different alkyl halides were revealed by considering the example of excited mercury.

$$Hg^* + EtCl \rightarrow HCl + C_2H_4 + Hg$$
$$Hg^* + EtF \rightarrow H_2 + CH_2 = CHF + Hg$$

It was found that the metal atom is inserted into C–Cl and C–Br bonds and is not inserted into the C–F bond. The interaction of Zn, Cd, and Hg (metals with $3p$ orbitals) with ethyl halides is a spectacular example of a solid-phase reaction at superlow temperatures. The resulting hydrogen halide (HX) was bound in a complex with ethylene at low temperatures, whereas the products of the first reaction could be obtained only upon heating. The unusual reactivity of ethyl halides was explained based on the energy diagrams of the metal–ethyl halide system.

The reactions with ethylene and propylene at 77 and 4 K were studied by considering the example of copper, silver, and gold. In hydrocarbon matrices at 77 K, the formation of complexes $Cu(C_2H_4)$, $Cu(C_2H_4)_2$, $Cu(C_2H_4)_3$, $Ag(C_2H_4)$, and $Ag(C_2H_4)_2$ was observed.[141–144] At 4 K, complexes with propylene $Cu(C_3H_6)$, $Cu(C_3H_6)_2$, $Au(C_3H_6)$, and $Au(C_3H_6)_2$ were found. The fact that no silver complexes were observed deserves interest. This suggests that the formation of complexes is extremely sensitive to the energy levels of the interacting orbitals. Thus, gold with acetylene yields a vinyl radical, rather than a complex.[145]

$$Au + HC \equiv CH \rightarrow Au-CH=CH$$

The interaction of gold anions Au_n^- ($n = 1 - 3$) with CO and O_2 at 100 K was studied,[146] particularly, for Au_6^- anions.[147]

Benzene layers formed on Au{111} surface were studied at 4 K, using the STM technique. At low coverage degrees, benzene was preferentially adsorbed on the vertexes of the monoatomic steps, while its adsorption on the terraces was relatively weak. The microscope probe induced the concert-cascade movements of benzene molecules, rather than random motion.[148]

It was found that gold nanoparticles measuring 2–4 nm deposited on TiO_2 exhibit catalytic activity and can promote selective catalytic reactions such as epoxidation of propylene.[149] The unexpectedness of these results stimulated experimental and, particularly, theoretical studies. In analyzing the catalytic properties of gold, the key role was played by the cluster structure. Thus, the question of planar–nonplanar structure of gold particles Au_6 and Au_8 was analyzed based on several theoretical methods.[150]

A method for synthesizing polyethylene-stabilized copper nanoparticles was developed.[151] According to the EXAFS and electron microscopy studies, the average size of copper particles was 17 nm, and their structure largely corresponded to the crystalline structure of metal copper. A highly dispersed phase was formed by the decomposition of copper diacetate. It was found that copper nanoparticles in polyethylene matrix resisted oxidation in air.

The high catalytic activity of copper nanoparticles incorporated into p-xylylene toward isomerization of 3,4-dichlorbutene-1 into *trans*-1,4-dichlorbutene-2 was described.[152] The activity of cryochemically synthesized copper in poly-p-xylylene exceeded that of highly dispersed copper in silica gel by two orders of magnitude. The catalytic activity depended on the copper content in poly-p-xylylene and correlated with the conductivity of samples.

The reactivity of nanopowders of copper and some other metals obtained by electric explosion was studied.[153,154] The chemical activity of nanopowders was investigated by the example of photoreduction of o-, p-, and m-nitroanilines, the formation of copper phthalocyanine, and by the oxidation of isopropylbenzene. It was proposed to use the oxidation of isopropylbenzene as a model reaction for characterizing the properties of electroexplosive powders.

Nanorods of CdS were synthesized by the decomposition of molecular precursors in the absence of surfactants.[155] Using the solvothermal method, tetrapod-shaped CdSe and CdTe nanocrystals were synthesized.[156–158] Tetrapod-shaped ZnO[159] and ZnS[160] nanoparticles were synthesized by combining the evaporation and vapor-condensation methods. Nanowires, nanorods, and nanoribbons were obtained for ZnSe.[161,162]

Zinc oxide nanoparticles are a promising material for electronic and optoelectronic applications. Attention was focused on one-dimensional structures of the nanorod type. In addition to high-temperature vapor-phase synthesis, a great attention was paid to low-temperature processes of synthesizing zinc oxide nanorods.[163–165] Among these reactions, the processes based on the growth of ZnO nanocolumns 50–2000 nm long and 50–100 nm wide around zinc oxide grains deposited on substrates were most successful.[166] A simple method of preparation of ZnO nanorods and nanotubes was proposed.[167] This was based on magnetron sputtering to yield different surfaces coated with ZnO films. To form zinc oxide nanorods, zinc plates or foils were immersed in 5% formamide solution. The average diameter of grown nanorods was ~100 nm.

Recently, it was demonstrated that zinc oxide nanorods can be formed in the microstructures of self-assembling organic templates.[168] The templates were prepared by microcontact printing of self-assembled monolayers of 11-mercaptoundecanoic acid ($HSC_{10}H_{20}COOH$) on electrobeam evaporated silver films. Then, the chemically patterned Ag surfaces were immersed for several hours in an aqueous solution of zinc nitrate and hexamethylenetetraamine (($CH_2)_6N_4$) at 50–60 °C to produce nanorods of 2-mm diameters. The assumed mechanism of selective nucleation is based on the combination of the pH-regulated interaction of charged particles and the changes in supersaturation conditions near the film–solution interface.

Nanostructures based on gold and silver find various applications in science and technology. Mention should be made of their most interesting applications in optics and catalysis,[169–171] optoelectronics and electronics,[172–174] chemical and biological sensing, and clinical diagnostics.[175–179] Silver particles are actively used in detecting surface-enhanced Raman scattering.[180]

Various structural combinations based on gold and silver particles and, particularly, core–shell nanostructures are finding ever-increasing application. In contrast to individual particles, such structures are characterized by the presence of two surface-plasmon resonance bands. For gold nanoshells deposited on dielectric cores, viz. silica beads or latex polymers, these resonance bands could be shifted from 500 to 1200 nm by varying the core diameter and/or the shell thickness.[181]

It was shown that gold particles effectively adsorb on the surfaces of silica beads preliminarily modified by monolayers of 2-amino-propyl-trimethoxysilane. The adsorbed gold particles could then serve as nucleation sites for the deposition of more gold via the electroless plating process, which resulted in the formation of a complete gold shell on each silica bead.[182] Sonochemical deposition of gold and silver on silica beads from aqueous solutions of $HAuCl_4$ and $AgNO_3$ resulted in the formation of shell-like structures.[183] By repeated gold and silica deposition steps, nanoparticles with multiple, concentric gold and silica shells were prepared.[184] It was also demonstrated that gold particles could be covered with a double shell of silica and polymer. The removal of silica sandwiched between gold and polymer by etching it out with an HF solution gave rise to rattle-like nanostructures (i.e. core–shell particles with cores movable inside shells).[185] The core size, the shell thickness, and the gap between the core and the shell could be separately controlled by changing the experimental conditions.

For gold, palladium, and platinum, the preparation of hollow structures with well-defined void sizes and walls of definite thickness and controlled porosity was described.[186–188] The synthesis of nanotubes with multiple layers of gold and silver was realized.[189] The properties of gold nanoparticles have been reviewed.[190]

Preparation of nanorattles built of Ag/Au alloy cores and the shells of the same alloy was described.[191] The same publication reports preparation of nanoscale multiple-walled "Matrioshka" structures, the formation of which involved a repeated cycle of electroless deposition of a silver layer followed by galvanic replacement reaction between Ag and $HAuCl_4$. The resulting nanosystems displayed interesting optical properties with well-separated absorption bands, one of which corresponded to the core alloy, while the other pertained to the shell alloy.

6.6 SUBGROUP OF BORON AND ARSENIC

The boron subgroup includes B, Al, Ga, In, and Tl. All these elements are important; however, aluminum was studied in more detail. For the results of experimental studies of aluminum, see Refs. 192 and 193; theoretical studies can be found in Refs. 194 and 195. The active development of calculation methods for analyzing clusters was associated with the fact that ligand-free clusters cannot be prepared in amounts sufficient for experimental studies. Moreover, small particles are as a rule metastable.

A comprehensive theoretical analysis of aluminum clusters can be found.[195] Using the density-functional method, the neutral and charged particles of Al_n ($n \leq 15$) were studied. For aluminum clusters, the binding energy, the relative stability, fragmentation channels of particles, the ionization potential, and the vertical and adiabatic electron affinity were analyzed as a function of cluster size. As shown, the particles containing less than six atoms are characterized by two-dimensional structures, whereas those including more than six atoms have three-dimensional structures. The changes in the geometry were accompanied by the corresponding transformations in the electronic structure, namely, in the concentration of s- and p-electrons in HOMO. The binding energy changed monotonically with an increase in particle size; however, Al_7, Al_7^+, Al_7^-, Al_{11}^-, and Al_{13}^- were more stable as compared with their neighbors. The authors of this study[529] explained this by the existence of mixed valence states. The univalent state was typical of particles containing less than seven atoms, while the trivalent state characterized the particles with more than seven atoms.

The bandgap in the Al_{13}^- anion was found to be 1.5-eV wide, which far exceeds 0.3 and 0.4 eV for Al_{12} and Al_{14}, respectively. Number 13 is far less a "magic" number for neutral and positively charged clusters. Another magic number is 7 but only for the positively charged cluster Al_7^+. The calculated bandgaps were 1.6, 0.4, and 0.45 eV for particles Al_7^+, Al_6^+, and Al_8^+, respectively.

A series of particles containing five atoms were also studied.[196,197] Clusters of Al_5^- and Al_5 were studied in more detail.[532] The study employed quantum-chemical calculations of MP2/6-311+G* level and a hybrid Hartree–Fock and density-functional method used in combination with the photoelectron spectrum of Al_5^- anion. A planar structure C_{2v} was shown to be in best agreement with experimental and calculation results. An analysis of the electronic structure and molecular orbitals of clusters made it possible to conclude that the appearance of planar five-component structures is caused by the four-center peripheral bonds.

For Al_5^- and Al_5 clusters, the following structure provided the best agreement with experimental results:

$$\begin{array}{c} Al \xrightarrow{2.54} Al \\ 2.60 / \backslash 2.71 / \backslash \\ Al \underset{2.51}{} Al Al \end{array}$$

where bond distances are shown in angstroms. From the viewpoint of the authors,[198] the reasons for the planar structure of such systems are still unclear.

For aluminum, reactions of its cluster ions Al_n^+ with oxygen[199] and deuterium[200] were studied. In the case of deuterium, both the chemical reactions and the formation of metastable adducts were observed. The formation of metastable particles was associated with the absence of stabilizing collisions. The activation energy E_{act} of the adduct formation increased with the increase in the cluster size, which was accompanied by oscillations depending on the even–odd nature of the cluster. Odd clusters exhibited higher E_{act}. A decrease in E_{act} for clusters with even numbers of atoms is probably associated with the weaker repulsion in the transition state because of the presence of an unpaired electron in the higher occupied electron orbital. The main products of the chemical reaction were $Al_{n-1}D^+$, Al_{n-2}^+, and Al^+ for reactions of small clusters and Al_nD^+ and $Al_{n-1}D^+$ for large clusters.

By analyzing the effects of the particle size and the energy of collisions for a reaction of Al_n with oxygen, it was shown that the reaction cross-section increases with the particle size and the O_2 chemisorption on a cluster is accompanied by a quick loss of two Al_2O_3 molecules. The remaining Al_{n-4}^+ particles have excessive energy, sufficient to liberate one or several aluminum atoms.

Neutral Al_n and charged particles react in different ways with O_2. A neutral adduct Al_nO_2 can merely be considered as an aluminum cluster with weakly bound oxygen.

In contrast to oxygen, ammonium was shown to weakly adsorb on aluminum clusters.[201]

Clusters, containing two crystals, such as Ni_xAl_y, NbAl, and CoAl, were obtained.[202] Here, we only touch upon the new chemistry field that deals with low-valence compounds of aluminum.[203,204]

The interaction of fused aluminum with gaseous HCl leads to the following reaction:

$$Al_{liquid} + HCl_{gas} \xrightarrow{1200\,K} AlCl_{gas} + \tfrac{1}{2} H_{2gas}$$

This reaction proceeds with 90% efficiency. Gaseous AlCl formed is condensed on a cold cryostate surface. Heating to 180 K results in the disproportionation reaction:

$$3AlCl_{solid} \xrightarrow{180\,K} AlCl_{3solid} + 2Al_{solid}$$

Thus, active solid AlCl can exist for a long time only at 77 K. Co-condensation of AlCl with butine-2 yields a dimer, namely, 1,4-dialuminum cyclohexadiene, which exists in the crisscross sandwich form:

$$AlCl + H_3C-C\equiv C-CH_3 \rightarrow$$

[Structure: 1,4-dialuminum cyclohexadiene with Cl–Al groups at positions 1 and 4]

Similar chemical transformations were mentioned for gallium as well.[205] Much attention was drawn to gallium compounds with arsenic,[206] which provided the grounds for the development of semiconductor devices. Aluminum was also actively used in the preparation of thin films.[207]

The reactivity of the majority of active clusters was studied as a function of the nature of molecules involved in reactions. Table 6.4 shows the relative chemical reactivity of aluminum particles Al_n ($n=1-30$) with different molecules.[208] Aluminum particles were fabricated by laser evaporation, and the products were analyzed on a time-of-flight mass spectrometer. As seen, the reactivity of aluminum particles decreases in a series $O_2 > CH_3OH > CO > D_2O > D_2$. Methane reacted with none of the aluminum particles studied. It was assumed that all reactions represent chemisorption of molecules, the activity of which was studied based on the variations in the interaction energy between molecular orbitals by using a proposed method.[209] The central idea of this study was to promote the electron transfer from a metal cluster to the antibinding orbital of a

TABLE 6.4 Chemical Reactivity of Aluminum Particles with Different Molecules[208]

Molecule	Relative reactivity	Most active clusters	Reaction products
CH_4	No reaction	–	–
D_2	4	Al_{16}	Al_nD_2 ($6 \leq n \leq 15$)
D_2O	200	Al_{10}, Al_{17-20}	$Al_n(D_2O)_y$ ($n \geq 8$, $y \geq 1-3$)
CO	400	Al_6	–
CH_3OH	2000	Al_{10}, Al_{16-17}	$Al_n(CH_3OH)_y$ ($n \geq 3$, $y \geq 1-2$)
O_2	6000	Al_2, Al, Al_n ($n > 25$)	Al_2O_3, $Al_n(O_2)y$ ($n \geq 7$, $y \geq 1-2$)

molecule to be added. The reactivity of particles was explained on the basis of the reactions of cluster growth such as $Al_n + Al \rightarrow Al_{n+1}$ or $Al_n + Al_2 \rightarrow Al_{n+2}$.

Such reactions could transform active particles into inactive and vice versa.

Interesting studies on the activity of aluminum particles were carried out.[210] The synthesis of aluminum particles involved using the condensation method developed at the Institute of Chemical Physics, Academy of Sciences of the USSR, in the late 1950s.[211] Its modernized version named "levitation-jet" method employs the contact-free confinement and heating of an evaporated metal droplet in a high-frequency inductor field and an inert-gas flow.[212] This method allows one, by varying the inert-gas pressure and the velocity of its flow, to regulate the size of metal nanoparticles. Its great advantage is the absence of contact between the droplet and any possible impurities, which ensures the absence of foreign admixtures in metal nanoparticles.

Aluminum particles measuring 5.6, 37, and 100 nm and synthesized by the levitation-jet method were studied in the adsorption processes and reactions with ordinary and deuterated water, carbon oxides, acetic acid vapors, hexamethylsiloxane, and acrylic acid.[210,213] The water vapor was adsorbed at room temperature, while the desorption was studied under the programmed linear temperature variations; the reaction products were identified by mass spectrometry. The smaller particles were found to have wider distributions of active centers in energy. The water adsorption on the fresh surface involved strong chemical bonding and was independent of the average particle size. A single adsorption–desorption cycle resulted in the complete loss of active water-adsorption sites.

The experiments with acetic acid vapors led to a conclusion that three types of active centers of different activities were present on the aluminum surface. These centers formed chemical bonds and promoted the decomposition of the acids by the following scheme:

$$CH_3 - C(O) - OH + Al_n \rightarrow CH_3 - C(O) - OAl_n + \frac{1}{2}H_2$$

It was found that the 6-nm particles have additional active centers and form stronger bonds with adsorbed molecules, which did not prevent the aluminum particle surface from interacting with water vapor and atmospheric oxygen.

The arsenic subgroup includes the elements As, Sb, and Bi, all of which can form nanosemiconductors. Evaporation in a Knudsen cell of individual metals and their mixtures allowed quite a number of intermetallic polar compounds such as Cs_2Sb_2 and Cs_2Sb_4 to be identified. Compounds Cs_6Sb_4, Cs_6Sb_4Bi, $Cs_6Sb_2Bi_2$, $Cs_2Sn_2Sb_3$, and $CsInSb_3$ were also formed.[214,215] Presumably, Cs atoms and Sb_2 and Sb_4 molecules were the precursors of compounds containing cesium and antimony. (Compounds Na_2Te_2, Na_2Te_3, Ce_2Te_2, and Ce_2Te_3 were synthesized by a similar method.[216])

In the gas phase, particles In_xP_y containing 5–14 atoms were obtained.[217] As compared with even clusters, odd clusters were characterized by more intensive light absorption. The excitation energy changed from 0.84 to 1.84 eV. Even clusters were more abundant as compared with odd ones and had a higher dissociation energy. Even clusters were assumed to have closed shells and a singlet ground state, whereas odd clusters were characterized by the open multiplet ground state. The fact that the absorption spectra corresponding to the adsorption of clusters were comparable with those of the semiconductor junction deserves attention.

Indium nitride attracts attention as a promising material for optoelectronics. Nanocrystals of InN were synthesized in a reaction of $InCl_3$ with Li_3N at 250 °C. The synthesized material represented a mixture of cubic and hexagonal phases.[218] Hexagonal InN nanocrystals with a diameter of 10–30 nm were obtained by reacting In_2S_3 with $NaNH_2$.[219] CVD of indium and ammonium vapors on Si(100) yielded hexagonal indium nitride crystals of micrometer sizes.[220] Indium nitride nanowires grown on gold-coated silicon by the thermal evaporation of In in the presence of NH_3 had a diameter of 40–80 nm.[221] By employing a vapor–liquid–solid route, InN nanowires with diameters of 10–100 nm were synthesized at 700 °C in the reaction of In_2O_3, metal In, and NH_3.[222] CVD of a mixture of In_2O_3, In, and NH_3 onto Si/SiO_2 surfaces coated with 20-nm-thick gold layers allowed InN nanowires with a diameter of 15–30 nm to be obtained.[223]

REFERENCES

1. Moskovits, M.; Ozin, G. Eds.; *Cryochemistry (Russ.)*; Wiley: New York, 1976, p 594.
2. Sergeev, G. B.; Batyuk, V. A. *Cryochemistry*; Khimiya: Moscow, 1978, p 296, 167.
3. Sergeev, G. B.; Batyuk, V. A. *Cryochemistry*; Mir, 1986, p 326.
4. Sergeev, G. B.; Smirnov, V. V. *Molecular Halogenating of Olefines*; Izd. MGU: Moscow, 1985, p 240.
5. Wang, Y. A.; Li, J. J.; Chen, H.; Peng, X. *J. Am. Chem. Soc.* **2002**, *124*, 2293–2298.
6. Guo, W.; Li, J. J.; Wang, Y. A.; Peng, X. *J. Am. Chem. Soc.* **2003**, *125*, 3901–3909.
7. Gerion, D.; Parak, W. J.; Williams, S. C.; Zanchet, D.; Micheel, C. M.; Alivisatos, A. P. *J. Am. Chem. Soc.* **2002**, *124*, 7070–7074.
8. Dubertret, B.; Skourides, P.; Norris, D. J.; Noireaux, V.; Brivanlou, A. H.; Libchaber, A. *Science* **2002**, *298*, 1759–1762.
9. Wu, X.; Liu, H.; Liu, J.; Haley, K. N.; Treadway, J. A.; Larson, J. P.; et al. *Nat. Biotechnol.* **2002**, *21*, 41–46.
10. Mattoussi, H.; Mauro, J. M.; Goldman, E. R.; Anderson, G. P.; Vikram, C. S.; Mikulec, F. V.; et al. *J. Am. Chem. Soc.* **2000**, *122*, 12142–12150.
11. Derfus, A. M.; Chan, W. C. W.; Bhatia, S. N. *Nano Lett.* **2004**, *4*, 11–18.
12. Hawker, P. N.; Timms, P. L. *J. Chem. Soc. Dalton Trans.* **1983**, *6*, 1123–1126.
13. Garland, D. A.; Lindsay, D. M. *J. Chem. Phys.* **1983**, *78*, 2813–2816.
14. Howard, J. A.; Sutcliffe, R.; Mile, B. *Chem. Phys. Lett.* **1984**, *112*, 84–86.
15. Wang, Y.; George, T. F.; Lindsay, D. M.; Beri, A. C. *J. Chem. Phys.* **1987**, *86*, 3493–3499.
16. Lindsay, D. M.; Chu, L.; Wang, Y.; George, T. F. *J. Chem. Phys.* **1987**, *87*, 1685–1689.
17. Carrol, S. J.; Pratontep, S.; Strein, M.; Palmer, R. E. *J. Chem. Phys.* **2000**, *113*, 7723–7727.
18. McCaffrey, J. G.; Ozin, G. A. *J. Chem. Phys.* **1994**, *101*, 10354–10365.
19. Miller, J. C.; Ault, B. S.; Andrews, L. *J. Chem. Phys.* **1977**, *67*, 2478–2487.

20. Duley, W. W.; Graham, W. R. M. *J. Chem. Phys.* **1971**, *55*, 2527–2532.
21. Francis, J. E.; Webber, S. E. *J. Chem. Phys.* **1972**, *56*, 5879–5886.
22. Kargin, V. A.; Fodiman, E. V. *J. Phys. Chem. (Russ.)* **1934**, *5*, 423.
23. Skell, P. S.; Girard, J. E. *J. Am. Chem. Soc.* **1972**, *94*, 5518–5519.
24. Richey, H. G., Jr. Ed.; *Grinhard Reagents. New Developments*; Wiley: Chichester, 2000, p 418.
25. Tjurina, L. A.; Smirnov, V. V.; Barkovskii, G. B.; Nikolaev, E. N.; Esipov, S. E.; Beletskaya, I. P. *Organometallics* **2001**, *20*, 2449–2450.
26. McCaffrey, J. G.; Parnis, J. M.; Ozin, G. A.; Breckenridge, W. H. *J. Phys. Chem.* **1985**, *89*, 4945–4950.
27. McCaffrey, J. G.; Ozin, G. A. *J. Chem. Phys.* **1988**, *89*, 1844–1857.
28. Greene, T. M.; Lanzisera, D. V.; Andrews, L.; Downs, A. J. *J. Am. Chem. Soc.* **1998**, *120*, 6097–6104.
29. Thompson, C. A.; Andrews, L. *J. Am. Chem. Soc.* **1996**, *118*, 10242–10249.
30. Andrews, L.; Yunstein, J. *J. Phys. Chem.* **1993**, *97*, 12700–12704.
31. Imizu, Y.; Klabunde, K. J. *Inorg. Chem.* **1984**, *23*, 3602–3605.
32. Klabunde, K. J.; Whetten, A. *J. Am. Chem. Soc.* **1986**, *108*, 6529–6534.
33. Jasien, P. G.; Dykstra, C. E. *J. Am. Chem. Soc.* **1983**, *105*, 2089–2090.
34. Jasien, P. G.; Dykstra, C. E. *J. Am. Chem. Soc.* **1985**, *107*, 1891–1895.
35. Li, W.; Yang, S. *J. Phys. Chem. A* **1998**, *102*, 825–840.
36. Watanabe, H.; Iwata, S.; Hashimoto, K.; Misaizu, F.; Fuke, K. *J. Am. Chem. Soc.* **1995**, *117*, 755–763.
37. Sun, N.; Klabunde, K. J. *J. Catal.* **1999**, *185*, 506–512.
38. Sun, N.; Klabunde, K. J. *J. Am. Chem. Soc.* **1999**, *121*, 5587–5588.
39. Russel, G. A.; Brown, H. C. *J. Am. Chem. Soc.* **1955**, *77*, 4578–4582.
40. Stark, J. V.; Park, D. J.; Lagadic, I.; Klabunde, K. J. *Chem. Mater.* **1996**, *8*, 1904–1912.
41. Stark, J. V.; Klabunde, K. J. *Chem. Mater.* **1996**, *8*, 1913–1918.
42. Khaleel, A.; Kapoor, P. N.; Klabunde, K. J. *Nanostr. Mater.* **1999**, *11*, 459–468.
43. Stoimenov, P. K.; Klinger, R. L.; Marchin, G. L.; Klabunde, K. J. *Langmuir* **2002**, *18*, 6679–6686.
44. Kohn, A.; Weigend, F.; Ahrichs, R. *Phys. Chem. Chem. Phys.* **2001**, *3*, 711–719.
45. Bauschlicher, C. W., Jr.; Partridge, H. *Chem. Phys. Lett.* **1999**, *300*, 364.
46. Ehlers, A. W.; van Klink, G. P. M.; van Eis, M. J.; Bickelhaupt, F.; Nederkoorn, P. H. J.; et al. *J. Molec. Model.* **2000**, *6*, 186–194.
47. Mascetti, J.; Tranquille, M. *J. Phys. Chem.* **1988**, *92*, 2177–2184.
48. Arnold, P. Ph.D. thesis, Univ. of Sussex, 1997, p 209.
49. Cloke, F. G. N.; Gibson, V. C.; Green, M. L. H.; Mtetwa, V. S. B.; Prout, K. *J. Chem. Soc. Dalton Trans.* **1988**, *8*, 2227–2229.
50. Brown, P. R.; Cloke, F. G. N.; Green, M. L. H.; Tovey, R. C. *J. Chem. Soc. Chem. Commun.* **1982**, *9*, 519–520.
51. Van Zee, R. J.; Baumann, C. A., Jr.; Weltner, W. *J. Chem. Phys.* **1985**, *82*, 3912–3920.
52. El-Sayed, M. A. *J. Phys. Chem.* **1991**, *95*, 3898–3906.
53. Berces, A.; Hackett, P. A.; Lian, L.; Mitchell, S. A.; Rayner, D. M. *J. Chem. Phys.* **1998**, *108*, 5476–5490.
54. Mitchell, S. A.; Lian, L.; Rayner, D. M.; Haskeff, P. A. *J. Chem. Phys.* **1995**, *103*, 5539–5544.
55. Whetten, R. L.; Zakin, M. R.; Cox, D. M.; Trevor, D. J.; Kaldor, A. *J. Chem. Phys.* **1986**, *85*, 1697.
56. Elkind, J. L.; Weiss, F. D.; Alford, J. M.; Laaksonen, R. J.; Smalley, R. E. *J. Chem. Phys.* **1988**, *88*, 5215–5224.
57. Morse, M. D.; Geusic, M. E.; Heath, J. R.; Smalley, R. E. *J. Chem. Phys.* **1985**, *83*, 2293–2304.
58. Hamrick, Y. M.; Morse, M. D. *J. Phys. Chem.* **1989**, *93*, 6494–6501.

59. Zakin, M. R.; Brickman, R. O.; Cox, D. M.; Kaldor, A. *J. Chem. Phys.* **1988**, *88*, 3555–3560.
60. Nayak, S. K.; Rao, B. K.; Khanna, S. N.; Jena, P. *Chem. Phys. Lett.* **1996**, *259*, 588–592.
61. Kietzmann, H.; Morenzin, J.; Bechthold, P. S.; Gantefor, G.; Eberhardt, W. *J. Chem. Phys.* **1998**, *109*, 2275–2278.
62. Athanassenas, K.; Kreisle, D.; Collings, B. A.; Rayner, D. M.; Hackett, P. A. *Chem. Phys. Lett.* **1993**, *213*, 105–110.
63. Holmgren, L.; Andersson, M.; Rosen, A. *J. Chem. Phys.* **1998**, *109*, 3232–3239.
64. Holmgren, L.; Rosen, A. *J. Chem. Phys.* **1999**, *110*, 2629–2636.
65. Zemski, K. A.; Justes, D. R., Jr.; Castleman, A. W. *J. Phys. Chem. A* **2001**, *105*, 10237–10245.
66. Lan, J. T.; Achleither, A.; Wurth, W. *Chem. Phys. Lett.* **2000**, *317*, 269–275.
67. Wang, D.; Chang, Y. -L.; Wang, Q.; Cao, J.; Farmer, D. B.; Gordon, R. G.; et al. *J. Am. Chem. Soc.* **2004**, *126*, 11602–11611.
68. Pinna, N.; Garnweitner, G.; Beato, P.; Niederberger, M.; Antonietti, M. *Small* **2005**, *1*, 112–121.
69. Vargaftik, M. N.; Zagorodnikov, V. P.; Stolyarov, I. P.; Likholobov, V. A.; Chuvilin, A. L.; Zaikovskii, V. I.; et al. *Doklady AN SSSR* **1985**, *284*, 896–899.
70. Volokitin, Y.; Sinzig, J.; Schmid, G.; de Jongh, L. J.; Vargaftik, M. N.; et al. *Nature* **1996**, *384*, 621–623.
71. Bul'enkov, N. A.; Tytik, D. L. *Russ. Chem. Bull. (Russ.)* **2001**, *1*, 1–19.
72. Locke, S. A.; Shevlin, P. B. *Organometallics* **1984**, *3*, 217–222.
73. Gardenas, G.; Shelvin, P. B. *J. Org. Chem.* **1984**, *49*, 4726–4728.
74. Berry, A. D. *Organometallics* **1983**, *2*, 895–898.
75. Rohlfing, E. A.; Cox, D. M.; Kaldor, A. *Bull. Am. Phys. Soc.* **1983**, 28.
76. Rohlfing, E. A.; Cox, D. M.; Kaldor, A. *J. Phys. Chem.* **1984**, *88*, 4497–4502.
77. Knickelbein, M.; Yang, S.; Riley, S. J. *J. Chem. Phys.* **1990**, *93*, 94–104.
78. Richtsmeier, M.; Parks, E. K.; Lin, K.; Pobo, L. G.; Riley, S. J. *J. Chem. Phys.* **1985**, *82*, 3659–3665.
79. Parks, E. K.; Nieman, G. S.; Pobo, L. G.; Riley, S. J. *J. Chem. Phys.* **1988**, *88*, 6260–6272.
80. Parks, E. K.; Weiller, B. N.; Bechthold, P. S.; Hoffman, W. F.; Nieman, G. S.; Pobo, L. G.; et al. *J. Chem. Phys.* **1988**, *88*, 1622–1632.
81. Whetten, R. L.; Cox, D. M.; Trevor, D. J.; Kaldor, A. *J. Phys. Chem.* **1985**, *89*, 566–569.
82. Zakin, M. R.; Brickman, R. O.; Cox, D. M.; Kaldor, A. *J. Chem. Phys.* **1988**, *88*, 6605–6610.
83. Teranishi, T.; Inoue, Y.; Nakaya, M.; Oumi, Y.; Sano, T. *J. Am. Chem. Soc.* **2004**, *126*, 9914–9915.
84. Gu, H.; Zheng, R.; Zhang, X.; Xu, B. *J. Am. Chem. Soc.* **2004**, *126*, 5664–5665.
85. Gu, H.; Yang, Z.; Gao, J.; Chang, C. K.; Xu, B. *J. Am. Chem. Soc.* **2005**, *127*, 34–35.
86. Nayak, S. K.; Khanna, S. N.; Rao, B. K.; Jena, P. *J. Phys. Chem. A* **1997**, *101*, 1072–1080.
87. Billas, I. M.L.; Chatelain, A.; de Heer, W. A. *Science* **1994**, *265*, 353–359.
88. Fayet, P.; McGlinchey, M. J.; Woeste, L. H. *J. Am. Chem. Soc.* **1987**, *109*, 1733–1738.
89. Cox, D. M.; Reichmann, K. C.; Trevor, D. J.; Kaldor, A. *J. Chem. Phys.* **1988**, *88*, 111–119.
90. Fayet, P.; Kaldor, A.; Cox, D. M. *J. Chem. Phys.* **1990**, *92*, 254–261.
91. Blomberg, M. R. A.; Siegbahn, P. E. M.; Svensson, M. *J. Phys. Chem.* **1992**, *96*, 5783–5789.
92. Trevor, D. J.; Cox, D. M.; Kaldor, A. *J. Am. Chem. Soc.* **1990**, *112*, 3742–3749.
93. Trevor, D. J.; Whetten, R. L.; Cox, D. M.; Kaldor, A. *J. Am. Chem. Soc.* **1985**, *107*, 518–519.
94. Klabunde, K. J. *Free Atoms, Clusters and Nanosized Particles*; Academic Press: San Diego, New York, Boston, London, Sydney, Tokyo, 1994, p 311.
95. Klabunde, K. J.; Jeong, G. H.; Olsen, A. W. In *Selective Hydrocarbon Activation: Principles and Progress*; VCH: NY, 1990, p 433.
96. Sergeev, G. B.; Zagorskii, V. V.; Badaev, F. Z. *Chem. Physics (Russ.)* **1984**, *3*, 169–175.
97. Sergeev, G. B. *Mend. Chem. J. (Russ.)* **1990**, *35*, 566–575.

98. Tan, B. J.; Klabunde, K. J.; Sherwood, P. M. A. *J. Am. Chem. Soc.* **1991**, *113*, 855–861.
99. Klabunde, K. J.; Li, Y. X.; Tan, B. J. *Chem. Mater.* **1991**, *3*, 30–39.
100. Ozin, G. A.; McCaffrey, J. G. *Inorg. Chem.* **1983**, *22*, 1397–1399.
101. Ozin, G. A.; McIntosh, D. F.; Mitchell, S. A. *J. Am. Chem. Soc.* **1981**, *104*, 7351.
102. Ozin, G. A.; McCaffrey, J. G. *J. Am. Chem. Soc.* **1982**, *104*, 7351–7352.
103. Kafafi, Z. H.; Hauge, R. H.; Fredin, L.; Billups, W. E.; Margrave, J. L. *J. Chem. Soc. Chem. Commun.* **1983**, *21*, 1230.
104. Ball, D. W.; Kafafi, Z. H.; Hauge, R. H.; Margrave, J. L. *J. Am. Chem. Soc.* **1986**, *108*, 6621–6626.
105. Chang, S. C.; Hauge, R. H.; Kafafi, Z. H.; Margrave, J. L.; Billups, W. E. *J. Am. Chem. Soc.* **1988**, *110*, 7975–7980.
106. Chang, S. C.; Kafafi, Z. H.; Hauge, R. H.; Margrave, J. L. *Inorg. Chem.* **1990**, *29*, 4373.
107. Kline, E. S.; Kafafi, Z. H.; Hauge, R. H.; Margrave, J. L. *J. Am. Chem. Soc.* **1985**, *107*, 7559–7562.
108. Kline, E. S.; Kafafi, Z. H.; Hauge, R. H.; Margrave, J. L. *J. Am. Chem. Soc.* **1987**, *109*, 2402–2409.
109. Simons, L. H.; Lagowski, J. J. *J. Organometal. Chem.* **1983**, *249*, 195–203.
110. Schneider, J. J.; Goddard, R.; Krueger, C. *Organometallics* **1991**, *10*, 665–670.
111. Fan, H. -J.; Lin, C. -W.; Liao, M. -S. *Chem. Phys. Lett.* **1997**, *273*, 353–359.
112. Andrews, M. P.; Ozin, G. A. *Chem. Mater.* **1989**, *1*, 174–187.
113. Yuasa, M.; Oyaizu, K.; Yamaguchi, A.; Kuwakado, M. *J. Am. Chem. Soc.* **2004**, *126*, 11128–11129.
114. Bryan, J. D.; Heald, S. M.; Chambers, S. A.; Gamelin, D. R. *J. Am. Chem. Soc.* **2004**, *126*, 11640–11647.
115. Glavee, G. N.; Easom, K.; Klabunde, K. J.; Sorensen, C. M.; Hadjipanayis, G. C. *Chem. Mater.* **1992**, *4*, 1360–1363.
116. Bunker, Ch. E.; Karnes, J. J. *J. Am. Chem. Soc.* **2004**, *126*, 10852–10853.
117. Sobal, N. S.; Hilgendorff, M.; Moewald, H.; Giersig, M.; Spasova, M.; Radetic, T.; et al. *Nano Lett.* **2002**, *2*, 621–624.
118. Kim, H.; Achermann, M.; Balet, L. P.; Hollingsworth, J. A.; Klimov, V. I. *J. Am. Chem. Soc.* **2005**, *127*, 544–546.
119. Lu, Y.; Yin, Y.; Mayers, B. T.; Xia, Y. *Nano Lett.* **2002**, *2*, 183–186.
120. Roucoux, A.; Schulz, J.; Patin, H. *Chem. Rev.* **2002**, *102*, 3757–3778.
121. Sakamoto, Y.; Fukuoka, A.; Higuchi, T.; Shimomura, N.; Inagaki, Sh.; Ichikawa, M. *J. Phys. Chem. B* **2004**, *108*, 853–858.
122. Chen, J.; Herricks, T.; Geissler, M.; Xia, Y. *J. Am. Chem. Soc.* **2004**, *126*, 10854–10855.
123. Gu, H.; Ho, P. -L.; Tsang, K. W. T.; Wang, L.; Xu, B. *J. Am. Chem. Soc.* **2003**, *125*, 15702–15703.
124. Xu, C.; Xu, K.; Gu, H.; Zhong, X.; Guo, Z.; Zheng, R.; et al. *J. Am. Chem. Soc.* **2004**, *126*, 3392–3393.
125. Klemmer, T. J.; Shukla, N.; Liu, C.; Wu, X. W.; Svedberg, E. B.; Mryasov, O.; et al. *Appl. Phys. Lett.* **2002**, *81*, 2220–2222.
126. Takahashi, Y. K.; Koyama, T.; Ohnuma, M.; Ohkubo, T.; Hono, K. *J. Appl. Phys.* **2004**, *95*, 2690–2696.
127. Chen, M.; Liu, J. P.; Sun, S. *J. Am. Chem. Soc.* **2004**, *126*, 8394–8395.
128. Harbich, W.; Fedrigo, S.; Buttet, J.; Lindsay, D. M. *J. Chem. Phys.* **1992**, *96*, 8104–8108.
129. Balbuena, P. B.; Derosa, P. A.; Seminario, J. M. *J. Phys. Chem. B* **1999**, *103*, 2830–2840.
130. Musolino, V.; Selloni, A.; Car, R. *Surf. Sci.* **1998**, *402–404*, 413–417.
131. Konopka, M.; Rousseau, R.; Stich, I.; Marx, D. *J. Am. Chem. Soc.* **2004**, *126*, 12103–12111.
132. Nygren, M. A.; Siegbahn, P. E. M.; Jin, C.; Guo, T.; Smalley, R. E. *J. Chem. Phys.* **1991**, *95*, 6181–6184.
133. Liu, H.; Zhao, Q.; Li, Y.; Liu, Y.; Lu, F.; Zhuang, J.; et al. *J. Am. Chem. Soc.* **2005**, *127*, 1120–1121.

134. Wang, J.; Tian, M.; Mallouk, T. E.; Chan, M. H. W. *J. Phys. Chem. B* **2004**, *108*, 841–845.
135. Allard, M.; Sargent, E. H.; Lewis, P. C.; Kumacheva, E. *Adv. Mater.* **2004**, *16*, 1360–1364.
136. Lin, S. T.; Frenklin, M. T.; Klabunde, K. J. *Langmuir* **1986**, *2*, 259–260.
137. Pengo, P.; Polizzi, S.; Battagliarin, M.; Pasquato, L.; Scrimin, P. *J. Mater. Chem.* **2003**, *13*, 2471–2478.
138. Chang, S. C.; Kafafi, Z. H.; Hauge, R. H.; Billups, W. E.; Margrave, J. L. *J. Am. Chem. Soc.* **1987**, *109*, 4508–4513.
139. Cartland, H. E.; Pimentel, G. C. *J. Phys. Chem.* **1990**, *94*, 536–540.
140. Cartland, H. E.; Pimentel, G. C. *J. Phys. Chem.* **1989**, *93*, 8021–8025.
141. Howard, J. A.; Joly, H. A.; Mile, B. *J. Phys. Chem.* **1990**, *94*, 1275–1281.
142. Kasai, P. H. *J. Am. Chem. Soc.* **1984**, *106*, 3069–3075.
143. Kasai, P. H.; McLeod, D., Jr.; Watanabe, T. *J. Am. Chem. Soc.* **1980**, *102*, 179–190.
144. Howard, J. A.; Joly, H. A.; Mile, B. *J. Phys. Chem.* **1990**, *94*, 6627–6631.
145. Kasai, P. H. *J. Phys. Chem.* **1988**, *92*, 2161–2165.
146. Hagen, J.; Socaciu, L. D.; Elijazyfer, M.; Heiz, U.; Bernhard, T. M.; Woeste, L. *Phys. Chem. Chem. Phys.* **2002**, *4*, 1707–1709.
147. Wallace, W. T.; Whetten, R. L. *J. Am. Chem. Soc.* **2002**, *124*, 7499–7505.
148. Yang, H.; Holloway, P. H. *Appl. Phys. Lett.* **2003**, *82*, 1965–1967.
149. Hayashi, T.; Tanaka, K.; Haruta, M. *J. Catal.* **1998**, *178*, 566–575.
150. Olson, R. M.; Varganov, S.; Gordon, M. S.; Metiu, H.; Chretien, S.; Piecuch, P.; et al. *J. Am. Chem. Soc.* **2005**, *127*, 1049–1052.
151. Yurkov, G. Yu.; Kozinkin, A. V.; Nedoseykina, T. I.; Shuvaev, A. T.; Vlasenko, V. G.; Gubin, S. P.; et al. *Inorg. Mater.* **2001**, *37*, 997–1001.
152. Vorontsov, P. S.; Grigor'ev, E. I.; Zav'yalov, S. A.; Zav'yalova, P. M.; Rostovshikova, T. N.; Zagorskaya, O. V. *Khim. Fizika* **2002**, *21*, 45–49.
153. Fedushak, T. A.; Il'in, A. P.; Pisareva, S. I.; Shiyan, L. N. *Physical Chemistry of Ultra-Dispersed Systems*; Book of Abstracts, Part 2 (Russ.); UrO RAN: Ekaterinburg, 2001; pp 225–230.
154. Fedushak, T. A.; Il'in, A. P. *Appl. Chem. J. (Russ.)* **2002**, *75*, 359–363.
155. Jun, Y.; Lee, S. -M.; Kang, N. J.; Cheon, J. *J. Am. Chem. Soc.* **2001**, *123*, 5150–5151.
156. Yang, Q.; Tang, K.; Wang, C.; Qian, Y.; Zhang, S. *J. Phys. Chem. B* **2002**, *106*, 9227–9230.
157. Chen, M.; Xie, Y.; Xiong, Y.; Zhang, S.; Qian, Y.; Liu, X. *J. Mater. Chem.* **2002**, *12*, 748–753.
158. Gao, F.; Lu, Q.; Xie, S.; Zhao, D. *Adv. Mater.* **2002**, *14*, 1537–1540.
159. Yan, H.; He, R.; Pham, J.; Yang, P. *Adv. Mater.* **2003**, *15*, 402–405.
160. Zhu, Y. -C.; Bando, Y.; Xue, D. -F.; Goldberg, D. *J. Am. Chem. Soc.* **2003**, *125*, 16196–16197.
161. Jiang, Y.; Meng, X. -M.; Yiu, W. -C.; Liu, J.; Ding, J. -X.; Lee, C. -S.; et al. *J. Phys. Chem. B* **2004**, *108*, 2784–2787.
162. Zhang, X. T.; Ip, K. M.; Liu, Z.; Leung, Y. P.; Li, Q.; Hark, S. K. *Appl. Phys. Lett.* **2004**, *84*, 2641–2643.
163. Liu, B.; Zeng, H. C. *J. Am. Chem. Soc.* **2003**, *125*, 4430–4431.
164. Feng, X.; Feng, L.; Jin, M.; Zhai, J.; Jiang, L.; Zhu, D. *J. Am. Chem. Soc.* **2004**, *126*, 62–63.
165. Liu, B.; Zeng, H. C. *Langmuir* **2004**, *20*, 4196–4204.
166. Peterson, R. B.; Fields, C. L.; Gregg, B. A. *Langmuir* **2004**, *20*, 5114–5118.
167. Yu, H.; Zhang, Z.; Han, M.; Hao, X.; Zhu, F. *J. Am. Chem. Soc.* **2005**, *127*, 2378–2379.
168. Hsu, J. W. P.; Tian, Z. R.; Simmons, N. C.; Matzke, C. M.; Voigt, J. A.; Liu, J. *Nano Lett.* **2005**, *5*, 83–86.
169. Murphy, C. J.; Jana, N. R. *Adv. Mater.* **2002**, *14*, 80–82.
170. Kim, F.; Song, J. H.; Yang, P. *J. Am. Chem. Soc.* **2002**, *124*, 14316–14317.

171. Teng, X.; Black, D.; Watkins, N. J.; Gao, Y.; Yang, H. *Nano Lett.* **2003**, *3,* 261–264.
172. Chen, S.; Yang, Y. *J. Am. Chem. Soc.* **2002,** *124,* 5280–5281.
173. Kamat, P. V. *J. Phys. Chem. B* **2002**, *106,* 7729–7744.
174. Maier, S. A.; Brongersma, M. L.; Kik, P. G.; Meltzer, S.; Requicha, A. A. G.; Atwater, H. A. *Adv. Mater.* **2001,** *13,* 1501–1505.
175. Tkachenko, A. G.; Xie, H.; Coleman, D.; Glomm, W.; Ryan, J.; Anderson, M. F.; et al. *J. Am. Chem. Soc.* **2003,** *125,* 4700–4701.
176. Roll, D.; Malicka, J.; Gryczynski, I.; Lakowicz, J. R. *Anal. Chem.* **2003,** *75,* 3440–3445.
177. Nath, N.; Chilkoti, A. *Anal. Chem.* **2002,** *74,* 504–509.
178. Thanh, N. T. K.; Rosenzweig, Z. *Anal. Chem.* **2002,** *74,* 1624–1828.
179. Kim, Y.; Johnson, R. C.; Hupp, J. T. *Nano Lett.* **2001,** *1,* 165–167.
180. Dick, L. A.; McFarland, A. D.; Haynes, C. L.; Van Duyune, R. P. *J. Phys. Chem. B* **2002,** *106,* 853–860.
181. Prodan, E.; Nordlander, P. *Nano Lett.* **2003,** *3,* 543–547.
182. Jackson, J. B.; Halas, N. J. *J. Phys. Chem. B* **2001,** *105,* 2743–2746.
183. Pol, V. G.; Gedanken, A.; Calderon-Moreno, J. *Chem. Mater.* **2003,** *15,* 1111–1118.
184. Caruso, F.; Spasova, M.; Salgueirino-Maceira, V.; Liz-Marzan, L. M. *Adv. Mater.* **2001,** *13,* 1090–1094.
185. Kamata, K.; Lu, Y.; Xia, Y. *J. Am. Chem. Soc.* **2003,** *125,* 2384–2385.
186. Sun, Y.; Mayers, B.; Xia, Y. *Adv. Mater.* **2003,** *15,* 641–646.
187. Sun, Y.; Xia, Y. *Nano Lett.* **2003,** *3,* 1569–1572.
188. Sun, Y.; Xia, Y. *J. Am. Chem. Soc.* **2004,** *126,* 3892–3901.
189. Sun, Y.; Xia, Y. *Adv. Mater.* **2004,** *16,* 264–268.
190. Daniel, M. -C.; Astruc, D. *Chem. Rev.* **2004,** *104,* 293–346.
191. Sun, Y.; Wiley, B.; Li, Z. -Y.; Xia, Y. *J. Am. Chem. Soc.* **2004,** *126,* 9399–9406.
192. Taylor, K. J.; Pettiette, C. L.; Craycraft, M. J.; Chesnovsky, O.; Smalley, R. E. *Chem. Phys. Lett.* **1988,** *152,* 347–352.
193. Gantefoer, G.; Eberhardt, W. *Chem. Phys. Lett.* **1994,** *217,* 600–604.
194. Dolgounitcheva, O.; Zakrewski, V. G.; Oritz, J. V. *J. Chem. Phys.* **1999,** *111,* 10762–10765.
195. Rao, B. K.; Jena, P. *J. Chem. Phys.* **1999,** *111,* 1890–1904.
196. Li, X.; Wang, L. -S.; Boldyrev, A. I.; Simons, J. *J. Am. Chem. Soc.* **1999,** *121,* 6033–6038.
197. Wang, L. S.; Boldyrev, A. I.; Li, X.; Simons, J. *J. Am. Chem. Soc.* **2000,** *122,* 7681–7687.
198. Geske, G. D.; Boldyrev, A. I.; Li, X.; Wang, L. S. *J. Chem. Phys.* **2000,** *113,* 5130–5133.
199. Jarrold, M. F.; Bower, J. E. *J. Chem. Phys.* **1987,** *87,* 5728–5738.
200. Jarrold, M. F.; Bower, J. E. *J. Am. Chem. Soc.* **1988,** *110,* 70–78.
201. Fuke, K.; Nonose, S.; Kikuchi, N.; Kaya, K. *Chem. Phys. Lett.* **1988,** *147,* 479–483.
202. Rohlfing, E. A.; Cox, D. M.; Petkovic-Luton, R.; Kaldor, A. *J. Phys. Chem.* **1984,** *88,* 6227–6231.
203. Tacke, M.; Schnoeckel, H. *Inorg. Chem.* **1989,** *28,* 2895–2896.
204. Dohmeier, C.; Mattes, R.; Schnoeckel, H. *J. Chem. Soc. Chem. Commun.* **1990,** *5,* 358–359.
205. Tacke, M.; Kreinkamp, H.; Plaggenborg, L.; Schnoeckel, H. *Z. Anorg. Allg. Chem.* **1991,** *604,* 35–40.
206. Van Zee, R. J.; Li, S., Jr.; Weltner, W. *J. Chem. Phys.* **1993,** *98,* 4335–4338.
207. Yamada, I. *Appl. Surf. Sci.* **1989,** *43,* 23.
208. Cox, D. M.; Trevor, D. J.; Whetten, R. L.; Kaldor, A. *J. Phys. Chem.* **1988,** *92,* 421–429.
209. Saillard, J. -Y.; Hoffmann, R. *J. Am. Chem. Soc.* **1984,** *106,* 2006–2026.
210. Kuskov, M. L. Synthesis and Study of Physical–Chemical Properties of Ultradispersed Particles Al, Abstract of PhD thesis, Moscow (Russ.), 2000, p 23.
211. Gen, M. Ya.; Ziskin, M. C.; Petrov, Yu. N. *Doklady AN SSSR* **1959,** *127,* 366.

212. Gen, M. Ya.; Miller, A. V. *Surf. Phys. Chem. Mech. (Russ.)* **1983**, *2*, 150–154.
213. Zhigach, A. N.; Leypunskii, I. O.; Kuskov, M. L.; Verzhbitskaya, T. M. *Physical Chemistry of Ultra-dispersed Systems*; Book of Abstracts Part 2. (Russ.); UrO RAN: Ekaterinburg, 2001, pp 39–47.
214. Hartmann, A.; Weil, K. G. *Z. Phys. D.: At. Mol. Clusters* **1989,** *12*, 11.
215. Hartmann, A.; Weil, K. G. *High Temp. Sci.* **1990,** *27*, 31.
216. Hartmann, A.; Poth, L.; Weil, K. G. *High Temp. Sci.* **1991,** *31*, 121.
217. Rinnen, K. -D.; Kolenbrander, K. D.; De Santolo, A. M.; Mandich, M. L. *J. Chem. Phys.* **1992,** *96*, 4088–4101.
218. Bai, Y. -J.; Liu, Z. -G.; Xu, X. -G.; Gui, D. -L.; Hao, X. -P.; Feng, X.; et al. *J. Crystal Growth* **2002,** *241*, 189–192.
219. Xiao, J.; Xie, Y.; Tang, R.; Luo, W. *Inorg. Chem.* **2003,** *42*, 107–111.
220. Takahashi, N.; Niwa, A.; Sugiura, H.; Nakamura, T. *Chem. Commun.* **2003,** *3*, 318–319.
221. Liang, C. H.; Chen, L. C.; Hwang, J. S.; Chen, K. H.; Hung, Y. T.; Chen, Y. F. *Appl. Phys. Lett.* **2002,** *81*, 22–24.
222. Zhang, J.; Zhang, L.; Peng, X.; Wang, X. *J. Mater. Chem.* **2002,** *12*, 802–804.
223. Sardar, K.; Deepak, F. L.; Govindaraj, A.; Seikh, M. M.; Rao, C. N. R. *Small* **2005,** *1*, 91–94.

Chapter 7

Assemblies Involving Nanoparticles

Chapter Outline
7.1 Assemblies Involving
    Nanoparticles   209
7.2 Forces between
    Nanoparticles   215
    7.2.1 Attraction Forces   215
7.2.2 Theory of NP Interaction
      Potentials   215
7.2.3 Nanocrystal
      Superlattices   216

7.1 ASSEMBLIES INVOLVING NANOPARTICLES

In different versions of the nanoparticle synthesis that involve the condensation procedure, the process starts from the formation of nanoparticles (NP) from individual metal atoms, which can be considered as self-assembling or self-organization of atoms to give their assemblies. Self-organization is determined as a series of concurrent multicomponent processes of self-association. This is a complex spontaneous process resulting in the appearance of an order with respect to space and/or time,[1] which, in principle, can give rise to different size effects. This also includes the structural and dynamic kinds of order in equilibrium and nonequilibrium structures.

As seen in metal NP, a common practice is to consider the assemblies of both particles themselves and stabilizing protective layers. Attention is focused on the effect of the chemical nature of stabilized compounds on the self-organization processes. The analysis of studies carried out before 2000 can be found elsewhere.[2] Preparation of thiol-covered metal particles and their self-organization into one-, two-, and three-dimensional superlattices were surveyed.

A sufficiently general approach to controlling self-organization of assemblies of metal NP and composite materials based on carbon nanotubes and metals was proposed.[3–5] The central idea of the authors was to employ molecular films based on multidentate thioethers such as $Me_{4-n}Si(CH_2SMe)_n$ ($n = 2, 3, 4$) as the mediators and tetraalkylammonium bromide as the template.[5] For gold particles measuring from 4.8 to 6.4 nm and covered with a tetraalkyl bromide monolayer, it was shown that the mediator provides coordination properties of the resulting assemblies, while the template controls its geometry. By varying the mediator–template ratio, it is possible to regulate the size and shape of particles in the synthesized assemblies.

Many branches of nanotechnology employ materials based on NP. These applications realize the ability of NP to form assemblies of controlled size and shape and with special interparticle interactions that provide a possibility of utilizing unique nanosize properties. Much attention was attracted to the problems of synthesizing NP of different sizes and shapes and studying the peculiarities of their organization and self-assembling.[6–10]

Various applications of metal and semiconductor NP are largely determined by the chemical properties of their surfaces. For 2-nm particles, more than half of the atoms are localized on the surface and can influence the nanocrystal behavior. Uncompensated surface atoms can act as a sort of trap, e.g. for photogenerated charge carriers, thus reducing the efficiency of semiconductor emission.

By changing the nature of ligands that interact with a nanoparticle, one can govern its synthesis, stabilization, and chemical reactivity. Surface ligands prevent aggregation of individual NP. At the same time, they can facilitate dispersion of nanocrystals in various solvents, which is of special importance for aqueous solutions in view of biological labeling applications. Surface ligands containing appropriate functional groups can serve as "bridging" units for coupling of molecules or macromolecules to a nanocrystal, that leads to the development of new hybrid materials. In many cases, it was shown that thiols containing two thiol groups or combinations of several ligands can determine the size and behavior of NP.[11,12] At the same time, for CdSe particles, it was shown that thiols that coat a particle are easily oxidized to yield disulfides, which result in deposition of crystals.[13]

Processes of organization and self-organization of NP were studied most extensively considering the examples of silver and gold particles. Monodispersed silver nanocrystals were stabilized by the chemisorption of dodecane thiol on the receptor sites of the crystal surface.[14] The kinetics of aggregation and the structure of the formed ensembles depended on the number of receptor sites on the nanocrystal surface.

The synthesis of alternate positively charged gold particles and negatively charged silver particles and their self-organization in the presence of 4-mercaptoaniline and 4-mercaptobenzoic acid on a glass surface was described.[15] The formation of one-dimensional layers of gold particles on silicon substrates was also considered.[16] This process was performed using an AFM for the surface treatment and the chemical deposition of gold particles.

The effect of gold particles on the self-organization of dodecane thiol molecules was described.[17] Particles measuring from 1.5 to 5.2 nm were utilized, which corresponded to the presence of about 1100–4800 gold atoms and from 53 to 520 alkane thiol chains. The properties of thiols adsorbed on NP were studied using NMR, IR Fourier spectroscopy, differential scanning calorimetry, mass spectrometry, and thermal analysis techniques. The conformation of alkyl chains in thiols was more ordered on coarse particles. A changeover in the packing type was observed starting from a size of 4.4 nm. On larger particles, a two-dimensional layer was formed, whereas for smaller particles the elements of three-dimensional packing were observed.

The self-organization of 4.6 nm gold particles resulted in the formation of particle aggregates with a diameter of ca. 30 nm.[18] The original particles were formed by the reduction of $HAuCl_4$ with $NaBH_4$ in the presence of $HSCH_2COONa$. The removal of excessive metal ions from the solution by dialysis initiated the formation of coarser aggregates. The electron spectra measured before the dialysis revealed the absence of the plasmon peak, typical of gold particles.

The strategy of production of materials based on a "building-blocks and binders" principle was put forward.[19] The adhesion between a polymer and gold particles, which were bound with thymine as the protector, was achieved because of the formation of hydrogen bonds. In the course of self-association, the size and morphology of aggregates were temperature-controlled. The original 2 nm gold NP formed spherical aggregates measuring 97 ± 17 nm at 23 °C and 0.5–1 μm at −20 °C, which in turn consisted of finer aggregates as the individual subunits. At 10 °C, spherical NP formed an extended chain 50 nm long. Such self-association processes were studied using small-angle X-ray scattering and TEM techniques. As assumed, the formation of nonspherical aggregates at 10 °C is an intermediate step in the formation of coarser ensembles at −20 °C; hence, these aggregates can be used as the precursors of nanosize associates of different shapes and sizes.

The self-association of rod-like gold particles, 12 nm in diameter and 50–60 nm long, was studied by high-resolution electron microscopy.[20] By selecting the concentration of these particles, their size distribution, the conditions of solvent evaporation, and the surfactant ionic strength and concentration, it was possible to obtain one-, two-, and three-dimensional structures by electrolysis. A mixture of hexadecyltrimethylammonium bromide and tetraoctylammonium bromide was used as the water-soluble electrolyte, where the second surfactant defined the formation of cylindrical particles, and the ratio of the two surfactants determined the diameter-to-length ratio of the resulting gold nanorods. The electrolysis time was 4.5 min. Gold nanorods were formed on a copper wire net partly immersed in solution, exactly along the line of its contact with solution.[20] In the opinion of the authors of this study, the following two factors were important for the further self-assembling into rods: the water evaporation-generated convective transfer of particles from solution to the thin film and the interaction between particles within the film, which defined the formation of various structures. A possible reason for the parallel arrangement of rod-like particles was associated with the capillary forces operating between the rods arranged in parallel to one another, although the authors[20] stressed that a satisfactory explanation of the observed anisotropy in assemblies requires a more detailed theoretical analysis.

The self-assembling of particles gives rise to the appearance of order not only of the translation kind but also with respect to orientation. The latter kind of ordering is typical of particles with well-defined shapes.[21] High-resolution electron microscopy allowed the structural changes in nanosize gold rods, which were generated by femto- and nanosecond laser pulses, to be observed.[22] The pulse energy was insufficient for melting of the gold rods but quite sufficient to initiate the deformation processes that could affect their shape: the nanorods were observed to transform into nanospheres. The results of the statistical analysis of the distribution of nanorods over sizes and shapes, which was performed before and after treating them with a laser, can be found in Ref. 23.

In connection with the development of new electronic nanodevices, the attention of scientists was attracted to the procedures of synthesizing NP protected by monolayers. New procedures were developed for synthesizing such layers on gold particles, which involve using solutions of hexanethiols and hexanedithiols in organic[24] and aqueous[25] media.

Stable gold films, which consisted of particles measuring nearly 4 nm, were obtained by the reaction of low-molecular-weight ($M \leq 4000$ u) polymers such as polyethylene imine and poly-L-lysine with gold particles covered with carboxylic acids.[26] The morphology of films was found to depend on the nature of the acid that covered the gold particles. The use of mercaptododecanoic and mercaptosuccinic acids resulted in the formation of more ordered films.

Thin porous gold films on glass substrates were fabricated by employing colloid crystals as the templates. The films were studied using STM and served as a model substance in the Raman scattering measurements.[27,28]

Gold particles enabled the study of the microstructure, wettability, and thermal stability of self-assembled monolayers of partly fluorinated alkane thiols such as $F(CF_2)_{10}(CH_2)_nSH$ ($n = 2, 6, 11, 17,$ and 33).[29] The introduction of alkane fluoride chains as the terminal groups into the alkane thiols enhanced the stability of self-assembled monolayers. A monolayer of nonfluorinated alkane thiols on gold lost its ordering at a temperature about 100 °C. Fluorinated compounds with the number of methylene units in a chain $n = 11, 17,$ and 33 formed well-ordered monolayers, whereas for smaller numbers of units, the degree of ordering decreased. The wettability of monolayers also depended on the number of methylene units. The thermal stability of self-assembled monolayers increased with an increase in n; thus the monolayer films with $n=33$ were stable in air at 150 °C for 1 h.

The effect of the substrate nature on the morphology of two- and three-dimensional superlattices formed by dodecane thiol-covered silver sulfide nanocrystals measuring 5.8 nm was studied. As substrates, oriented pyrolytic graphite and molybdenum disulfide were used.[30] The self-assembling process was observed by the SEM and AFM techniques. It was shown that self-organization into superlattices depends on the substrate nature, being determined by the particle–particle and particle–substrate van der Waals interactions.

Self-organized monolayers formed by 4′-hydroxy-4-mercaptobiphenyl, 4-(4-mercaptophenyl) pyridine, and their mixtures with 4′-methyl-4-mercaptobiphenyl were deposited on the gold surface and used as the templates for growing glycine crystals.[31] The morphology of the resulting glycine crystals depended on the properties of surfaces used, being determined by the hydrogen bonds between the glycine molecules in the growing layer and the functional groups on the surface of self-assembling monolayers. Moreover, it was assumed that the hydrogen bond-defined interaction between CO_2^- and NH_3^+ groups of glycine with the hydroxyl groups on the surface of a monolayer built of $HOC_6H_4C_6H_4SH$ molecules is stronger than the interaction between the NH_3^+ group of glycine and the nitrogen atom of pyridine on the surface of a monolayer built of $NC_5H_4C_6H_4SH$ molecules.

As a rule, the self-assembling of monolayers formed by various substances on the surfaces of nanosize metal particles and films was studied using X-ray photoelectron spectroscopy, ellipsometry, SEM, and diffraction techniques. Electron microscopy was successfully employed for studying self-assembling of monolayers formed by metalloporphyrins and metallophthallocyanins on ultrathin gold films.[32] Gold films, 1.3–10 nm thick, on mica were prepared by means of a high-vacuum evaporator at low temperatures. The deposition rate was varied in a range of 0.2–0.4 Å/s. The deposited films were annealed at 250 °C for 2.5–4 h. The resulting films were examined using scanning AFM, electron spectroscopy, and XRD techniques. The appearance of separate structurized islets, the size and optical properties of which were controlled by the conditions of evaporation and subsequent annealing was observed on the mica surface. The electron spectra of these films demonstrated a peak of gold surface plasmon at a wavelength shifting from 606 to 530 nm with a decrease in the film thickness from 10 to 1.3 nm. The results of the light absorption kinetics were used for semiquantitative estimation of the nature of the chemical bonding and the chemical and structural properties of monolayers. As shown, the absorption in the vicinity of the gold plasmon band, which was caused by the assembling of molecules, can be used for monitoring molecules containing no chromophore groups.

Not only AOT salts but also their derivatives with other metals were used as templates for self-assembling. By correctly interpret the effect of templates on the self-assembling of metal particles, the knowledge of phase diagrams for these systems is necessary.[33] Considering the example of a system Cu(AOT)$_2$–isooctane–water used as the template, it was shown that the nanocrystalline form of particles can undergo profound changes under the effect of small concentrations of various salts.[34]

Currently, when preparing and stabilizing metal NP, attention was focused on the development of methods that would produce particles of definite controllable sizes. The majority of studies in this direction were apparently devoted to the use of thiolates and the synthesis of gold NP. The use of alkane thiol monolayers made it possible to synthesize gold particles that are stable both in solution and in the dry state. This, in turn, extended the possibilities of chemical transformations by introducing new functional groups.

When alkane thiols are used, the formation of gold NP and the appearance of a protective layer proceed in two stages:[35]

$$AuCl_{4(toluene)}^- + RSH \rightarrow (Au^ISR)_n(polymer)$$
$$(Au^ISR)_n + BH_4^- \rightarrow Au_x(SR)_y$$

These reactions combined the processes of nucleation and passivation. On average, an increase in the RSH–Au molar ratio and the acceleration of the reducing agent addition favored the formation of particles with smaller metal cores. A brief interruption of the reaction caused the formation of thicker coatings and very small cores (<2 nm).

The dynamics of gold core growth in the presence of multilayer protective coatings was studied.[35] The use of TEM made it possible to observe the slow changes in the sizes of gold cores, which followed the active initial stage of this reaction. Thus, when hexane thiol was used as the stabilizing coating, the cores built of gold atoms grew up to 3 nm during the first 60 h of the reaction, after which they were stabilized. An insight into the mechanisms of growth and annealing of NP would allow refining the synthetic methods with the aim of synthesizing finer particles with narrower size distributions.

Organic solvents are preferred for preparation of NP. They perform stabilizing functions. Such solvents or surfactants play the key role in the synthesis of NP. They are bound to the surface of growing nanocrystals via polar groups, form complexes with species in solutions, and control their chemical reactivity and diffusion to the surface of a growing particle. All the mentioned processes depend on the temperature and on quantities such as surface energy of a nanocrystal, concentration of free particles in solution and their sizes, and the surface-to-volume ratio of a particle.[36]

The formation of NP from metal atoms and their subsequent self-assembling into a functional system play a decisive role in the fabrication of chemical nanoreactors. In actual practice, the assembling of nanosystems is controlled by the interplay of the aggregation and fragmentation processes, which in the liquid phase are additionally complicated by the presence of stabilizing ligands.

Studying kinetics and thermodynamics of formation and subsequent self-assembling of NP is the most challenging direction of nanochemistry.

7.2 FORCES BETWEEN NANOPARTICLES

7.2.1 Attraction Forces

Once NP are made, they almost always have a strong attraction for one another. If this attraction is strong enough, and if the lattice energy of the particles in question is low enough, two NP will hit each other and coalesce to form one large particle (like two water droplets coming together to form one).

Why are NP attracted to each other? In the absence of like charges on the two particles, Van der Waals interactions and Hammaker constants are important. If the NP have surface ligands on them, the two particles may attract each other due to the ligands. For example, if long-chain alkyl groups are present, the chains from the two particles could intermingle, especially if the solvent medium they are in is hydrophilic (e.g. water).

Reversible temperature and solvent dependent solubility of uniform NP draws strong analogy to the phase behavior of molecular solutions, and it is reasonable to ask the following: does the classical nucleation theory apply to NP nucleation, or is the pathway more complex? Does a two-step model (proposed for protein crystallization) where a stable cluster forms as a dense liquid, and then the particles in the cluster reorganize into an ordered structure, apply to NP? Is the shape of the fluctuating pre-nucleating clusters spherical, as assumed in classical nucleation theory, or otherwise? Further studies of NP solution nucleation to the solid phase would greatly aid an understanding of this science and enable control of self-assembly of NPs inter superlattices, ramified aggregates, gels, or films on surfaces. To this end, a combined theoretical and experimental approach will be needed.

The NP cluster growth process is determined by the NP pair interaction potential, while the solubility is related to the NP–NP adhesion surface energy λ_{ad}. The long-range behavior is determined approximately by the van der Waals interaction with effective Hamaker constant A, while λ_{ad} is the NP–NP interaction potential at equilibrium-separation distance in the condensed NP phase, i.e. the minimum value of $V_{(x)}$.

7.2.2 Theory of NP Interaction Potentials

A phenomenological model for the interactions between ligated gold NPs was used to study self-assembly and superlattice formation in Brownian Dynamics simulations.[37,38] Further theoretical studies need to be directed in accurately determining *effective* NP–NP pair potentials by placing two NP particles at some fixed distance from each other and carrying out a Monte Carlo simulation of the ligand monomers (*i.e.* $-CH_2$ *units as a whole*) while including both bending and dihedral potentials among monomers. From Monte Carlo simulations, ligand configurations, and the NP positions, *effective* NP–NP pair potential can be determined. Systematic variation in the NP–NP separation would provide a numerical table for the NP–NP interaction for any separation. The parameters in the MC simulations are changed so that the model NP–NP potential is consistent with the effective Hamaker constant A, derived from the AFM colloidal probe experiments. Next, Brownian dynamics studies with the refined pair potential yield kinetics and morphology of cluster growth, which could be compared against light-scattering results. With such a dynamic relationship between theory and experiments, the most important questions about NP supercluster nucleation could be answered.[39,40]

7.2.3 Nanocrystal Superlattices

Nanocrystals that are soluble in a chosen solvent, and are monodisperse, can be considered as very large molecules.[40] This was shown by measurements of solubility curves of a thiol-ligated gold nanoparticle (dodecanethiol with 4.9 nm gold). Ligand length is 1.7 nm, and dynamic light scattering measurements showed a hydrodynamic diameter of 8.4 ± 1.0 nm, which is in good agreement with the TEM measurements (4.9 + 1.7 + 1.7 nm = 8.3 nm). Temperature induced quenches from a chosen solvent system (t-butyltoluene plus 2-butanone) led to a two-phase regime where superclusters of NP of several hundred nanometers were observed by dynamic light scattering with a red shift in the plasmon absorption band. Figure 7.1 shows the size of the superclusters with deeper temperature quenches (down from 60 °C). This is a classic result known for both ionic and molecular solutions, that is, deeper quenches yield smaller molecular clusters. Thus, classical nucleation theory can be used to predict the size of the nanoparticle clusters as a function of quench depth with a very small, but reasonable, surface tension (adhesion strength) of the nanoparticle solid phase. Thus, it was concluded that these huge species such as $Au_{4000}(HSR)_{350}$ behave as molecules in solution.

The forces that compel ligated NP to nucleate and grow into superlattices have been discussed in detail.[41] If certain conditions of concentration, Hammaker constant, particle diameter, center interparticle separation, ligand chain length, and area of head group on the particle surface are met, a nanoparticle solution will stay a solution indefinitely. However, if concentration increases to supersaturation, nucleation and self-assembly begins and, over time, this crystallization of nanocrystals can lead to perfectly ordered nanocrystal superlattices. (Figure 7.2)[42] which shows a superlattice with face-centered-cubic solid-state structure. Other examples have shown hexagonal-close-packed structures. Numerous variations are possible depending on the nanocrystal chemical makeup, ligand head group, and length, solvent, and conditions of crystallization.[41,42]

Chapter | 7 Assemblies Involving Nanoparticles

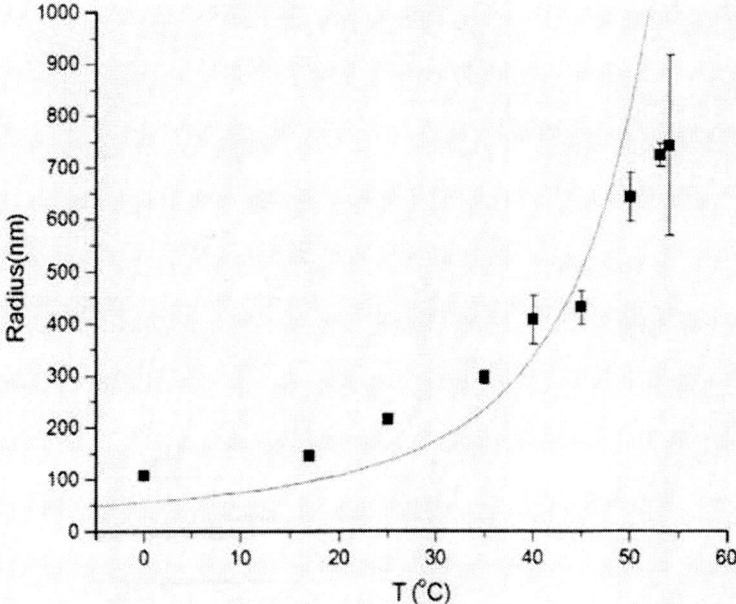

FIGURE 7.1 The sizes of superclusters formed with different quench temperatures from 60 °C. The line is the best fit to the nucleation theory described in the text with a single-fit variable and the surface tension of the solid phase with a value of 0.042 erg/cm^2.[40]

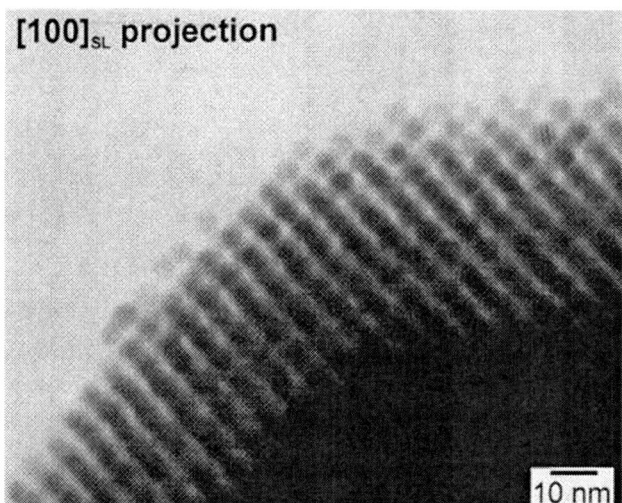

FIGURE 7.2 TEM micrograph of a nanocrystal superlattice composed of AU NP obtained by the inverse micelle method viewed along the $(100)_{th}$ direction.[42]

REFERENCES

1. Lehn, J. -M. *Supramolecular Chemistry. Concepts and Perspectives*; Wiley: Weinheim, 1995.
2. Roldugin, V. I. *Russ. Chem. Rev.* **2000,** *69,* 821–844.
3. Maye, M. M.; Luo, J.; Lim, I. -I.; Han, L.; Kariuki, N. N.; Rabinovich, D.; et al. *J. Am. Chem. Soc.* **2003,** *125,* 9906–9907.
4. Han, L.; Wu, W.; Kirk, F. L.; Luo, J.; Maye, M. M.; Kariuki, N. N., et al. *Langmuir* **2004,** *20,* 6019–6025.
5. Maye, M. M.; Lim, I. -I. S.; Luo, J.; Rab, Z.; Rabinovich, D.; Liu, T.; et al. *J. Am. Chem. Soc.* **2005,** *127,* 1519–1529.
6. Puntes, V. F.; Zanchet, D.; Erdonmez, C. K.; Alivisatos, A. P. *J. Am. Chem. Soc.* **2002,** *124,* 12874–12880.
7. Wei, Z.; Mieszawska, A. J.; Zamborini, F. P. *Langmuir* **2004,** *20,* 4322–4326.
8. Milliron, D. J.; Hughes, S. M.; Cui, Y.; Manna, L.; Li, J.; Wang, L. -W.; et al. *Nature* **2004,** *430,* 190–195.
9. Kim, F.; Connor, S.; Song, H.; Kuyukendall, T.; Yang, P. *Angew. Chem. Int. Ed.* **2004,** *43,* 3673–3677.
10. Thomas, K. G.; Barazzouk, S.; Ipe, B. I.; Joseph, S. T. S.; Kamat, P. V. *J. Phys. Chem. B* **2004,** *108,* 13066–13068.
11. Pathak, S.; Choi, S. -K.; Arnheim, N.; Thompson, M. E. *J. Am. Chem. Soc.* **2001,** *123,* 4103–4104.
12. Wuister, S. F.; Swart, I.; van Driel, F.; Hickey, S. G.; de Mello Donega, C. *Nano Lett.* **2003,** *3,* 503–507.
13. Aldana, J.; Wang, Y. A.; Peng, X. *J. Am. Chem. Soc.* **2001,** *123,* 8844–8850.
14. Fullam, S.; Rao, S. N.; Fitzmaurice, D. *J. Phys. Chem. B* **2000,** *104,* 6164–6173.
15. Kumar, A.; Mandale, A. B.; Sastry, M. *Langmuir* **2000,** *16,* 6921–6926.
16. Zheng, J.; Zhu, Z.; Chen, H.; Liu, Z. *Langmuir* **2000,** *16,* 4409–4412.
17. Hostetler, M. J.; Wingate, J. E.; Zhong, C. -J.; Harris, J. E.; Vachet, R. W.; Clark, M. R.; et al. *Langmuir* **1998,** *14,* 17–30.
18. Adachi, E. *Langmuir* **2000,** *16,* 6460–6463.
19. Boal, A. K.; Ilhan, F.; DeRouchey, J. E.; Thurn-Albrecht, T.; Russel, P.; Rotello, V. M. *Nature (London)* **2000,** *404,* 746–748.
20. Nikoobakht, B.; Wang, Z. L.; El-Sayed, M. A. *J. Phys. Chem. B* **2000,** *104,* 8635–8640.
21. Wang, Y.; Ren, J.; Deng, K.; Gui, L.; Tang, Y. *Chem. Mater.* **2000,** *12,* 1622–1627.
22. Link, S.; Wang, Z. L.; El-Sayed, M. A. *J. Phys. Chem. B* **2000,** *104,* 7867–7870.
23. Link, S.; Burda, C.; Nikoobakht, B.; El-Sayed, M. A. *J. Phys. Chem. B* **2000,** *104,* 6152–6163.
24. Chen, S. *J. Phys. Chem. B* **2000,** *104,* 663–667.
25. Chen, S. *J. Am. Chem. Soc.* **2000,** *122,* 7420–7421.
26. Maya, L.; Muralidharan, G.; Thundat, T. G.; Kenik, E. A. *Langmuir* **2000,** *16,* 9151–9154.
27. Velev, O. D.; Kaler, E. W. *Adv. Mater.* **2000,** *12,* 531–534.
28. Tessier, P. M.; Velev, O. D.; Kalambur, A. T.; Rabolt, J. F.; Lenhoff, A. M.; Kaler, E. W. *J. Am. Chem. Soc.* **2000,** *122,* 9554–9555.
29. Fukushima, H.; Seki, S.; Nishikawa, T.; Takiguchi, H.; Tamada, K.; Abe, K.; et al. *J. Phys. Chem. B* **2000,** *104,* 7417–7423.
30. Motte, L.; Lacaze, E.; Maillard, M.; Pileni, M. P. *Langmuir* **2000,** *16,* 3803–3812.
31. Kang, J. F.; Zaccaro, J.; Ulman, A.; Myerson, A. *Langmuir* **2000,** *16,* 3791–3796.
32. Kalyzhny, G.; Vaskevich, A.; Ashkenasy, G.; Shanzer, A.; Rubinshtein, I. *J. Phys. Chem. B* **2000,** *104,* 8238–8248.

33. Andre, P.; Filankembo, A.; Lisiecki, I.; Petit, C.; Gulik-Krzywicki, T.; Ninham, B. W.; et al. *Adv. Mater.* **2000**, *12*, 119–123.
34. Filankembo, A.; Pileni, M. -P. *J. Phys. Chem. B* **2000**, *104*, 5865–5868.
35. Chen, S.; Templeton, A. C.; Murray, R. W. *Langmuir* **2000**, *16*, 3543–3548.
36. Roldugin, V. I. *Russ. Chem. Rev.* **2004**, *73*, 115–146.
37. Lopez-Acevedo, O.; Tsunoyama, T.; Tsukuda, T.; Hakkinen, H.; Aikens, C. M. *J. Am. Chem. Soc.* **2010**, *132*, 8210–8218.
38. Quian, H.; Eckenhoff, W. T.; Zhu, Y.; Pintauer, T.; Jin, R. *J. Am. Chem. Soc.* **2010**, *132*, 8280–8281.
39. Khan, S. J.; Pierce, F.; Sorensen, C. M.; Chakrabarti, A. *Langmuir* **2009**, *25*, 13861–13868.
40. Yan, H.; Cingarapu, S.; Klabunde, K. J.; Chakrabarti, A.; Sorensen, C. M. *Phys. Rev. Lett.* **2009**, *102*, 095501.
41. Prasad, B. L. V.; Sorensen, C. M.; Klabunde, K. J. *Chem. Soc. Rev.* **2008**, *37*, 1871–1883.
42. Stoeva, S. I.; Prasad, B. L. V.; Sitharaman, U.; Stoimenov, P.; Zaikovski, V.; Sorensen, C. M.; et al. *J. Phys. Chem. B* **2003**, *107*, 7441–7448. (invited paper for A. Henglein Special Issue).

Chapter 8

Group of Carbon

Chapter Outline

8.1 Fine Particles of Carbon and Silicon 221
8.2 Fullerenes 223
8.3 Carbon Nanotubes 225
  8.3.1 Filling of Tubes 226
  8.3.2 Grafting of Functional Groups. Tubes as Matrices 227
  8.3.3 Intercalation of Atoms and Molecules into Multiwalled Tubes 229
8.4 Graphene 230
8.5 Carbon Aerosol Gels/ Turbstratic Graphite/ Graphene 231

The role of carbon in nanochemistry has grown in importance since the discovery of fullerenes and nanotubes; thus we devote a separate chapter to the elements of this group. In addition to carbon materials, silicon particles will be considered.

8.1 FINE PARTICLES OF CARBON AND SILICON

An adduct of the reaction between a particle built of three carbon atoms and water was obtained at 10 K, using the matrix isolation method.[1] Under the effect of light, this adduct undergoes a reaction

$$C_3(H_2O) \xrightarrow{h\nu} HC\equiv C-C-OH$$

to yield hydroxyacetenylcarbene.

In matrices, the interaction of carbon vapors with carbon monoxide produced C_4O and C_6O molecules, which were in the triplet state.[2] They were studied by EPR technique; their identification involved labeling with carbon ^{13}C and oxygen ^{17}O. Presumably, these molecules had linear structures.

It was found that C_4O and C_6O molecules are formed only during photolysis, i.e. only excited C_3 molecules can react with CO and C_3O:

$$C_3 + h\nu \to C_3^* \xrightarrow{CO} \begin{array}{l} C_4O \\ C_3O \\ C_6O \end{array}$$

Note that the transition from the linear to the cyclic form occurs for carbon particles containing 9 or 10 atoms.[3]

Along with carbon clusters, silicon clusters were synthesized. Clusters Si_n^+ were shown to chemisorb ammonium at the rates varying by three orders of magnitude. Silicon particles with $n = 21, 25, 33, 39$, and 45 were poorly active, while an Si_{43} cluster turned out to have the highest activity.[4] It was assumed that some clusters form several structural isomers. The low activity of Si_{39} and Si_{45} particles was associated with their stable crystalline structures. Chemisorption of ammonium on cluster ions Si_n^+ ($n < 70$) was also studied.[5] At room temperature, all clusters reacted at a rate virtually approaching the collision rate. At 400 K, equilibrium was established and the energy of ammonium bonding was ca. 1 eV, which apparently suggests molecular adsorption. At elevated temperatures, dissociative adsorption occurred. The fact that at 700 K the probability of attachment of ammonium to cluster ions was lower by 2–4 orders of magnitude as compared with compact silicon deserves interest.

In the gas phase, reactions of silicon clusters Si_n^+ ($n = 10$–65) with water resulted in the formation of a series of $Si_n(D_2O)_m^+$ adducts.[6] Wide variations in the reactivities of silicon clusters containing less than 40 atoms were observed. Clusters Si_n^+ with $n = 11, 13, 14, 19$, and 23 exhibited low reactivity. Large clusters were also less active than compact silicon. These results made it possible to put forward a two-step mechanism, in which the fast initial formation of the adduct was followed by the slow regrouping to give a strongly bound particle.

Similar results were also obtained for the reaction of oxygen with silicon clusters Si_n^+, i.e. in this reaction too, the clusters exhibited a weaker activity as compared with compact silicon.[7] In the reaction with oxygen, small clusters yielded Si_{n-2}^+ and two additional molecules of silicon oxide. Large clusters formed adducts such as $Si_n^+ O_2^+$. A demarcation line between the two reaction pathways corresponded to clusters containing from 29 to 36 silicon atoms. It is likely that a two-step process occurs too, when the molecular chemisorption is accompanied by a slow dissociative chemisorption.

The low activity of clusters as compared with compact silicon is probably caused by the presence of more active sites in compact silicon. This means that the clusters retain a close-packed structure containing several isolated bonds, and such structures most resemble closed nuclei rather than the compact silicon surface. These peculiarities made it possible to liken silicon clusters to fullerenes. It is probable that only carbon and silicon, in contrast to the majority of elements, exhibit an unusual behavior that shows up in the lower activity of their clusters as compared with the surfaces of compact elements.

8.2 FULLERENES

Fullerenes, which were discovered in mid-1980s, attract much attention nowadays. A large number of books and reviews have been devoted to them.[8]

A fullerene containing 60 carbon atoms is considered as classical. It represents a spherical structure, in which six-membered rings are bound with one another by five-membered cycles. Without going into details of physico-chemical properties of fullerenes, we consider only certain examples of their chemical behavior.

At room temperatures, fullerenes C_{40}–C_{80} do not react with active molecules such as nitrogen oxide, oxygen, and sulfur oxide. Nonetheless, several new chemical reactions with fullerenes were discovered.

The simplest fullerene compound with hydrogen, $C_{60}H_2$, was obtained and characterized.[9] Halogen derivatives of fullerenes were synthesized. By direct addition of fluorine, the series of compounds $C_{60}F_x$ and $C_{70}F_y$, where $x \leq 48$ and $y \leq 56$, were obtained.[10] Fluorination of fullerenes has been analyzed.[11] Chlorination and bromination of fullerenes were also carried out.[12] Chlorination was performed in tubes heated up to 250 °C. As a rule, 24 chlorine atoms were added. At 400 °C, polychlorfullerenes were dechlorinated to original fullerenes.

Chlorine in a fullerene can be substituted. For example, the following reactions were carried out:

$$C_{60}Cl_n \xrightarrow{MeOH, KOH} C_{60}(OMe)_n$$

$$C_{60}Cl_n \xrightarrow{C_6H_6, AlCl_3} C_{60}(Ph)_n$$

Chemistry of fullerenes is extensive and diverse; hence we mention here only some aspects. The insertion of C_60 into a polymer as a compound with covalent bonds was accomplished.[13] This was the reaction with xylylene biradical. The resulting polymer was nonsoluble and had a net structure with a xylylene/C_{60} ratio equal to 3.4:1.

The highly selective synthesis of dimers of fullerenes C_{60} and C_{70} was accomplished in neutron beams.[14] The obtained compounds were analyzed using chromatography mass spectrometry.

Incorporation of metal particles into fullerenes was carried out. A special symbolics was proposed for particles such as M@C_n for metal atoms inside a fullerene and MC_n for external metal atoms.[15] The first synthesis of such substances was based on the laser evaporation of a mixture of lanthanum salts and graphite.[16] In high-temperature plasma, lanthanum ions were reduced to atoms and got incorporated into fullerene cages during their formation.

A particle Sc_3@C_{82} was studied; particularly, by using the EPR technique, it was shown that Sc_3 represents an isosceles triangle similar to that found in inert matrices.[17]

Several fullerene adducts with metals such as M_xC_{60} were synthesized.[18] The interest in such compounds was arisen by the fact that one of the first compounds K_xC_{60} exhibited superconduction. For K_3C_{60}, superconduction was observed at a temperature $T_c = 19.3$ K.[19] Films $Ca_xRb_yC_{60}$ exhibited superconduction at $T_c = 33$ K.[20] It was concluded that superconduction of such compounds is determined by the density of states in the Fermi level. Another interpretation of this phenomenon was based on experiments with Ca_xC_{60}, Sr_xC_{60}, K_6C_{60}, and Ca_3C_{60} and was concerned with the electron transfer from a metal to a fullerene.[21,22]

Fullerenes formed the basis for not only synthesizing superconductive compounds but also for fabricating substances that surpass diamonds with respect to their bulk modulus of elasticity and hardness.[23] Superhard materials were synthesized from fullerenes C_{60} and C_{70} at a pressure up to 13 GPa and a temperature of 1600 °C.

In addition to insertion of metals into internal cavities of fullerene C_{60} cages, noble gases and small molecules can be put inside fullerenes by using elevated temperatures (650 °C) and pressures (3000 atm).[24] Moreover, chemical methods were developed for opening different-sized windows in fullerenes.[25–27] The use of open fullerenes made it possible to determine not only the rate of ^{3}He escape from this fullerene[28,29] but also the activation barrier and the equilibrium constant of incorporation–extraction of helium. The activation energy of helium extraction from a fullerene modified by a chemical method 617 was found to be 22.8 Kcal/mol. In the temperature range 50–60 °C, the equilibrium constant weakly depended on the temperature, which pointed to equivalence of the barriers of ^{3}He entry and escape from the modified fullerene. Presumably, the activation energy should depend on the orifice size of the fullerene.

An $La_2@C_{78}$ compound was synthesized in a DC arc-discharge followed by extraction with 1,2,4-trichlorobenzene.[30]

For fullerenes with metal particles put inside their hollow carbon cages $M@C_{82}(M=Y, La, Ce)$, the chemical reactivities of anions and cations formed upon fullerene reduction and oxidation were compared.[31] $M@C_{82}$ particles were prepared by arc evaporation. Their oxidation and reduction were carried out in 1,2-dichlorobenzene. In the chemical-activity studies, 1,1,2,2-tetrakis(2,4,6-trimethylphenyl)-1,2-disilirane was used as the reagent. It was shown that positively charged fullerenes $[M@C_{82}]^-$ easily react with disilirane, whereas negatively charged $[M@C_{82}]^+$ do not enter into this reaction. The difference in reactivity was associated with the electrophilicity of metallofullerenes, which with regard to disilirane increases with oxidation and decreases with reduction, i.e. the chemical reactivity of $M@C_{82}$ can be controlled by ionization of the fullerene.

Three fullerenepyrrolidine derivatives were entrapped in the interlayer spacing of lamellar aluminum silicate materials.[32] It was shown that a neutral derivative can be incorporated between clay layers together with solvent, producing stable compositions. The incorporation of a charged fullerene was easier due to its water solubility. The presence of negatively charged molecules induced no pronounced changes in the electronic structure of fullerenes. New hybrid nanocompositions in which the properties of C_{60} deviated from those in crystals or solutions were synthesized. It was assumed that a sizable amount of charge is transferred between the host and guest.

8.3 CARBON NANOTUBES

The discovery of fullerenes made a substantial contribution to the development of nanochemistry of nonmetals. At present, the discovery of nanotubes is believed to mark a transition to real nanotechnologies.[33,34] Methods of nanotube synthesis, their structure, and physicochemical properties can be found.[35–38]

Single- and multiwalled coaxial nanotubes are formed as a result of rolling up strips of graphene sheets to form seamless cylinders. The inner diameter of carbon nanotubes can vary from 0.4 to several nanometers, and foreign substances can enter their inner spaces. Single-walled tubes contain a smaller amount of defects; their annealing at high temperatures in inert atmospheres allows obtaining defect-free tubes. The tube's structure type affects its chemical, electronic, and mechanical properties. Individual tubes aggregate to form different types of bundles, containing slots.

Without going into details of the carbon nanotube synthesis, we only comment on the dynamics of this process. The first method consisted in the arc-discharge evaporation of graphite in an inert gas flow. This method is still actively used. It allowed single-walled carbon nanotubes with a diameter of 0.79 nm to be obtained in the presence of CeO_2 and nanosize nickel.[39] The arc method gave way to the evaporation of a graphite target in a hot furnace by a scanning laser beam. Nowadays, the catalytic pyrolysis of methane, acetylene, and carbon monoxide is gaining acceptance.[38] Nanotubes with diameters from 20 to 60 nm were obtained in the methane flame on an Ni–Cr wire.[40] The pyrolysis of an aerosol prepared from a benzene and ferrocene solution, which was carried out at 800–950 °C, turned out to be a highly efficient method for synthesizing multiwalled nanotubes 30–130-μm long with inner diameters from 10 to 200 nm.[41] The latter method is based on employing hydrocarbon solutions and catalysts. The preparation of nanotubes involves difficulty in controlling; they are usually accompanied by the formation of different carbon forms, which can be eliminated by cleaning procedures.

The use of single-walled carbon nanotubes is often defined by the uniformity of their distribution; certain applications require tubes 20–100 nm long. For shortening and purifying freshly prepared nanotubes, various oxidation procedures with participation of nitric acid and its mixtures with other oxidants (H_2SO_4, $KMnO_4$) were employed.[42,43] Chemical oxidation resulted in the appearance of different oxygen-containing groups on the tube's ends and often on its sidewalls.[44] When using chemical oxidation for cutting the tubes, it is important to regulate the origination of various defects and take into account the decrease in the mass of the treated material. In contrast, the use of a mixture of 96% H_2SO_4 and 30% H_2O_2 (piranha solutions) with the volume ratio of 1:4 made it possible to cut nanotubes at room temperature without creating defects on their walls.[45] At 70 °C, the appearance of defects and selective etching of the nanotube diameter were observed.

A controlled multistep procedure was proposed for the removal of iron impurities and nonnanotube carbon materials from raw single-walled carbon nanotubes.[46] The tubes were synthesized at high CO pressures in the presence of iron pentacarbonyl as a catalyst. The earlier most popular method used strong acids as oxidants for the removal of impurities.

A more perfect procedure involves two processes: oxidation and deactivation of the metal oxide. In the oxidation process, the coating on the metal catalyst that consists of a nonnanotube carbon material is oxidized with oxygen to form carbon dioxide, and the metal is transformed into oxide. In the next step, the metal oxide is deactivated by reacting with $C_2H_2F_4$ or SF_6, to avoid the oxide involvement in the oxidation of carbon nanotubes.

Apparently, the term "chemistry of nanotubes" was used for the first time by Cook et al., 1996.[47] Currently, this means the synthesis of tubes, their purification, and the different types of chemicals that modify both the external and internal surfaces of tubes. The intercalation of foreign particles into the intratube space of bundles and the use of nanotubes as the matrices for synthesizing various materials such as adsorbents, sensors, and catalysts can also be assigned to the chemistry of nanotubes.

The peculiarities in the structure of carbon nanotubes distinguish their chemical behavior as compared with the behavior of fullerenes and graphite. The internal cavities in fullerenes are so small that they can house only several atoms of foreign elements, while the nanotube's inner spaces are greater in volume. Fullerenes can form molecular crystals, e.g. graphite is a lamellar polymeric crystal. Nanotubes represent an intermediate state. Single-walled tubes closely resemble molecules, while the behavior of multiwalled tubes approaches that of carbon fibers. A separate tube is usually considered as a one-dimensional crystal; their bundles form two-dimensional crystals.[48]

Chemistry of carbon nanotubes has been comprehensively discussed in reviews.[38,49,50]

8.3.1 Filling of Tubes

Filling of carbon nanotubes can be accomplished either during or after their synthesis.

For filling tubes during the synthesis, of prime importance are additives, which prevent the closure of the tube channel. Boron is among such additives.[51] Tubes with inner spaces filled with fullerenes C_{60} and C_{70} are of interest as the composite materials.[52] A remarkable fact is that "nanopod"-like structures were observed in the products of their laser–thermal synthesis upon their annealing in vacuum at 1100 °C.[53] In such structures, the tube diameter (1.4 nm) exceeds twice the diameter of a C_{60} molecule (0.7 nm) so that fullerene molecules can move and form pairs.

The ends of nanotubes are usually closed by five- or six-carbon cycles, where five-membered cycles are less resistant to oxidation. To fill already synthesized tubes, their ends should be opened, e.g. via selective oxidation, which can be realized using gaseous reagents such as oxygen, air, carbon dioxide, or their aqueous solutions. The acid treatment is also effective, where nitric acid is the conventional reagent. The oxidation mechanism is still not entirely clear.[54]

The tube's inner spaces can be filled in liquid media, particularly, in fused oxides of various metals. In the tubes with inner diameters <3 nm, a glassy phase was formed, rather than a crystalline phase.[55] Interesting results were obtained by filling nanotubes with potassium iodide crystals, which was performed in a fused salt medium.[56,57] In a tube of 1.6-nm diameter, a KI crystal comprising only nine atoms was compressed along the (0 0 1) axis by 0.695–0.705 nm as compared with the compact substance. A compressed crystal had the coordination number 5 for face atoms and 4 for edge atoms. Insofar that the ratio of such atoms was high, it could be expected that the deviations in the geometry would affect the electronic properties of the substances. In the opinion of the authors,[56,57] typical metals can be transformed into dielectrics.

Substances that are incorporated into the channels of carbon nanotubes can take part in chemical reactions. By thermal decomposition of oxides and their reduction, metal-containing nanotubes were obtained and a transformation of potassium oxide into its sulfide was accomplished in the intratube space.[58] The inner spaces of tubes could be filled by chemical deposition from the gas phase by utilizing, e.g. volatile metal compounds. During the catalytic pyrolysis of hydrocarbons in porous $AlPO_4$, nanotubes could be filled without being preliminarily opened.[59]

8.3.2 Grafting of Functional Groups. Tubes as Matrices

Planting various functional groups on nanotube walls forms an extensive and important branch of nanochemistry of carbon tubes. Such a process can be performed by long-term treatment of tubes in acids, where the behavior of single-walled tubes was shown to depend on the method of their synthesis.[60] Functional groups can be removed from the tube's walls by heating to temperatures above 623 K.[61]

An assumption that protonation of single-walled carbon nanotubes is associated with the formation around a tube of an acidic layer that promotes its subsequent dissolution was confirmed.[62,63] XDS studies have shown that carbon nanotubes can serve as templates for sulfuric acid crystallization. This fact is considered as a direct evidence of its protonation.[64]

The use of carbon nanotubes involves preparation of their uniform dispersions. For single-walled nanotubes, this can be achieved by either wrapping the tubes with polymers, e.g. polyvinylpyrrolidone[65] and poly(arylene ethylene),[66] or by protonation of nanotubes with superacids.[62] Moreover, it was demonstrated that the reduction of single-walled nanotubes with alkali metals produced polyelectrolyte salts soluble in polar organic solvents such as dimethylsulfoxide.[67]

Nanosize vibrations of single-walled carbon nanotubes synthesized by arc discharge and chemical vapor deposition were studied with spatial resolution of ~15 nm.[68]

The method of jet linear dichroism (differential absorption of polarized light oriented in parallel and normally to the jet direction) was used for studying the structural interactions of anthracene, naphthalene, and DNA molecules with carbon nanotubes. The differences in interactions of aromatic molecules oriented in parallel and normally to the nanotube were revealed.[69]

Attachment of functional groups to the walls of carbon nanotubes serves for conferring certain functions to AFM probes. The best results were obtained in gases.[70] Figure 8.1 illustrates this process. Modification was performed via a discharge in such gases as O_2, H_2, N_2, and mixtures of H_2 and N_2. In the nitrogen atmosphere, nitrogen atoms entered into the composition of heterocycles at the end of a nanotube. Probes thus modified can be used for studying the surfaces of layers containing hydroxyl groups.

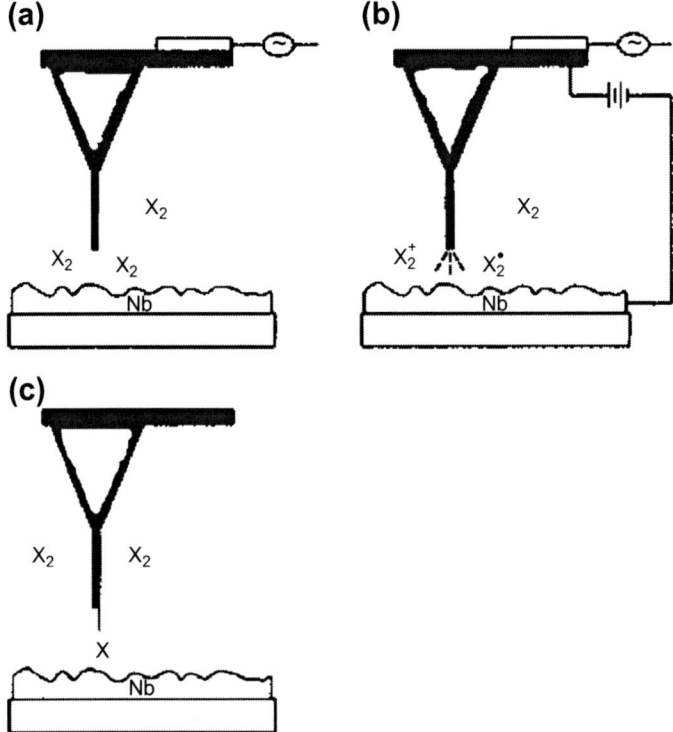

FIGURE 8.1 Scheme of performance of an AFM cantilever: (a) scanning of a surface with sputtered niobium by a vibrating probe made of nanotubes in the atmosphere of X2 gas; (b) discharge between a nanotube and the surface; (c) probe with a nanotube containing X.[70]

Fluorination was actively used for modifying the tube's walls. The process was found to be reversible at $T \leq 325$ K. When fluorinated tubes were acted on by anhydrous hydrazine, fluoride atoms were removed and the original tube structure was recovered. At 400 °C, the fluorination was irreversible. A partial recovery of the structure can be associated with the reaction

$$CF_{n(Tp)} + n/4N_2H_4 \rightarrow C_{(Tp)} + nHF + n/4N_2$$

As a result of long-term (up to seven days) fluorination with BrF_3 vapors, the arc-synthesized multiwalled tubes and nanoparticles formed tubular or quasi-spherical C_2F particles.[71] With an increase in fluorination degree, the tube diameter and interlayer gaps increased and, at a certain crucial degree, tubes were observed to unroll to form multilayer planar particles.

The use of carbon nanotubes as the matrices made it possible to obtain copper particles with narrow size distributions.[72] Original nanotubes with diameters from 5–10 to 25–35 nm were synthesized via the catalytic pyrolysis of methane. By varying both the copper salt concentration in aqueous solution and the copper–tube ratio, the reduction by hydrogen could produce either copper nanoparticles or nanowires. The smallest copper particles of 5–10 nm were obtained for the low concentrations of copper ions in solution. An increase in the salt concentration favored the formation of copper nanowires with diameters from 100 nm to 5 mm and a length up to several hundred microns.

8.3.3 Intercalation of Atoms and Molecules into Multiwalled Tubes

The intercalation differs for single- and multiwalled tubes. In multiwalled tubes intercalated particles are arranged between separate layers, whereas in single-walled tubes, intercalated particles penetrate into the intratube space of bundles.

Intercalation of nanotubes differs from the corresponding process in fullerenes. Fullerenes, e.g. C_{60}, form charge–transfer complexes only with electron donors.[8] According to the results obtained by using Raman spectroscopy and conductivity measurements, bundles of single-walled tubes behave in a dual way; they can interact both with donors and acceptors.[73,74] Crystalline bundles of single-walled tubes exhibit metallic properties. In these tubes, a positive temperature coefficient was observed. The intercalation of bromine or potassium resulted in a 30-fold decrease in the conductivity at 300 K and in an extension of the positive temperature coefficient range. This fact suggests that tubes doped with bromine or potassium can be classified with synthetic metals.[74]

An interesting result was achieved by incorporating electron acceptors. An attempt was undertaken to transform tubes into diamonds by synthesizing additional amount of carbon simultaneously with intercalation of potassium into tubes.[75] It was assumed that, via the reaction $4K + CCl_4 \rightarrow 4KCl + C$ in the intratube space of multiwalled tubes, the adjusting graphene sheets could be bound, and the sp^2 bonds would change to the sp^3 bonds. The reaction was carried out in an autoclave at 200 °C. Disordering, amorphization, and the formation of potassium chloride crystals between graphene sheets were observed. The absence of KCl crystals was observed for the nanotubes with an outer diameter below 10 nm.

The sorption of gases by nanotubes can proceed on both external and internal walls and also in the intratube space. Thus, experimental studies of nitrogen adsorption at 77 K on multiwalled tubes with mesopores (4.0 ± 0.8)-nm wide showed that adsorption occurs on both internal and external walls of nanotubes.[76] In this case, the external surfaces adsorbed five times more particles than the internal surfaces, and these processes were described by different isotherms. Adsorption in mesopores occurred in agreement with classical condensation in capillaries; the pore diameter was estimated as 4.5 nm. Certain peculiarities of this process were associated with the fact that the tubes were only one-end open.

Bundles of nanotubes adsorb nitrogen well. The internal specific surface area of raw nonpurified tubes was 233 m^2/g, and their external specific surface area was 143 m^2/g.[77]

Their treatment with hydrochloric and nitric acids increased the total specific surface and enhanced their adsorbability with respect to benzene and methanol.

Applications of carbon nanotubes are considered in Section 8.5. The examples of their use in nanotechnology clearly demonstrate that in contemporary research the borders between fundamental and applied studies disappear.

8.4 GRAPHENE

Graphite is a form of carbon with sp^2 bonding made up of flat six-membered rings. Each layer is independent in that it is not bound by covalent bonds to the other layers. However, they are weakly bound to each other by van der Waals and π–π interactions.

Energetically speaking, it is not difficult to slide these layers across each other, and graphite is, therefore, a good lubricant. It is also not energetically costly to separate these layers, although separation and purification are rather difficult.

In 2004, Novoselov and coworkers[78] used a Scotch Tape method to pull apart top layers, a mechanical method.

Soon other methods were developed, such as liquid phase exfoliation.[79] Chemical vapor deposition methods have also been developed, where single layers of carbon were prepared on flat metal surfaces. For example, methane and inert carrier gases are passed over high-temperature solid surfaces of copper or other metal. It is believed that methane decomposition yields hydrogen and carbon atoms that traverse into the metal surface forming carbides. By cooling the high-temperature sample, carbon comes to the surface and, under the right cooling conditions, forms graphene on the metal surface.[80]

Characterization of graphene is rather difficult.[81] Microscopy methods (TEM, SEM, and AFM) are the primary tools, and high resolution images are needed for a convincing characterization. Another technique is Raman Spectroscopy, which can indicate purity and quality of graphene samples. With Raman, two absorbance bands are observed: D-band at 2700 cm^{-1} and G-band at 1600 cm^{-1}. In graphite, the G-band is larger, while in graphene, the D-band is predominant.

Potential applications of graphene are many. It possesses unique electrical properties and could be used as a component in integrated circuits for solar cells and other devices.

8.5 CARBON AEROSOL GELS/TURBSTRATIC GRAPHITE/GRAPHENE

In 1998, it was demonstrated that a spontaneous super aggregation/gelation of soot particles took place in a laminar diffusion flame.[82] This work led to a closed-chamber system where, instead of a laminar flame, a detonation of hydrocarbon-oxygen mixtures (lean in oxygen) led to aerosol gels.[83] The densities of these carbon aerosol gels are very low, as little as 2.0 mg/cc. Microscopy studies by TEM showed graphite-like layers, and XRD spectra resembled that reported for graphene.[84] Raman spectroscopic studies indicated large amounts of sp^2 hybridized carbon, as well as many defects and disorder.[85-87]

It appears that this carbon aerosol gel is similar to turbostratic graphite, where layers are arranged in a rotationally symmetric fashion, and may be a good starting point for efficient exfoliation to form single graphene layers.

REFERENCES

1. Ortman, B. J.; Hauge, R. H.; Margrave, J. L.; Kafafi, Z. H. *J. Phys. Chem.* **1990,** *94,* 973–7977.
2. Van Zee, R. J.; Smith, G. R., Jr.; Weltner, W. *J. Am. Chem. Soc.* **1988,** *110,* 609–610.
3. McElvany, S. W. *J. Chem. Phys.* **1988,** *89,* 2063–2075.
4. Alford, J. M.; Laaksonen, R. T.; Smalley, R. E. *J. Chem. Phys.* **1991,** *94,* 2618–2630.
5. Jarrold, M. F.; Ijiri, Y.; Ray, U. *J. Chem. Phys.* **1991,** *94,* 3607–3618.
6. Ray, U.; Jarrold, M. F. *J. Chem. Phys.* **1991,** *94,* 2631–2639.
7. Jarrold, M. F.; Ray, U.; Creegan, K. M. *J. Chem. Phys.* **1990,** *93,* 224–229.
8. Sokolov, V. I.; Stankevich, I. V. *Russ. Chem. Rev.* **1993,** *62,* 419–436.
9. Matsuzawa, N.; Dixon, D. A.; Fukunaga, T. *J. Phys. Chem.* **1992,** *96,* 7594–7604.
10. Tuinman, A. A.; Mukherjee, P.; Adcock, J. L.; Hettich, R. L.; Compton, R. N. *J. Phys. Chem.* **1992,** *96,* 7584–7589.
11. Boltalina, O. V.; Galeva, N. A. *Russ. Chem. Rev.* **2001,** *69,* 609–621.
12. Olah, G. A.; Bucsi, I.; Lambert, C.; Aniszfeld, R.; Trivedi, N. J.; Sensharma, D. K.; et al. *J. Am. Chem. Soc.* **1991,** *113,* 9387–9388.
13. Loy, D. A.; Assink, R. A. *J. Am. Chem. Soc.* **1992,** *114,* 3977.
14. Yuasa, M.; Oyaizu, K.; Yamaguchi, A.; Kuwakado, M. *J. Am. Chem. Soc.* **2004,** *126,* 11128–11129.
15. Curl, R. F.; Smalley, R. E. *Science* **1988,** *242,* 1017.

16. Chai, Y.; Guo, T.; Jin, C.; Haufler, R. E.; Chibante, P. F.; Fare, J.; et al. *J. Phys. Chem.* **1991**, *95*, 7564–7568.
17. Yannoi, C. S.; Hoinkis, M.; deVries, M. S.; Bethune, D. S.; Salem, J. R.; Crowder, M. S.; et al. *Science* **1992**, *256*, 1191.
18. Hebard, A. F.; Rosseinsky, M. J.; Haddon, R. C.; Murphy, D. W.; Glarum, S. H. G.; Palstra, T. T. M.; et al. *Nature* **1991**, *350*, 600.
19. Holczer, K.; Klein, O.; Huang, S. M.; Kaner, R. B.; Fu, K. J.; Whetten, R. L.; et al. *Science* **1991**, *252*, 1154.
20. Tankigaki, K.; Ebbesen, T. W.; Saito, S.; Mizuki, J.; Tsai, J. S.; Kubo, Y.; et al. *Nature* **1991**, *352*, 222.
21. Hadden, R. C.; Kochanski, G. P.; Hebard, A. F.; Fiory, A. T.; Morris, R. C. *Science* **1992**, *258*, 1636.
22. Wertheim, G. K.; Buchanan, D. N. E.; Rowe, J. E. *Science* **1992**, *258*, 1638.
23. Blank, V. D., Buga, S. G., Dubitsky, G. A. Fizika i Tehnika vysokih davlenii, 2000; Vol. 10, p 127.
24. Cross, R. J.; Khong, A.; Saunders, M. *J. Org. Chem.* **2003**, *68*, 8281–8283.
25. Schick, G.; Jarrosson, T.; Runbin, Y. *Angew. Chem. Int. Ed.* **1999**, *38*, 2360–2363.
26. Murata, Y.; Murata, M.; Komatsu, K. *J. Am. Chem. Soc.* **2003**, *125*, 7152–7153.
27. Murata, Y.; Murata, M.; Komatsu, K. *Chem. – A Eur. J.* **2003**, *9*, 1600–1609.
28. Stanisky, C. M.; Cross, R. J.; Saunders, M.; Murata, M.; Murata, Y.; Komatsu, K. *J. Am. Chem. Soc.* **2005**, *127*, 299–302.
29. Rubin, Y.; Jarrosson, T.; Wang, G.-W.; Bartberger, M. D.; Houk, K. N.; Schick, G.; et al. *Angew. Chem. Int. Ed.* **2001**, *40*, 1543–1546.
30. Yang, H.; Holloway, P. H. *Adv. Funct. Mater.* **2004**, *14*, 152–156.
31. Maeda, Y.; Miyashita, J.; Hasegawa, T.; Wakahara, T.; Tsuchiya, T.; Feng, L.; et al. *J. Am. Chem. Soc.* **2005**, *127*, 2143–2146.
32. Wang, Y.; Jiang, X.; Herricks, T.; Xia, Y. *J. Phys. Chem. B* **2004**, *108*, 8631–8640.
33. Iijima, S. *Nature (London)* **1991**, *354*, 56.
34. Service, R. F. *Science* **1998**, *281*, 940–943.
35. Eletskii, A. V. *Physics-Uspekhi* **1997**, *40*, 899–924.
36. Ivanovskii, A. L. *Russ. Chem. Rev.* **1999**, *68*, 103–118.
37. Rakov, E. G. *Russ. J. Inorg. Chem.* **1999**, *44*, 1736–1748.
38. Rakov, E. G. *Russ. Chem. Rev.* **2000**, *69*, 35–52.
39. Liu, B.; Wagberg, T.; Olsson, E.; Yang, R.; Li, H.; Zhang, S.; et al. *Chem. Phys. Lett.* **2000**, *320*, 365–372.
40. Yuan, L.; Saito, K.; Pan, C.; Williams, F. A.; Gordon, A. S. *Chem. Phys. Lett.* **2001**, *340*, 237–241.
41. Mayne, M.; Grobert, N.; Kumalakaran, R.; Ruehle, M.; Kroto, H. W.; Walton, D. R. M. *Chem. Phys. Lett.* **2001**, *338*, 101–107.
42. Hu, H.; Zhao, B.; Itkis, M. E.; Haddon, R. C. *J. Phys. Chem. B* **2003**, *107*, 13838–13842.
43. Zhang, J.; Zou, H.; Quing, Q.; Yang, Y.; Li, Q.; Liu, Z.; et al. *J. Phys. Chem. B* **2003**, *107*, 3712–3718.
44. Mawhinney, D. B.; Naumenko, V.; Kuznetsova, A.; Yates, J. T., Jr.; Liu, J.; Smalley, R. E. *J. Am. Chem. Soc.* **2000**, *122*, 2383–2384.
45. Ziegler, K. J.; Gu, Z.; Peng, H.; Flor, E. L.; Hauge, R. H.; Smalley, R. E. *J. Am. Chem. Soc.* **2005**, *127*, 1541–1547.
46. Meziani, M. J.; Pathak, P.; Hurezeanu, R.; Thies, M. C.; Enick, R. M.; Sun, Y.-P. *Angew. Chem. Int. Ed.* **2004**, *43*, 704–707.

47. Cook, J.; Sloan, J.; Green, M. L. H. *Chem. Ind.* **1996,** 600.
48. Sung, K. -M.; Mosley, D. W.; Peelle, B. R.; Zhang, S.; Jacobson, J. M. *J. Am. Chem. Soc.* **2004,** *126,* 5064–5065.
49. Rakov, E. G. *Russ. Chem. Rev.* **2001,** *70,* 827–864.
50. Dai, H. *Surf. Sci.* **2002,** *500,* 218–241.
51. Hernandez, E.; Ordejon, P.; Boustani, I.; Rubio, A.; Alonso, J. A. *J. Chem. Phys.* **2000,** *113,* 3814–3821.
52. Smith, B. W.; Monthioux, M.; Luzzi, D. E. *Chem. Phys. Lett.* **1999,** *315,* 31–36.
53. Smith, B. W.; Monthioux, M.; Luzzi, D. E. *Nature (London)* **1998,** *396,* 323–324.
54. Duesberg, G. S.; Muster, J.; Byrne, H. J.; Roth, S.; Burghard, M. *Appl. Phys. A* **1999,** *69,* 269–274.
55. Ajayan, P. M.; Stephan, O.; Redlich, Ph.; Colliex, C. *Nature (London)* **1995,** *375,* 564.
56. Meyer, R. R.; Sloan, J.; Dunin-Borkovski, R. E.; Kirkland, A. I.; Novotny, M. C.; Bailey, S. R.; et al. *Science* **2000,** *289,* 1324.
57. Sloan, J.; Novotny, M. C.; Bailey, S. R.; Brown, G.; Xu, C.; Williams, V. C.; et al. *Chem. Phys. Lett.* **2000,** *329,* 61–65.
58. Cook, J.; Sloan, J.; Green, M. L. H. *Fuller. Sci. Technol.* **1997,** *5,* 695.
59. Sinha, A. K.; Hwang, D. W.; Hwang, L. -P. *Chem. Phys. Lett.* **2000,** *332,* 455–460.
60. Nagasava, S.; Yudasaka, M.; Hirahara, K.; Ichihashi, T.; Iijima, S. *Chem. Phys. Lett.* **2000,** *328,* 374–380.
61. Kuznetsova, A.; Mawhinney, D. B.; Naumenko, V.; Yates, J. T., Jr.; Liu, J.; Smalley, R. E. *Chem. Phys. Lett.* **2000,** *321,* 292–296.
62. Ramesh, S.; Ericson, L. M.; Davis, V. A.; Saini, R. K.; Kittrell, C.; Pasquali, M.; et al. *J. Phys. Chem. B* **2004,** *108,* 8794–8798.
63. Zhou, W.; Vavro, J.; Guthy, C.; Winey, K. I.; Fischer, J. E.; Ericson, L. M.; et al. *J. Appl. Phys.* **2004,** *95,* 649–655.
64. Zhou, W.; Heiney, P. A.; Fan, H.; Smalley, R. E.; Fischer, J. E. *J. Am. Chem. Soc.* **2005,** *127,* 1640–1641.
65. O'Connell, M. J.; Boul, P.; Ericson, L. M.; Huffman, C.; Wang, Y.; Haroz, E.; et al. *Chem. Phys. Lett.* **2001,** *342,* 265–271.
66. Chen, J.; Liu, H.; Weimer, W. A.; Halls, M. D.; Waldeck, D. H.; Walker, G. C. *J. Am. Chem. Soc.* **2002,** *124,* 9034–9035.
67. Penicaud, A.; Poulin, P.; Derre, A.; Anglaret, E.; Petit, P. *J. Am. Chem. Soc.* **2005,** *127,* 8–9.
68. Anderson, N.; Hartschuh, A.; Cronin, S.; Novotny, L. *J. Am. Chem. Soc.* **2005,** *127,* 2533–2537.
69. Derfus, A. M.; Chan, W. C. W.; Bhatia, S. N. *Nano Lett.* **2004,** *4,* 11–18.
70. Wong, S. S.; Woolley, A. T.; Joselevich, E.; Lieber, C. M. *Chem. Phys. Lett.* **1999,** *306,* 219–225.
71. Okotrub, A. V.; Yudanov, N. F.; Chuvilin, A. V.; Asanov, I. P.; Shubin, Yu. V.; Bulusheva, L. G.; et al. *Chem. Phys. Lett.* **2000,** *322,* 231–236.
72. Chen, P.; Wu, X.; Lin, J.; Tan, K. L. *J. Phys. Chem. B* **1999,** *103,* 4559–4561.
73. Rao, A. M.; Eklund, P. C.; Bandow, S.; Thess, A.; Smalley, R. E. *Nature (London)* **1997,** *388,* 257–259.
74. Lee, R. S.; Kim, H. J.; Fisher, J. E.; Thess, A.; Smalley, R. E. *Nature (London)* **1997,** *388,* 255.
75. Hsu, W. K.; Li, W. Z.; Zhu, Y. Q.; Grobert, N.; Terrones, M.; Terrones, H.; et al. *Chem. Phys. Lett.* **2000,** *317,* 77–82.
76. Inoue, S.; Ichikuni, N.; Suzuki, T.; Uematsu, T.; Kaneko, T. *J. Phys. Chem. B* **1998,** *102,* 4689–4692.
77. Sen, R.; Eswaramoorthy, M.; Rao, C. N. R. *Chem. Phys. Lett.* **1999,** *304,* 207–210.

78. Novoselov, K. S.; Geim, A. K.; Morozov, S. V.; Jiang, D.; Zhang, Y.; Dubonos, S. V.; et al. *Science* **2004**, *306*, 666.
79. Hernandez, Y.; Nicolosi, V.; Lotya, M.; Blighe, F. M.; Sun, Z.; De, S.; et al. *Nat. Nanotechnol.* **2008**, *3*, 563–568.
80. Yu, Q.; Lian, J.; Siriponglint, S.; Li, H.; Chen, Y. P.; Pei, S. S. *Appl. Phys. Lett.* **2008**, *93*, 113103.
81. Luo, Z.; Yu, T.; Shang, J.; Wang, Y.; Lim, S.; Gagik, L. L.; et al. *Adv. Funct. Mater.* **2011**, *21*, 911–917.
82. Sorensen, C. M.; Hageman, W. B.; Rush, T. J.; Huang, H.; Oh, C. *Phys. Rev. Lett.* **1998**, *80*, 1782–1785.
83. Dhaubadel, r.; Gerving, S. C.; Chakrabarti, A.; Sorensen, C. M. *Aerosol Sci. Tech.* **2007**, *41*, 804–810.
84. Wang, G.; Yang, J.; Park, J.; Gou, X.; Wang, B.; Lin, H.; et al. *J. Phys. Chem. C* **2008**, *112*, 8192–8195.
85. Malard, L. M.; Pimenta, M. A.; Dresselhaus, G.; Dresselhaus, M. S. *Phys. Rep.* **2009**, *473*, 51–87.
86. Dresselhaus, M. S.; Jorio, A.; Hofmann, M.; Dresselhaus, G.; Saito, R. *Nano Lett.* **2010**, *10*, 751–758.
87. Ferrari, A. C.; Meyer, J. C.; Scardaci, V.; Casiraghi, C.; Lazzeri, M.; Mauri, F.; et al. *Phys. Rev. Lett.* **2006**, *97*, 187401.

Chapter 9

Organic Nanoparticles

Chapter Outline

9.1 Introduction 235
9.2 Methods for the Preparation of Nanoparticles 237
    9.2.1 Physical Methods 237
        9.2.1.1 Mechanical Grinding of the Original Substance 237
        9.2.1.2 Laser Ablation 239
    9.2.2 Chemical Methods 242
        9.2.2.1 Solvent Replacement 242
        9.2.2.2 Antisolvents for Precipitation 244
        9.2.2.3 Chemical Reduction in Solution 245
        9.2.2.4 Ion Association 246
        9.2.2.5 Synthesis of Nanoparticles in Water–Oil Emulsion 247
        9.2.2.6 Photochemical Method 248
        9.2.2.7 The use of Supercritical Fluids 248
        9.2.2.8 Cryochemical Synthesis and Modification of Nanoparticles 251
9.3 Properties and Application of Organic Nanoparticles 257
    9.3.1 Spectral Properties 257
    9.3.2 Quasi-one-dimensional Systems 260
    9.3.3 Drugs and Nanoparticles 263
9.4 Conclusion 269

9.1 INTRODUCTION

Nanoscience is a multifunctional discipline that embraces such wide areas as nanotechnology and nanochemistry. Nanotechnologies are aimed at the preparation of nanoparticles with novel properties and the development of materials and devices on their basis. This is one of the topical directions of natural science in the twety-first century.

    The development of nanochemistry is associated with the pass from the microscale level to nanosizes. Studying nanosized particles resulted in the discovery of size effects, which can be considered as a new degree of freedom imparting new qualities to chemical compounds and resulting in earlier unknown chemical transformations.

The investigations evolved in such a way that the first efforts were directed at the studies of atoms, clusters, nanoparticles of different metals, semiconductors, fullerenes, carbon nanotubes as well as physicochemical properties of the respective materials. This has determined the first studies in the field of nanochemistry. For the most comprehensive analysis of the contribution made by nanochemistry into the development of nanotechnologies, see monographs,[1,2] and the size effect-based reactions and materials can be found in another monograph.[3]

The progress in biochemistry and nanomedicine is determined by the properties of complex organic compounds, which, in addition to carbon and hydrogen, can include atoms of oxygen, sulfur, halogens, and also by the mechanisms of processes involving these compounds. The methods for the preparation and physicochemical properties of nanoparticles containing complex organic molecules are studied to an insufficient extent yet. This is partly associated with the fact that the intermolecular interactions are weaker than the interatomic ones and the size dependence of physicochemical properties of organic nanoparticles is less pronounced as compared with inorganic nanoparticles.

Nanoparticles can be considered as objects in-between the compact organic compounds and single molecules.

From our point of view, the fact that the development of studies of organic nanoparticles, contrary to that of metal nanoparticles, has started only relatively recently can be explained by the following two main reasons. First, low melting temperatures and lesser thermal stability of organic nanostructures as compared with those of inorganic compounds limit the methods for their synthesis and applications. Second, it was believed that no size effect is manifested in organic nanomaterials. However, eventually it was shown that the size effect is typical of organic nanoparticles as well.

Inorganic and organic nanoparticles are both similar and dissimilar. The similarity is expressed in the size dependence of physicochemical (especially, spectral) properties. For nanoparticles of metals and inorganic semiconductors, the size effects are most strongly pronounced in the size range <10 nm. For organic compounds, the size effects were discovered in studying the spectral properties of particles with the sizes from several tens to several hundreds of nanometers. In metals, the size effect is associated with collective electrons that can hold within the nanobulk. In organic molecules, the electrons "pertain" to chemical bonds, the mobility of these electrons is limited, and hence, the interactions that result in spontaneous formation of ensembles and favor the transfer of electrons and charge are weak.

So far, the proneness for self-assembling that can be employed in designing devices for opto- and nanoelectronics has been observed only for the systems that comprise several cyclic and aromatic molecules. The structure and the interaction of these molecules favor the charge transfer and give rise to stacking—the appearance of excitons and shared electrons.

Yet another promising direction in the use of organic nanoparticles is associated with modification of drugs, particularly, preparation of their novel stable polymorphic forms.

The increased interest in the studies of organic nanoparticles has recently been manifested by the appearance of several reviews.[4-6] Each of these publications considers a single method of fabrication of nanoparticles and summarizes the results of studies by particular research teams. In the present review, we analyzed different approaches to the manufacture and the characterization of nanoparticles of organic compounds built of several cyclic and aromatic fragments.

Such molecules are the components of drugs and various dyes and determine the properties of numerous polyaromatic compounds. By the example of such substances, methods for the preparation of nanoparticles are analyzed and their physicochemical properties and certain application prospects are considered.

Undoubtedly, investigations into organic nanoparticles is a promising direction and its rapid progress in inevitable.

9.2 METHODS FOR THE PREPARATION OF NANOPARTICLES

Methods for the fabrication of nanoparticles can be divided into two groups:

- coarsening of original species (i.e. single atoms and molecules) to form nanosized particles (the "bottom-up" approach);
- grinding of coarse particles (molecular aggregates, powders, grains) to nanosized particles (the "top-down" approach).

The former approach includes mainly the chemical methods for the fabrication of nanoparticles, while the latter relies on physical methods. The fundamentally different technologies often produce nanomaterials with the same chemical composition but different properties.

The synthesis of organic nanoparticles is a laborious branch of chemistry, which is poorly explored so far. The use of physical methods, e.g. mechanical disintegration and laser ablation, is of limited application due to the low melting points and thermal instability of organic compounds. Only a small fraction of the known methods developed for the synthesis of nanoparticles of metals and inorganic semiconductors can be used for synthesizing organic nanoparticles. Below, the current methods for the preparation of organic nanomaterials are considered.

9.2.1 Physical Methods

9.2.1.1 Mechanical Grinding of the Original Substance

Depending on the nature of a substance and the required particle size, different kinds of mills and "dry" and "wet" grinding are used.[7] Dry grinding is the most popular technology that employs ball and jet mills. The surface of a mechanically ground substance is partly amorphous and has many defects, i.e. it is thermodynamically activated.[8]

Grinding of ibuprofen by the jet milling technology was described.[9] Figure 9.1 shows microimages of particles measuring from 1 to 7 mm, recorded using a scanning electron microscope (SEM). Nanoparticles of fluticasone propionate with the average size of 5 mm were fabricated by the same method.[10] Fairly unstable particles are normally produced upon dry grinding. Their stabilization was achieved by adding, for instance, cellulose ethers[11] or poly(vinylpyrrolidone),[12] which adsorbed on the newly formed surface of particles. The particles acquired a positive charge that kept them from agglomeration.

In wet grinding, antisolvents are used, which prevent dust formation and the agglomeration of particles. This method was used in the manufacturing of suspensions of nanosized drugs for intravenous injections.

In planetary ball mills, a drug can be ground to sizes <0.1 mm. Thus, a suspension of insulin nanoparticles (150 nm) was prepared.[13] To stabilize such suspensions, milling was carried out in the presence of special additives.[10,14]

The following drawbacks of the mechanical synthesis can be mentioned:

- partial dissolution of the organic substance and its uncontrolled recrystallization during drying are possible;
- impurities from the mill material can inevitably get into the final product (to suppress this effect, rubber-covered glass beads were proposed to use as the grinding tools)[15];
- destruction of thermally unstable substances due to heat liberated during grinding;
- the presence of a relatively large number of particles measuring >5 mm.

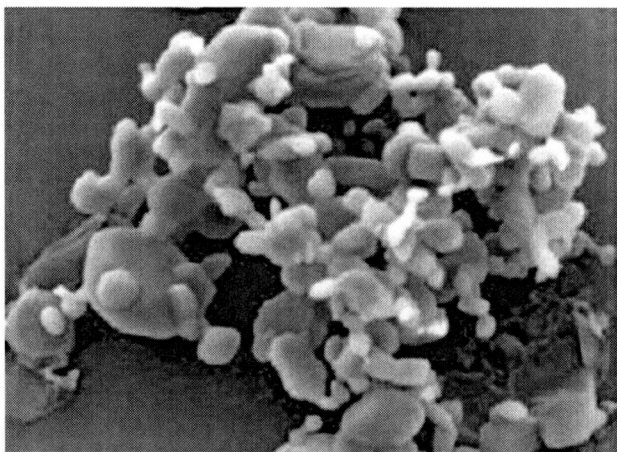

FIGURE 9.1   SEM image of ibuprofen particles produced by mechanical grinding.[9]

Suspensions of nanoparticles can be prepared by piston-gap high-pressure homogenization technology known under the "DissoCubes" trademark. The piston motion in the cylinder creates pressure up to 2000 bar. A flow of a particles-water mixture passes through a gap 3–5 mm wide, which occurs, according to the Bernoulli law, at a high rate.[16] A sharp decrease in the liquid flow section, i.e. from the cylinder section to the gap section, results in cavitation so that the vapor bubbles first appear and then burst at the exit from the gap. The shock wave induced due to the explosion of bubbles is the factor favoring the decrease in the particle size.[17] This technology of producing nanosuspensions can be effected with APV Gaulin (Germany), Avestin, Inc. (Canada), and GEA NiroSoavi (Italy) homogenizers.

A suspension of atovaquone (malorone), a drug for the prophylaxis and curing of malaria, was produced by high-pressure homogenization. The average particle size was 279 nm.[18] amphotericin B, itraconazole, and carbamazepine particles with the sizes of 493, 519 and 870 nm, respectively, were manufactured analogously.[19,20]

The peculiarities of mechanical milling of substances were surveyed in a review.[8] The main drawbacks of this method is the limited possibility of a control over the size and the shape of the synthesized particles, their wide size distribution, and also their irregular shape.[21]

The stabilization of particles can be achieved by introducing auxiliary agents. Mechanical grinding is a convenient method for the production of nanoparticle suspensions. A large number of pharmaceutical companies use and develop various mechanical grinding technologies to improve the bioavailability of poorly soluble substances. The ease of scaling such technologies for the industrial production should be mentioned.[19]

9.2.1.2 Laser Ablation

The first application of the laser ablation method in the formation of nanoparticles of organic compounds dates back to 2000.[22] This method is as follows: a suspension of the original microcrystalline powder is prepared in a solvent in which this substance is virtually insoluble; the thus prepared suspension is subjected to the impact of an intense laser pulse, which results in the fragmentation of grains and the transition of the original suspension to a transparent colloidal solution.[23] The method of laser ablation of microcrystalline powders of organic compounds in a solvent has opened new prospects for the synthesis of nanoparticles. The use of lasers allows preparation of stable colloidal solutions in the absence of stabilizers.[3] By tuning the excitation parameters such as the wavelength, the laser pulse width, and fluence, one can control the size and the degree of modification of the synthesized nanoparticles. Thus, the increase in the laser-radiation fluence results in a decrease in the nanoparticle size. The structure of formed nanoparticles, e.g. of quinacridone, depends on the laser pulse wavelength and fluence.

The method of laser ablation was used to synthesize nanoparticles of several dyes, namely, benzidine yellow (1), quinacridone (2), brilliant rose (3), and dark red (4).

The solvents used for these dyes were the 7:1 ethanol-water mixture for benzidine yellow, the[22] 39:1 ethanol-water mixture for brilliant rose, and dark red and distilled water for quinacridone. During the experiments, the solutions were thoroughly mixed with a magnetic stirrer. Laser ablation was employed for the preparation of 20 nm particles of indometacin and phenitoin, a drug used for treating epilepsy.[24] In the case of a laser with the pulse width of 8 ns (nanosecond laser), the nanoparticle size was 40–60 nm and did not depend on the nature of substance.

Dye end	1	2	3	4
Particle size, nm	59	50	40	43

With a laser with the pulse width of 150 fs (femtosecond laser), smaller particles could be synthesized.

The dependence of the nanoparticle size and structure on the laser pulse wavelength, width, and fluence is related to the laser ablation mechanism. In the nanosecond region, this process was of the photothermal nature,[25,26] whereas in the femtosecond region, it was of the photomechanical origin.[27] For nanosecond photothermal ablation in solution, the crystals of an organic compound absorbed laser radiation, which was accompanied by the local increase in temperature, compensated by cooling due to the heat transfer to the solution. The established equilibrium temperature determines the particle size. An increase in the laser radiation fluence increases the equilibrium temperature and hence favors fragmentation. For example, for a wavelength of 355 nm, the size of Quinacridone particles decreased from 120 to 50 nm as the fluence increased from 40 to 100 mJ cm^2. Figure 9.2 shows microimages of quinacridone particles prepared under the action of a laser pulse.

For femtosecond lasers, the multiphoton absorption induces a considerable increase in the amplitude of vibrations of lattice molecules and elevates the pressure in the irradiation zone. As a result, the mechanical fragmentation begins prior to establishment of the temperature equilibrium with the solvent, and the number of particles increases. Thus, quinacridone particles measuring 13 nm were prepared with the fluence of 40 mJ cm^{-2} and the pulse width of 150 fs.[3]

The effect of the laser radiation fluence on the formation of different modifications of a given compound depends on the thermodynamic stability of the latter. For example, the quinacridone β-phase is less stable than its γ-phase; therefore, the latter[10] formed at the higher fluence.[3]

Thus, regulation of the laser radiation parameters allows the control over the size of the formed nanoparticles and production of stable concentrated dispersions of organic nanoparticles with no stabilizers.

However, the following two drawbacks of the laser ablation method deserve mention:

- Nanoparticles are formed under the action of a narrow laser beam; hence, only a small amount of dispersion is formed. This implies that high power inputs are necessary for the production of considerable amounts of nanoparticles;
- Intense laser radiation may cause photochemical decomposition of certain organic substances.[28]

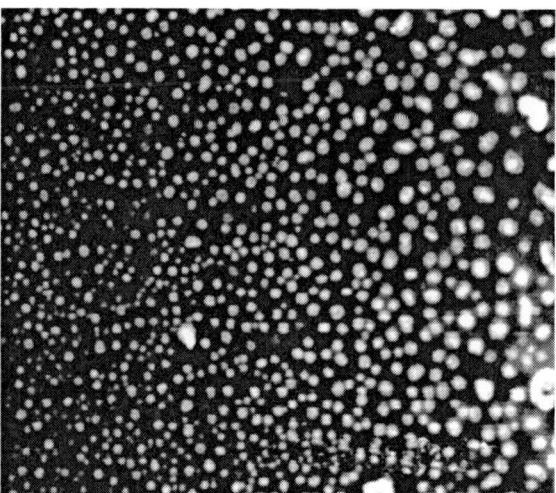

FIGURE 9.2 SEM image of quinacridone particles obtained upon laser radiation[4] of wavelength 355 nm, pulse duration 8 ns, and fluence 100 mJ cm.

9.2.2 Chemical Methods

9.2.2.1 Solvent Replacement

Preparation of an aqueous colloidal solution of β-carotene was described for the first time in 1965.[29] Carotene and poly(vinylpyrrolidone) were dissolved in chloroform, and then chloroform was evaporated in vacuo to form a solid mixture of β-carotene and polyvinylpyrrolidone. Upon the addition of water to the solid mixture, an aqueous colloidal solution formed, which was called "hydrosol" (by analogy with "aerosol").

The hydrosol contained mainly amorphous particles of water-dispersed carotene. The average particle size in the hydrosol was 200–400 nm, which allowed their use for intravenous injections. Such a technology of preparation of drug hydrosols had several features. First, the precipitation required a large amount of water, and the content of water in the product was 96–98%. Second, considerable amounts of stabilizers are added to prevent crystallization and the particle growth. For example, in the synthesis of a cyclosporine hydrosol with the particle size of 270 nm in the presence of gelatine, the gelatine-to-cyclosporine ratio was 20:1.25. The introduction of a large amount of viscous stabilizers can decrease the solubility; hence, the stabilizer should be accurately dosed. The final stabilization was accomplished by lyophilization or spray drying.[8]

A new procedure that involved solvent replacement, i.e. reprecipitation, was developed for the preparation of perylene nanoparticles.[30] This procedure consists of the fast introduction of a substance dissolved in an appropriate good solvent into a large volume of a miscible poor solvent. The sudden change in the medium induces the precipitation of organic molecules.[31] The reprecipitation method was widely used for manufacture of nanoparticles of different organic compounds.[32–35] For example, perylene particles with an average size of 100 nm (Fig. 9.3) were obtained by pouring the 0.25×10^{-3} M perylene solution in acetone to water with vigorous stirring.[36]

Nanoparticles of 5-(4-dimethylaminophenyl)-3-(4-dimethylaminostyryl)-1-phenyl-2-pyrazoline (5) were prepared by adding the ethanol solution of this compound (1.0×10^{-2} M) to water with vigorous stirring. Aggregation began immediately to form a dispersion of nanoparticles of compound 5 in water.

The size of the obtained nanoparticles can be controlled by changing the volume of the added ethanol solution of compound 5 and varying the temperature.

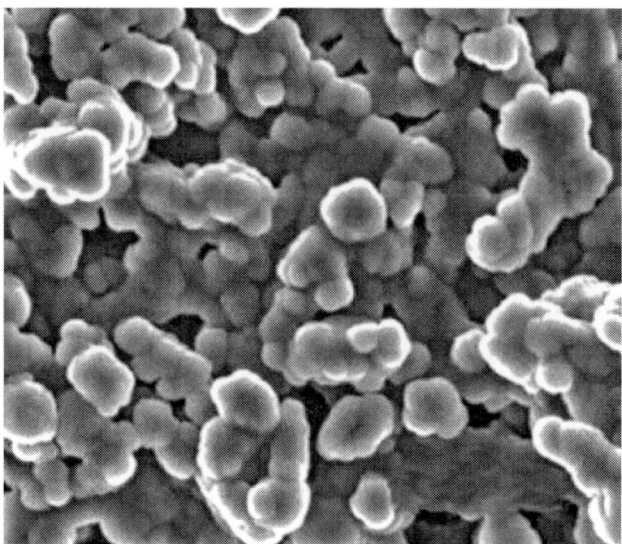

FIGURE 9.3 SEM image of perylene particles synthesized by the reprecipitation method.[36]

Thus, the addition of 40 and 100 mL of the ethanol solution to 10 mL of water resulted in 50- and 105-nm nanoparticles of compound 5 at 25 °C and 90- and 190-nm nanoparticles at 50 °C, respectively.[34]

Reprecipitation of 1×10^{-3} M acetone solution of 5-(4-dimethylaminophenyl)-3-naphthyl-1-phenyl-2-pyrazoline (100 and 50 mL) with water at 25 °C afforded nanoparticles measuring 65 and 40 nm, respectively.[37] In a similar way, nanoparticles of trans-1-cyano-1,2-bis(4′-methylbiphenyl-4-yl)ethylene (6) were prepared.

Tetrahydrofuran was used as the solvent and water served as the antisolvent. According to data from the emission scanning electron microscopy, nanoparticles of compound 6 represented spheres with the diameter of 30–40 nm. The suspension of nanoparticles was transparent without any sediment and stable for more than three months.[32] It was shown[38] that hydrophobic molecules containing the polar C=O groups were stable in water for more than a year with no surfactants.

Using the reprecipitation method, with dimethylsulfoxide as the solvent, particles of the dye indigo blue measuring 11–17 nm were obtained.[39]

The solvent substitution occurs under mild conditions, which allows fabrication of nanoparticles of thermally unstable compounds.[35] This method is relatively simple and accessible. However, its drawbacks include low-production performance.[40] Very dilute solutions of nanoparticles are obtained, because it is the millimolar concentrations of the organic compounds that are normally added to large volumes of antisolvents. Second, this method cannot be used for substances poorly soluble in organic solvents (e.g. for pentacene).

In the method of precipitation, keen attention was drawn to the use of antisolvents. Their application in preparation of drug nanoparticles is considered in an individual section.

9.2.2.2 Antisolvents for Precipitation

The rapid extension of pharmaceutical applications of nanoparticles may also simultaneously aggravate the problems associated with their use. For instance, the water-insoluble drugs should be mentioned. These compounds include a large number of modern drugs. The efforts of many scientists and manufacturers are aimed at increasing their solubility and, as a consequence, improving their bioavailability and the related therapeutic effects. In connection with this, attention should be focused on the general requirements to organic nanoparticles involved in the drug composition. As was mentioned above, suspensions prepared by different dispersing procedures are used most widely on the industrial scale. However, crystalline substances are of no less importance, being the most stable form for drug storage. For any method of synthesis, it is the particle size, the size distribution, and the stability of synthesized systems that are the most important factors. The decisive role pertains to kinetic peculiarities of the nucleation and the particle growth.

When considering the methods of preparation of organic nanoparticles, attention was drawn to a supersaturation that arises in the liquid phase upon the changeover of the solvent or in the gas phase at the cryosynthesis. Under conditions of the supersaturation, certain properties of synthesized drug nanoparticles can be regulated by varying the temperature or adding an antisolvent.[41] The dependence of the supersaturation on the particle radius and mass and also on the solvent density was also considered. The supersaturation and the supersaturation-determined nucleation kinetics were shown to be controlled by the local changes, for instance, proved to be more pronounced on cold walls as compared with the middle of the vessel. This affects the homogeneity and leads to heterogeneity, changing the kinetics of nucleation and particle growth steps at precipitation. To obtain coarse particles with a narrow size distribution, it is necessary to optimize the supersaturation and control it in the course of precipitation. For this purpose, the most advanced way is to use antisolvents, which are added to solutions of active drugs in order to lower down their solubility or affect their reversibility.[42]

For strongly hydrophobic compounds, the self-organization can occur as the result of the substitution of a solvent unsuitable for this compound, e.g. water, for the organic medium. For water-soluble compounds, the precipitation occurs as the result of addition of an organic solvent.[43]

The severe supersaturation simultaneously decreases the radius of the critical nucleus and increases the nucleation rate; hence, to obtain a reliable distribution of particles, it is necessary to use processes with short characteristic mixing times. The use of jet mixers allows the 10^4-fold supersaturation to be reached in several ms.[40] For deposition involving antisolvents, the temperature-controlled baths with magnetic stirrers and the continuous narrow impinging jets were used.[44] The dilution of an organic jet with water initiates precipitation. To reduce aggregation and retain nanoscale sizes of particles, aqueous dispersions of polymers are added to the aqueous phase. A copolymer, polypropylene oxide–polyethylene oxide, polyvinyl alcohol was used most often. Mixing in baths proved to be less efficient than mixing in jets. In a bath, it is more difficult to control the particle size so that coarser aggregates are formed. Supersaturation depends on the drug concentration in solution and its solubility in the mixture. The determination of the optimal solvent/antisolvent ratio is necessary for the subsequent precipitation.

In those cases where solubility in water limits the supersaturation, in order to attain the latter state in an aqueous medium, it is desirable to increase hydrophobicity of the substance, thus making easier the formation of nanoparticles and their stabilization.

The hydrophobicity of a substance can be changed by covalently binding it to a poorly soluble "anchor." The solubility is reduced as the result of both the anchor hydrophobicity and the overall increase in the molecular mass. For the case of fast precipitation, this favors the formation of nanoparticles in the aqueous media.[40]

An alternative to covalent binding is the formation of complexes with ion pairs, which allows the solubility of drugs to be changed. Hydrophobic ion pairs are formed in the specific interaction with an oppositely charged molecule, where the solubility decreases due to neutralization of the charge that promotes the water solubility.[40] Ion pairs can be formed either preliminarily or immediately in the reaction system, initiating the precipitation.

9.2.2.3 Chemical Reduction in Solution

Chemical reduction in solution is one of the most popular methods for the production of nanoparticles of different metals and semiconductors.[45] So far, this is not widely used for preparing organic nanoparticles; however, it was employed in some special cases.

For example, we consider the synthesis of perylene (Pe) nanoparticles. This is based on the reduction of perylenylium perchlorate with the bromide anions in the presence of cetylammonium bromide (CTA^+Br^-) in acetonitrile.[46]

Perylenylium perchlorate was synthesized by the reaction:

The reduction of perylenylium perchlorate proceeded according to the equation:

Perylenylium perchlorate was synthesized by the reaction:

$$2\,Pe + 2\,AgClO_4 + I_2 \longrightarrow 2\,Pe^+ ClO_4^- + 2\,AgI.$$

The reduction of perylenylium perchorate proceeded according to the equation:

$$2\,Pe^+ ClO_4^- + 2\,CTA^+ Br^- \longrightarrow 2\,Pe + Br_2 + 2\,CTA^+ ClO_4^-.$$

As a result, perylene nanoparticles measuring from 25 to 90 nm were obtained. It was noted that the changes in the cetylammonium bromide-to-perlenylium perchlorate molar ratio resulted in perylene nanoparticles of different sizes. Thus, for the ratio 0.2–0.3, particles of 25 nm were obtained. As the molar ratio increased, the particle size also increased.

The way of injection of the reducing agent strongly affected the shape of perylene nanoparticles. Its dropwise addition resulted in single spherical nanoparticles, whereas its addition in one portion led to self-assembling of nanoparticles to afford nanorods.

By an example of perylene, the following advantages of the method of reduction in a homogenous solution were revealed:

- separation of nucleation and particle growth steps;
- the possibility of control over the particle size by varying the substance concentration and the mode of its injection.

9.2.2.4 Ion Association

Ion association in aqueous solution was used for the preparation of nanoparticles of pseudocyanine dyes.[47] The essence of this method can be explained for the synthesis of a cationic styryl dye (7), 2-(4-dimethylaminostyryl)-1-ethylpyridinium tetraphenylborate.

Compound 7 tended to association with different hydrophobic borate anions in the presence of a neutral stabilizer, polyvinylpyrrolidone.[48] Fast injection of a solution of 2-(4-dimethylaminostyryl)-1-ethylpyridinium iodide to a solution of sodium tetraphenylborate afforded pure suspension of nanoparticles of compound 7 with the size from 30 to 100 nm.

Chapter | 9 Organic Nanoparticles

The method of ion association employs no organic solvents. The particle size can be controlled by varying the cation-to-anion molar ratio.[49]

9.2.2.5 Synthesis of Nanoparticles in Water–Oil Emulsion

Nanoparticles were synthesized in water–oil emulsions containing a solute in a volatile solvent. Evaporation of the solvent affords nanoparticles with sizes comparable with the emulsion drops. Emulsification required the use of special equipment to achieve the needed drop size. Nanodrops were prepared by high-pressure homogenization and sonication.

From microemulsions containing sodium dodecyl sulfate as the surfactant, nanoparticles of propylparaben with the sizes of 40–70 nm were obtained.[50]

A version of producing nanoparticles in a stabilizer-free emulsion of the oil-in-water type was proposed.[39] Figure 9.4 illustrates a possible mechanism of the organic nanocrystal fabrication, using the emulsion method. First, an emulsion of the target substance in an organic solvent was prepared at elevated temperature. Its subsequent cooling to room temperature brings about nucleation and the crystal growth inside the emulsion. The formed nanoparticles are transferred

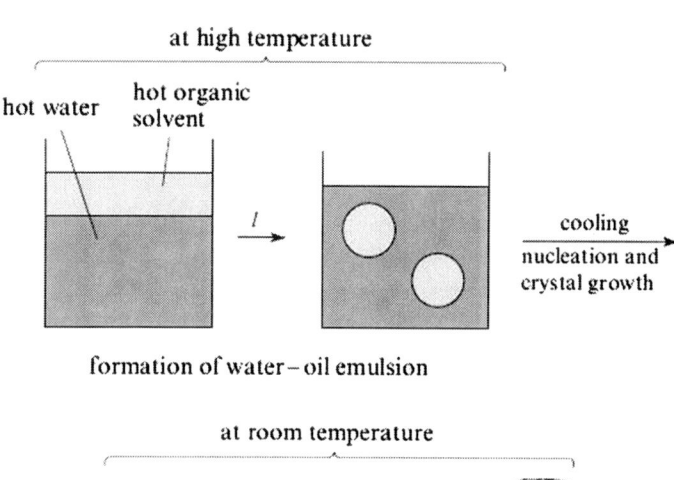

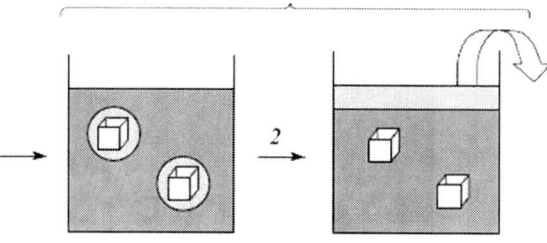

FIGURE 9.4 Schematic model of the organic nanocrystal fabrication by the emulsion method of nanoparticles: (1) preparation of the oil-in-water emulsion by stirring; (2) transfer of nanoparticles into the water phase, for example, by sonication.

into the water phase and dispersed by a mechanical method, e.g. by sonication. Thus, nanoparticles of tetracene measuring 85 nm and anthracene measuring 75 nm were prepared.

9.2.2.6 Photochemical Method

A new method was proposed for producing organic nanoparticles consisting of molecules of acridine dyes.[51] It is based on a photochemical reaction of tetrabromomethane and di-p-tolylamine in polystyrene films. The films were irradiated with light of different wavelengths and intensities in the presence of tetrabromomethane vapor. The reaction occurs at the polymer/vapor interface, which allows producing the dye nanoparticles in the fixed site on the polymer surface.

The size and the number of nanoparticles could be controlled by varying the intensity of the light flux, its spectral composition, and the duration of irradiation. Illumination with an incandescent lamp produced 100-nm particles. For the light flux of higher intensity, the argon laser was used. For the irradiation times of 20 s and 5 min, the average particle sizes were 40 and 80 nm, respectively.

An interesting experimental fact is that the micropores formed in the polymer film. The micropore size could be regulated by varying the parameters of the photon impact.

9.2.2.7 The use of Supercritical Fluids

At present, the use of different substances in subcritical and supercritical states for the production of nanoparticles of organic compounds is being intensively developed. The supercritical fluid (SCF) technologies have considerably extended the potential of preparation of organic nanoparticles. The classification of SCF technologies can be found in several reviews.[52-54]

A chemical compound in the supercritical state may be used as both the solvent and the precipitant.[55] Carbon dioxide was used most often as the supercritical solvent. It represents a noninflammable, cheap, nontoxic gas and is characterized by mild conditions of the transition to the critical state ($T = 304$ K and $P = 73.8$ atm).[56]

Supercritical fluids as the solvents were successfully used in a version of rapid expansion, where the saturated solution of a compound in an SCF is sprayed through a nozzle. As the pressure drops down, the solvent passes to the gaseous phase and the substance is precipitated to form fine aerosol. This method was used to produce, at a pressure of 220–450 atm and a temperature of 308–348 K, 100-nm spherical nanoparticles of a local anesthetic, lidocaine.[54] By a similar procedure, the nanoparticles of mefenamic acid (8) measuring 880 nm were fabricated.[57]

Several procedures for the rapid expansion of supercritical solutions were described. In one of them, the fluid with the target substance was sprayed into an organic solvent. Thus, ibuprofen particles measuring 40 nm were stabilized in a solution of sodium dodecyl sulfate.[58]

In another version, the SCF containing the dissolved compound was sprayed into water containing the surfactant Twin-80 to produce cyclosporine particles measuring 500 nm.[55]

Nanoparticles of the dye C545T (9) measuring 5–15 nm was produced using supercritical carbon dioxide.[59]

The drawback of the method under consideration is that the range of compounds well-soluble in supercritical carbon dioxide is limited.

Nanosizing of insoluble or poorly soluble substances can be effected, where the fluid performed the functions of the precipitant. For the dissolution of such substances, organic compounds are used that are soluble either completely or partly in the SCF. A saturated solution of the target compound in the organic solvent was sprayed under controlled conditions through a nozzle into a vessel containing the fluid, where the compound crystallized due to the abrupt decrease in the solubility to afford micro and nanoparticles. It was noted that the residual amounts of organic solvents could contaminate the final product.[54]

The following procedures were developed in which the supercritical solvents could be used as the medium for the precipitation of the target micro- and nanoparticles:

- precipitation in a gaseous antisolvent [Gas AntiSolvent (GAS)];[60]
- precipitation in a supercritical antisolvent (SAS); this procedure was used for producing tetracycline particles measuring 150 nm;[61]
- dissolution of a fine dispersion in an SCF [solution-enhanced dispersion by supercritical fluids (SEDS)]; Sulfathiazole particles measuring ~200 nm were prepared using this procedure;[62]
- supercritical extraction [aerosol solvent extraction system (ASES)].

As an example, let us consider supercritical extraction of a glucocorticosteroid, budesonide. The steroid solution in dichloromethane was sprayed into a large volume of SCF (carbon dioxide) for 1 h. Dichloromethane is soluble in SCF and hence is extracted, while the insoluble steroid forms the solid phase.[63] Figure 9.5 shows a microimage of budesonide nanoparticles.

When SCFs (e.g. carbon dioxide) are used with substances poorly soluble in these fluids, they represent antisolvents. The mechanism of accompanying

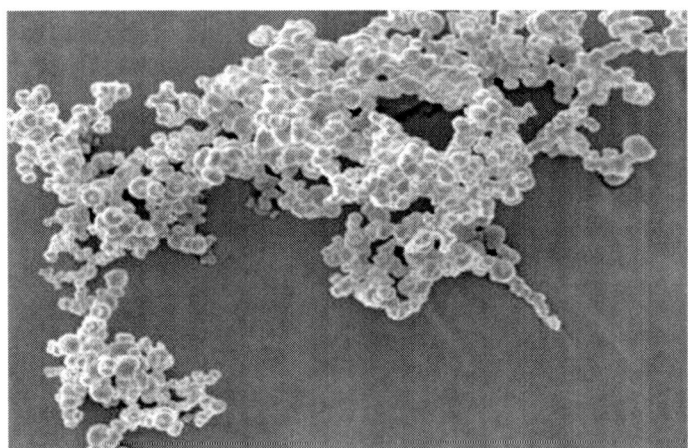

FIGURE 9.5 SEM image of budesonide particles, prepared by supercritical extraction.[63]

phenomena depends on the physical and chemical conditions of the process. The effect of pressure, temperature, and concentration on the size of synthesized particles was studied for precipitation in supercritical carbon dioxide. The model system represented the gadolinium acetate–dimethylsulfoxide system with carbon dioxide playing the roles of the SCF and the antisolvent.[64] According to scanning electron microscopic data, depending on the experimental conditions, one can obtain gadolinium acetate in the form of nanoparticles, microparticles, and extended microparticles.

At the fixed temperature (40 °C) and solution concentration (60 mg/mL), the pressure in the precipitator (500 mL) was varied in the range of 90–200 bar. Microparticles were synthesized at 90 bar and 120–180 bar. Nanoparticles could be obtained only at a pressure of 200 bar. According to XRD data, all particles were in the amorphous state. The effect of the solution concentration on the formation of particles was studied in the range of 20–300 mg/mL for the pressure of 150 bar and 40 °C. The average diameter of nanoparticles was found to vary from 86 nm at 20 mg/mL to 90 nm at 25 mg/mL. The particles represented irregular spheres. Both the particle size and their size distribution increased with the increase in the concentration. The effect of temperature was studied in the range of 35–60 °C, for the pressure of 150 bar and the solution concentration of 60 mg/mL. Under these conditions, nothing except microparticles and extended microparticles could be prepared. The authors associated this with the narrowness of the temperature interval. The formation of either nano- or microparticles was explained to be a result of changes in the jet dynamics and the surface tension. The competition between the two mechanisms was considered. In one mechanism, the jet breakup occurred, droplets were formed, and then dried, which gave rise to the formation of microparticles. In the other mechanism, the dynamic interface tension became weak, and a gas mixture was formed from which nanoparticles precipitated. These processes can be characterized by the time of the jet breakup (T) and the time of the changeover in the dynamic interface tension (t). The interplay of these mechanisms made it possible to explain the obtained results.[63] For $T<t$, the mechanism includes the formation of droplets to produce microparticles. For $T>t$, the process occurs via the formation of a gas mixture and results in precipitation of nanoparticles. Under different conditions, the important role is played by the critical mixing point of a binary system formed by a SCF and a solution added to the latter. The complex analysis of the effect of pressure, concentration and temperature has shown that the obtained results are well consistent with the earlier data.[65] Presumably, these results can be extended to other nanoparticles precipitated in the amorphous state.

For the case of precipitation of crystalline nanoparticles, different mechanisms were discussed.[64] For fluid CO_2 used as the antisolvent, the main problem is to pass from nanoparticles to microparticles. The process is determined by the dynamic properties of the system; when a fluid jet is used, its breakup can be organized near the nano-to-microparticle transition point in which the competition of different mechanisms occurs to afford a mixture of particles with different sizes.[63]

The merit of the SCF technologies is their environmental safety in the case of SCF-soluble substances; the drawbacks are the contamination of the final product with the solvent in the case of production of nanoparticles of substances insoluble in SCF as well as the sophisticated instrumentation required for this process and, as a consequence, high cost of the final product.[66] Certain progress was achieved in solving the problem of scaling the SCF technologies to make them suitable for industry.[67,68]

9.2.2.8 Cryochemical Synthesis and Modification of Nanoparticles

Cryochemistry studies the reactions proceeding at low (77 K) and ultralow (4 K) temperatures. In this temperature range, unusual chemical transformations involving free radicals, ions, and unstable molecular complexes were observed.[69] New mechanisms of traditional classic and model reactions of the halogen and hydrohalide addition to double bonds occurring at low temperatures were discovered.[70] The low-temperature limit for the chemical reaction rate was determined, and the presence of the tunneling mechanism was demonstrated for reactions at ultralow temperatures.[71]

The cryochemical synthesis of nanoparticles of organic compounds is based on a well-balanced combination of cold and heat; its peculiarities were considered in Ref. 72. Low temperatures are used to avoid uncontrolled changes in the intermediate and target products and to control the properties of the resulting materials. This method involves deep freezing of solutions of substances, which leads to fast solidification of both the solvent and the solute.[73] The obtained product in the form of cryogranules was subjected to freeze-drying (cryodrying) or cryoextraction. The cryodrying includes the following stages: freezing, primary drying, and secondary drying. The freezing rate affects the subsequent sublimation and the properties of the final product. The critical freezing rate was found, below which the particles tend to aggregate.[74] Cryochemical technology allows production of diverse materials (Fig. 9.6).

As a rule, the contemporary drugs represent complex organic compounds, of which many are virtually insoluble in water. The poor solubility limits the modes for drug administration into the organism and affects the kinetics and the mechanisms of their targeted delivery, metabolism, and excretion.

A new method was proposed for modifying and dispersing crystalline powders of drugs with recourse to low temperatures.[75,76] The method relies on the nonequilibrium–metastable states of systems formed as a result of vapor deposition onto a cold surface, which was followed by heating to room temperature (Fig. 9.7).

The fundamentals of this method were developed when studying different chemical transformations of organic compounds at low temperatures.[77,78]

By varying the condensation and heating conditions, one can control the parameters of processes and the properties of products. Two methods were developed for drug modification. One of them is based on vaporization and transfer of the substance with a flow of a heated gas, which was followed by its condensation on a cold surface. The other method is based on the vaporization and condensation in vacuum. The proposed methods were used for modifying and dispersion of the following drugs: carvedilol, moxonidine, glibenclamide, fluticasone propionate, gabapentin, androstenediol, and piroxicam. The majority of the drugs were virtually insoluble in water.

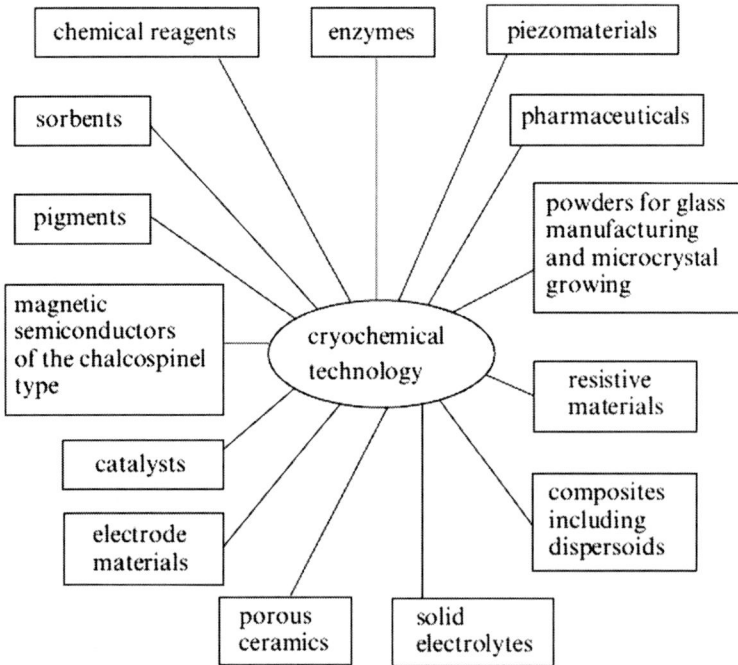

FIGURE 9.6 Materials as the cryochemical technology products.[72]

An amorphous form of a cardiac drug, carvedilol, was synthesized by the inflow modification.[79] According to data from the optical microscopy, the size of the original carvedilol particles was 10–15 mm, the average particle size of the cryomodified drug was 1 ± 0.1 mm, and the minimum size was 0.5 mm. The biological and medicinal trials pointed to an increase in the amorphous drug activity with the decrease in the particle size.

The jet method in an inert gas flow was used for studying the modification of the hormone D5-androstene-3b,17b-diol as a radioprotector. According to the IR spectroscopic data, the initial and the modified forms of androstenediol were similar. According to XRD studies, the modified drug represented a crystalline monohydrate of androstenediol. The particle size of the original drug varied from 3 to 300 mm. Figure 9.8 shows the size distributions of the original and the modified particles. The average size of androstenediol particles decreased to 200 nm and the minimum size decreased to 100 nm.

As the particles size decreased to nanometers, the mass rate of dissolution increased more than 1.5-fold for the same temperature and sample mass.

For the antiepileptic agent, gabapentin, the use of the method of vacuum sublimation–condensation made it possible to produce three forms of which one was obtained for the first time.[80] The particle size of the cryomodified drug was 0.5 mm.

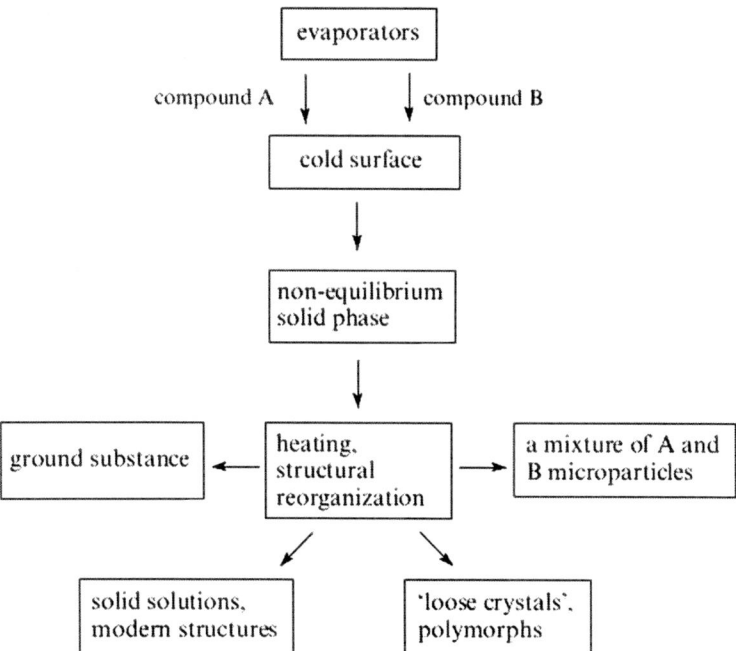

FIGURE 9.7 A scheme of possible transformations of one or two organic compounds in the condensation and heating.

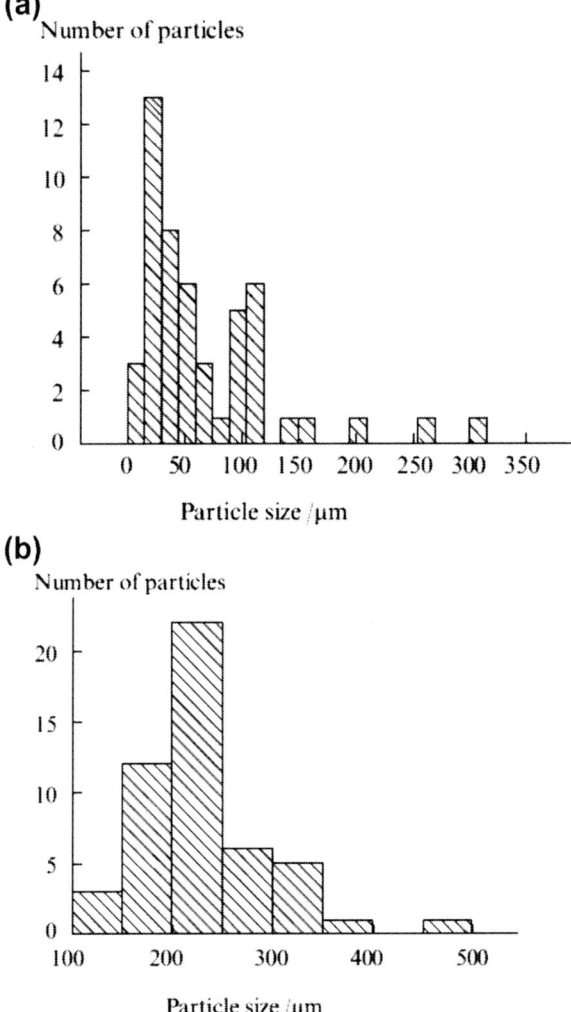

FIGURE 9.8 Histograms of the size distribution of D5-androstene-3b,17b-diol particles according to the data of scanning electron microscopy in the original (a) and modified (b) substances.

The method of vacuum cryomodification was developed for sublimation of compounds that are nonvolatile and chemically unstable at elevated temperatures.[81] The evaporation was carried out by pressing mechanically a thin layer of the substance to a heated metal grid. The condensation could be accomplished onto the walls of a polished copper cube cooled by liquid nitrogen; simultaneously, IR spectra of the condensate could be recorded.

The discussed methods of cryomodification had distinctive features. The transfer of the substance to the gas phase and the use of liquid nitrogen ensured that the final product was obtained from its molecular and nonequilibrium state. Using controlled heating, it was possible to change the size and the structure of modified particles and obtain their new forms.

Micronization of drugs was carried out in the absence of solvents. Cryochemical methods made it possible to modify the substances due to the introduction of auxiliary agents used in the drug synthesis.

At low temperatures, a new modification of 7-bromo-5-(2-chlorophenyl)-1,3-dihydro-2H-benzodiazepin-2-one (phenazepam),[82] one of the most efficient psychotropic agents used most widely in the medical practice, was obtained (Fig. 9.9).

A new modification of phenazepam was prepared in a unit that allowed transfer of a crystalline substance into the gas phase, which was followed by its condensation in an inert gas flow onto a surface cooled with liquid nitrogen. NMR-spectroscopic examination of solutions of the original phenazepam and its cryomodification in dimethylsulfoxide-d_6 has shown that the chemical composition of the phenazepam molecule remained unchanged.

A comparison of the results of XRD analysis of modified phenazepam with the data from the Cambridge database revealed the appearance of a new crystalline modification, β-polymorph. The lattice parameters of the cryomodified monoclinic phenazepam, $C_{15}H_{10}BrClN_2O$, were as follows: $a = 14.792(5)$, $b = 11.678(3)$, $c = 8.472(2)$ Å, $\beta = 93.677(19)8$, $V = 1460.4(7)$ Å^3, $\rho_{cal} = 1.59$ g cm^3, $Z = 4$, and space group P21/c. The average particle size of the β-polymorph of phenazepam was 55 ± 12 nm according to XRD data and 30–120 nm according to the transmission electron microscopy (TEM) data, i.e. the results obtained by these two methods were mutually consistent. In water, the β-polymorph dissolved faster by a factor of 3.9 as compared with the original phenazepam.

The particle size of drugs affected their solubility, dissolution rate, bioavailability, and metabolism. The mode of administration of a drug into the organism is chosen depending on the particle size.

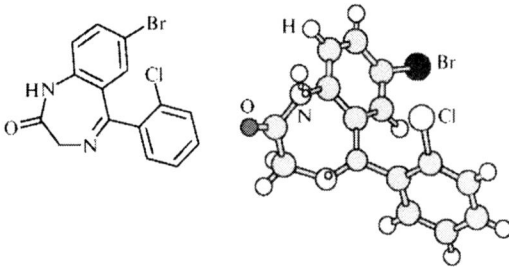

FIGURE 9.9 Phenazepam structural formula and a view of its molecule with the optimized structure.

Biopharmacological in vivo trials of cryochemically modified phenazepam have shown that the new modification retained the advantages of the original drug, while the side effects decreased by 40%.

In pharmacology, keen attention is paid to the synthesis and the use of new crystalline polymorphs of drugs. This is associated with the fact that different crystal systems with different unit cells (polymorphs) obtained from the same elements possess different physical and chemical properties and can have different therapeutic action. The state of the art in the studies of polymorphism and numerous examples of the use of this effect in pharmacology were described in a monograph.[83]

Sublimation of drugs followed by their condensation onto cold metal surfaces was employed[84] in the synthesis of polymorphic modifications of the antiepileptic drug, carbamazepin (10). The drug was condensed on the Au(111) and Cu(111) surfaces at 80 K, and its structure was studied at the same temperature by the scanning tunneling microscopy.

On the Au(111) surface, compound 10 formed the molecular structures different from those observed in three-dimensional crystals, namely, trimers were formed as a result of intermolecular hydrogen bonding and interaction with the metal (Fig. 9.10). In neither of the polymorphs known, such formations have ever been observed.

The nature of metals was shown to influence the chirality and the density of packing of two-dimensional molecular structures.

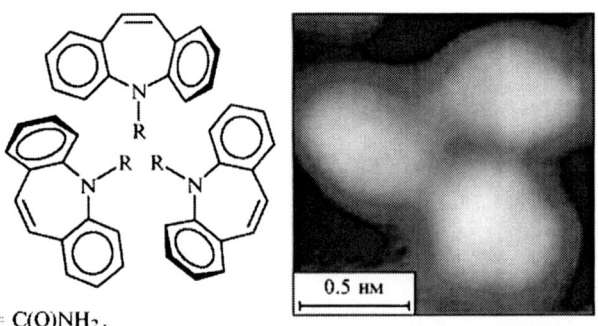

R = C(O)NH$_2$.

FIGURE 9.10 A scheme of formation of carbamazepin trimer and its STM image on the Au(111) surface.[84]

The obtained results can potentially be used for monitoring the polymorph formation. However, it is necessary to elucidate whether the synthesized two-dimensional structures can act as the templates in the formation of three-dimensional crystalline particles.

9.3 PROPERTIES AND APPLICATION OF ORGANIC NANOPARTICLES

In the Introduction to this chapter, the similar and dissimilar properties of nanoparticles of metals and organic compounds were outlined and certain application fields of organic nanoparticles were considered. In the present Section, the spectral and other properties of the latter are analyzed in greater detail.

9.3.1 Spectral Properties

The dependence of the Stokes shift in the molecular spectra on the size of organic particles was revealed.[85] The restriction for the exchange interactions in organic nanoparticles brings about an increase in the nonlinear optical characteristics.[86] The dependence of the melting temperature on the particle size was demonstrated.[87] A substantial difference in the optical and electronic properties of organic nanomaterials from the properties of nanomaterials based on metals and inorganic substances is associated with the presence in the former of weak intermolecular interactions such as the van der Waals forces and π–π conjugation.

The dependence of optical properties on the size of organic nanoparticles, which ranged from several tens to several hundreds of nanometers,[3] was considered by many authors. The optoelectronic properties of organic nanoparticles are governed by the charge-transfer excitons (P_{CT}) and the Frenkel excitons with small radii. The effect of exciton localization was rationalized.[34]

Figure 9.11 shows UV spectra of nanoparticles of compound 5 in water and in a dilute aqueous solution of ethanol. In solution, the following three absorption bands appear: the first is associated with the phenyl radical (PPh) and the other two are associated with the $n-\pi^*- (P_{n-\pi^*})$ and $\pi-\pi^*$-transitions ($P_{\pi-\pi^*}$) of the pyrazoline ring. As the particle size increased from several tens to a hundred of nanometers, the P_{Ph} and $P_{\pi-\pi^*}$ bands shifted in the long-wavelength direction. Presumably, the observed bathochromic shift is the result of the increased overlap of π-orbitals of the pyrazoline ring and the strengthening of interactions between the molecules of compound 5 with the increase in the particle size.[34]

The emerging absorption is explained by the appearance of charge-transfer excitons (P_{CT}) throughout the extended crystal structure of aggregates of molecules 5 closely stacked in nanoparticles.

An interesting phenomenon of multiple luminescence is manifested in nanoparticles of 1,3-diphenyl-5-pyrenyl-2-pyrazoline (11). This compound contains two sterically separated chromophore fragments, namely, pyrene and

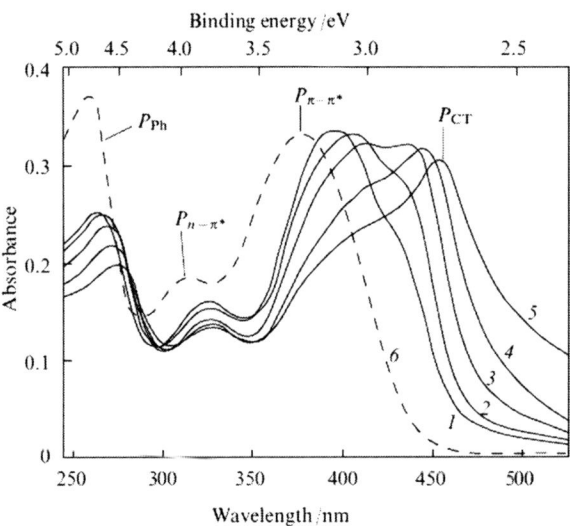

FIGURE 9.11 UV-Vis absorption spectra of dispersions of nanoparticles of compound 5 in water (1±5) and of a solution (C = 16105 M) in ethanol (6),[35] particle size/nm: (1) 20; (2) 50; (3) 105; (4) 190; (5) 310.

pyrazoline. In solution, the emission is associated only with the pyrene fragment, whereas in a solid crystal, this was exclusively due to the pyrazoline fragment. However, in nanoparticles of compound 11, the multiple emissions of radiation from pyrene, pyrazoline fragments and, additionally, from the charge-transfer complex between the pyrene and pyrazoline fragments were observed.[88] It was shown that with the increase in the particle size, the luminescence decreased.

Systematic studies of pyrazoline derivatives were carried out.[89] It was shown that the fluorescence spectra of 5-(2-anthryl)-1,3-diphenyl-2-pyrazoline nanoparticles differed from the spectra of solutions and crystals of this compound.

Guanosine or guanine (G) derivatives with oligo(p-phenylene-vinylene) substituents in position 8 prepared by a Pd-catalyzed cross-coupling were observed to form hydrogen-bonded self-associates.[90] In the absence of an alkali metal salt, they undergo a self-assembly with possible predominance of G-based tetramers (Fig. 9.12a and Fig. 9.13).

In the presence of alkali metal salts, the octamer can be formed as a result of cationic-dipolar interactions (Fig. 9.12b). The assembly of stable disk nanostructures with the diameter of 8.5 nm and the height of ~1.5–2.0 nm was observed.

Fluorescent organic nanoparticles were used, for instance, as the organic light-emitting diodes (OLED).[91] The efficiency of fluorescence of organic chromophores decreased in the solid state as a result of concentrational quenching. For certain fluorophores, the aggregation phenomenon was observed, which was

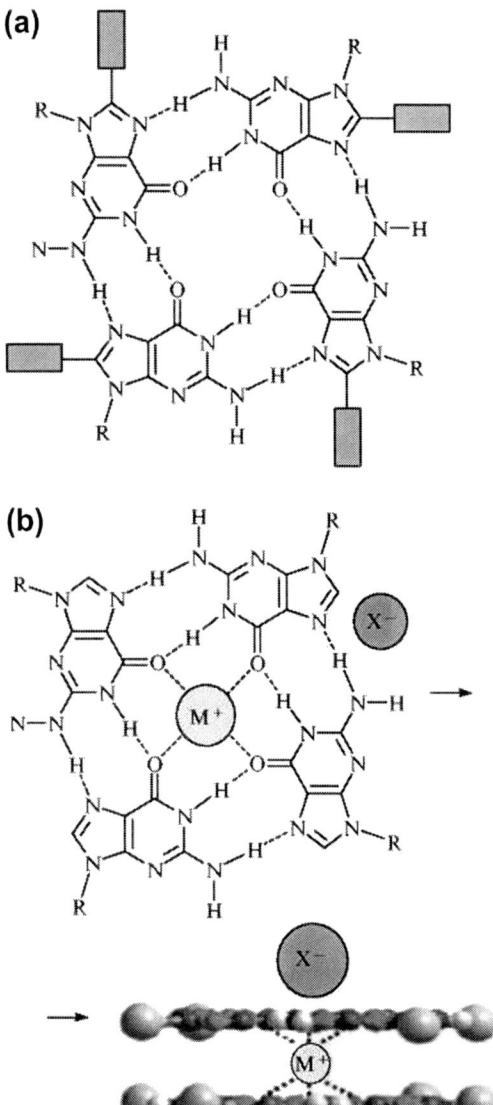

FIGURE 9.12 A model of self-assembly of guanine derivatives as a result of formation of hydrogen bonds (a) and in the presence of alkali metal cations (b).[90]

accompanied by intensification of fluorescence rather than its quenching.[92,93] For example, the fluorescence of solid trans-1,2-bis[4-(carbazolyl)phenyl]-1,2-dicyanoethylene nanoparticles was more intense as compared with a solution of this compound.[94] The photoluminescence of solid nanoparticles of compound 6 was 700 times (!) more intense than that of its solution.[32]

A new class of stable highly soluble substances, viz. donor-acceptor carbazole derivatives, was described in detail.[95] The dependences in their physicochemical properties on the change in the acceptor group were studied as well as their photophysical properties in different solvents. In tetrahydrofuran-water mixtures, organic nanoparticles are formed with the size of ~250 nm. Presumably, such particles could be used as OLED in dye-sensitized solar cells or the Gretzel cells.

Light-emitting diodes were mainly used in the design of devices for data representation in the production of thin full-color displays. The full-color displays based on light-emitting diodes are commercially available. However, no cheap displays have been developed so far, with the sizes comparable with those of the modern liquid-crystalline displays. The main problem the research encountered with was associated with the nonuniformity of light-emitting organic substances applied onto the screen surface. The difference in the organic film thicknesses led to nonuniform luminescence and had an adverse effect on the quality of color reproduction by these displays. An unexpected and interesting solution of this problem is based on the incorporation of organic field transistors into the displays.[96] Each of such transistors controls an individual pixel.[97] By controlling the current, it is possible to obtain the screen glow uniform over the whole surface, which required no substantial complications in the technological instrumentation for the selective deposition of light-emitting substances.

9.3.2 Quasi-one-dimensional Systems

When analyzing the properties and the application fields of organic nanoparticles, special attention should be drawn to quasi-one-dimensional (1D) organic materials in the form of nanotubes, nanoribbons, and nanowires. It is believed that such materials should have great influence on the development of microoptoelectronics of the next generation.[98]

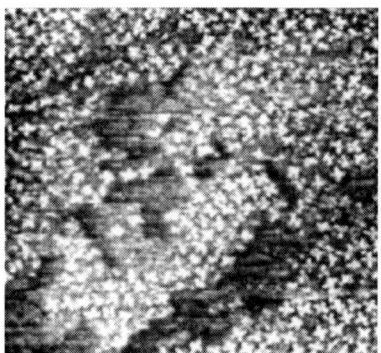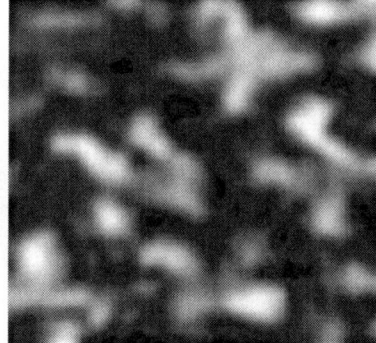

FIGURE 9.13 STM images of nanoparticles of a guanine derivative at different magnifications.[90]

Several methods were proposed for the preparation of 1D materials: by self-organization in solution,[99–102] template synthesis,[103–105] and physical vapor deposition.[106] Quasi-one-dimensional organic materials exhibit interesting optoelectronic properties. First of all, mention should be made of their inherent multicolor or the so-called red–green–blue (RBG) emission, which is applied in displays and screens. Such a color combination has usually been obtained by mixing different dyes, each emitting its own color. Recently, the multicolor emission has been observed for one substance, 1,2,3,4,5-pentaphenylcyclopenta-1,3-diene (12).[107]

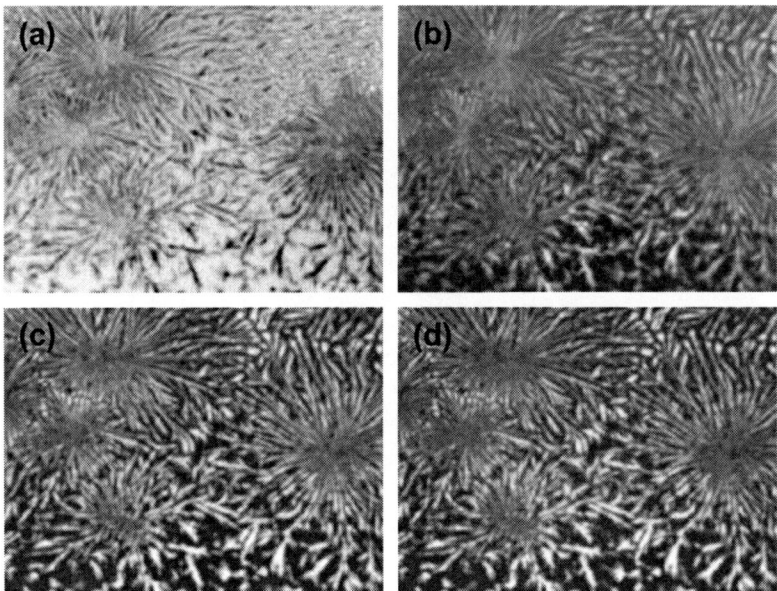

Under the action of light with the corresponding wavelengths, the RBG emission was observed for the 1D structure of compound 12 (Fig. 9.14).

FIGURE 9.14 Microphotographs (fluorescence microscopy) of nanoribbons of compound 12.[107] (a) bright-field image; irradiation with UV light with the wavelengths of 30 ± 380 (b), 460 ± 490 (c), and 520 ± 550 nm (d).

Yet another peculiarity of organic 1D nanostructures is associated with their properties of optical conductors. For example, 9,10-bis(phenylethynyl)anthracene nanotubes with the diameter of 690 nm and the length of 2.5 μm demonstrated the characteristic yellow emission with very bright luminescent sites at both ends of the nanotube and the relatively weak emission of the whole nanotube (Fig. 9.15).[108]

Finally, yet another significant property of organic 1D materials is the adjustable luminescence of binary nanowires. This effect was described in detail for nanowires of 1,3,5-triphenyl-2-pyrazoline with an addition of 9,10,11,12-tetraphenylnaphthacene (rubrene). It was observed that with the increase in the rubrene content, the emission color changed from blue to orange.[109]

Fluorescent nanoparticles of organic compounds can be used for detecting different substances. Thus, nanoparticles of 1-aminopyrene (13) are applicable for the quantitation of nitrites.

13

The fluorescence intensity is quenched by nitrites when involved in diazotization reaction, and the fluorescence extinction is proportional to the nitrite concentration.[110]

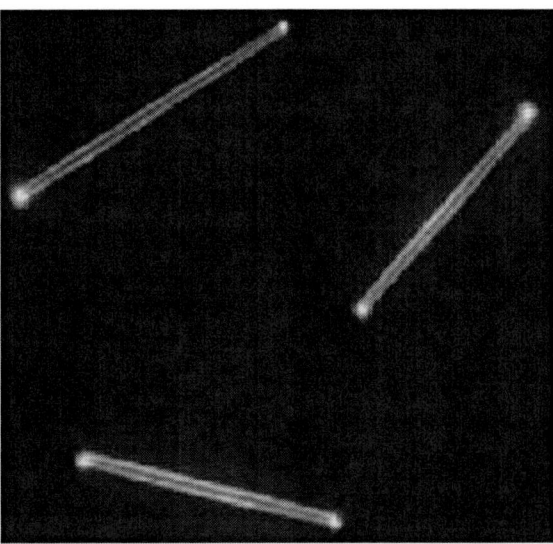

FIGURE 9.15 Microphotograph (fluorescence microcopy) of nanotubes of compound 12 irradiated with UV light of the wavelength 460 to 490 nm.[108]

9.3.3 Drugs and Nanoparticles

Nanoparticles of organic compounds demonstrate a great application potential in "green" chemistry and chemical biology. They easily form aqueous dispersions, which makes it possible to obviate toxic solvents in carrying out chemical reactions.[111] The preparation of aqueous solutions of organic substances as well as monitoring the reactions with living cells in these solutions required the use of detergents. Nanoparticles as such can react with cells and hence require no detergents.

Production of drug nanoparticles is one of the main directions in the preparation of new highly efficient formulations. Micro- and nanoforms of drugs exhibit unique properties and open up new promising approaches to the treatment of different diseases. Solubility of drugs is one of the properties that determine their efficacy.

The calculated and measured solubility values sometimes differed from 2–3-fold to 20-fold.[112] Such a difference can be associated with the fact that solubility depends on the particle size and shape and also on the presence of a charge that can affect both solubility and nucleation.

The properties determining the solubility can be divided into two groups, namely, intrinsic of coarse compact compounds and nanosized compounds. For coarse particles with the small surface-to-volume ratio (S/V), the important properties are the lattice enthalpy or the entropy and the properties pertaining to the solute–solvent interaction, such as the surface roughness and the charge. For nanoparticles with the larger S/V ratio, it is the properties affecting the nucleation and the particle growth, the state of particle surface, and its fractal dimensionality that are important.[113]

Theoretical estimates show that for particles measuring <100 nm, a considerable increase in solubility as compared with coarse particles could be expected. The classical Ostwald–Freundlich equation proved to be insufficient to comprehensively describe the nanoparticle solubilities.

The possibility of preparing nanosuspensions of poorly soluble substances is considered as an attractive merit of such drugs.[114]

The modification of the nanoparticle surface can be used in developing new methods of dissolution of poorly soluble substances. Thus, the dissolution rate of meloxicam nanoparticles is 6 times higher than that of the original drug.[115]

The drug celecoxib (14), poorly soluble in water, is used for treating rheumatoid arthritis, osteoarthritis, and psoriatic arthritis.

14

Production of amorphous nanoparticles of this drug measuring 17 nm increased its dissolution rate by 10-fold.[116]

The kinetics of nanoparticle nucleation by the example of four drugs insoluble in water was analyzed.[117] The effect of sonication and of hydroxypropylmethylcellulose (HPMC) on the particle deposition in the presence of an antisolvent (water) was studied. The antifungal drug itraconazole, the antibiotic griseofulvin, the nonsteroidal antiinflammatory drug ibuprofen, and the antibacterial drug sulfamethoxazole were tested, which made it possible to assess how the physicochemical properties affect the nucleation rate and the particle size determined by the light-scattering technique. The analysis of the results was performed within the framework of the theory of homogeneous nucleation, which allowed the contributions of kinetics and thermodynamics to be separated and the particle size to be taken into account. It was found that sonication and the addition of HPMC accelerate nucleation. Sonication favored the increase in the water molecule diffusivity. This increased the contribution of kinetic factors into the nucleation acceleration. The addition of HPMC reduced the solution viscosity and, hence, the contribution of the kinetic component.

However, the addition of HPMC also reduced the surface tension and decreased the contribution of thermodynamic factors. Hence, the use of HPMC favored a decrease in the energy barrier to nucleation and its acceleration. It was noted the decrease in this barrier was more pronounced for hydrophobic molecules such as Intraconazole as compared with other molecules, e.g. sulfamethoxazole. This is associated with the lower solubility of the former as compared with that of hydrophilic compounds such as sulfamethoxazole. The differences in the nucleation rates affect the particles involved in this process the sizes of which depend on the properties of the original substance such as the wettability for solid particles and the interface tension.

The therapeutic effect of drugs poorly soluble in water is apparently low. As a result, they have to be used in large doses, which entails undesirable side effects. This problem is especially important for drastic anticancer, hormonal, anti-inflammatory and antifungal drugs, and antibiotics. Nanotechnologies can provide a decrease in the therapeutic doses of drugs with the retention of their therapeutic effect, which would be favorable for reducing the size effects.

Stability plays the important role in the therapeutic effect of drugs. It is also of key importance for suspensions of many poorly soluble substances.

The physical and chemical stability of organic nanoparticles is one of the main characteristics determining their application limits.

The stability of nanocrystalline drugs was the subject of active discussions.[118] The size stability of drug nanoparticles is especially important for intravenous injections, because coarsening of particles to 5 μm leads to blocking of capillaries and arterial occlusion.[119] This is the control over the particle size and the size distribution should be provided during their storage as well as the thermodynamic and kinetic control over the stability of suspensions obtained from nanoparticles.

In contrast to microparticles, the properties and the stability of nanosystems depend on their proper dosage. The latter determines the size, the surface-to-volume ratio, and correspondingly, the properties of systems in vitro and in vivo.[120] Stability is one of the main aspects of drug efficacy and safety.

Stability of suspensions depends on physical processes such as sedimentation and agglomeration of particles, the growth of crystals, and the changes in their morphology.

The large surface of nanoparticles results in their high surface energy and thermodynamic instability, which is relieved with agglomeration. The agglomeration can change the stability of a suspension due to the fast precipitation of growing crystals. The concomitant problems are usually solved by the introduction of various stabilizers.

Two mechanisms of stabilization of suspensions in aqueous and nonaqueous media are traditionally considered, namely, the electrostatic repulsion and the steric stabilization.[121] The strong interaction between the stabilizer and the solvent is the decisive factor for the steric stability and prevention of agglomeration.[120] The main drawback of steric stabilization is associated with the necessity of tuning the stabilizer tails to a particular drug. The stabilization of indometacin nanoparticles involved the use of cyclodextrins and no surfactants.[122]

The latter gives rise to changes in the free energy $\Delta G = \Delta H - T\Delta S$. The case $\Delta G > 0$ corresponds to stabilization of the suspension; for $\Delta G < 0$, the agglomeration of particles occurs.

If a medium is a good solvent, then it partially favors stabilization because fine particles with adsorbed layers of the stabilizer cannot penetrate into one another. This decreases the number of configurations suitable for stabilization of tails and leads to changes in both the negative entropy and the positive free energy. On the other hand, if the medium is a poor solvent, the adsorption layers on particles can thermodynamically penetrate into one another, inducing agglomeration.

In the general case, poorly water-soluble pharmaceutical substances with the high molecular masses and the high melting points have a better chance to form nanosuspensions.[123]

The crystalline state is the most important factor that affects stability, solubility and efficacy. Under the effect of temperature, the high-energy amorphous state can transform into the crystalline state.

In formation of suspensions, one must take into account the possibility of chemical reactions that involve the solute and the solvent and can affect the stability of the system. The chemical stability depends on the properties of a particular pharmaceutical substance, its molecular structure, and the presence of different functional groups susceptible to hydrolysis or oxidation and their reactivity.[124] The general approach to using drug substances includes their storage in the crystalline state and the preparation of nanosuspensions immediately before their use.

Even more complicated problems arise for the case of peptides and proteins, for which besides the necessity of retaining the nanosize, one must control the secondary and other molecular structures.[125]

The particle size and the size distribution determine the physical stability of dispersions and are usually considered as its indicator or polydispersity index. The latter value is determined by the dynamic light scattering or laser diffraction techniques. For the polydispersity index in the interval of 0.1–0.25, the size distribution is considered as narrow, whereas the index of 0.5 points to a wide distribution.[118] The laser diffraction data give the relative size distribution. The morphology of nanoparticles can be determined by the scanning, transmission, and atomic-force microscopies. Indeed, the stability of a nanosuspension is higher for finer particles, because their sedimentation is slower. However, fine particles are not always desirable because they can strongly increase the dissolution rate, which may induce undesirable effects and weaken the therapeutic effect.

The wide use of stabilizers depends on understanding of their interaction with nanoparticles and the yet insufficiently developed procedure of screening of the selectivity of stabilizers themselves.

The use of the atomic-force microscopy allowed selective stabilizers for ibuprofen nanoparticles to be found.[126] It was shown that hydropropylcellulose (HPC) and hydropropylmethylcellulose (HPMC), in contrast to polyvinylpyrrolidone, can efficiency bind to the surface of ibuprofen nanoparticles. These results well correlate with the stability of suspensions and highlight the importance of the AFM in determining the selectivity of stabilizers and gaining insight into the mechanism of their action. However, the more efficient method of preparation of stable suspensions altogether excludes the use of stabilizers. This not only simplifies the procedure but also eliminates the problems associated with toxicity of stabilizers.

Despite the progress in pharmacology, the stability of drug nanoparticles remains a topical problem. It can be solved by gaining deeper insight into the interactions between nanoparticles and carrying out the quest for new highly efficient and selective stabilizers.

Morphology pertains to the fundamental characteristics or drugs. Nanoparticles of different morphology can be prepared by fast precipitation. By changing the kinetics of nucleation and growth, less stable highly energetic particles in amorphous or polymorphous states can be synthesized. The slow aggregation and precipitation usually lead to thermodynamically stable polymorphous crystalline structures. Hydrophobic substances in the amorphous state exhibit the higher dissolution rates and, correspondingly, the higher bioavailability. At the same time, substances in the form of crystals are more stable with respect to different external factors during the storage. The aforementioned processes should be taken into account when synthesizing nanoparticles.

A comparison of amorphous and crystalline itraconazole as regards its dissolution rate and bioavailability was carried out.[127] The amorphous state of this triazole antifungal agent was prepared by wet grinding and ultrafast freezing. The trials in mice have shown the 8-fold increase in the solubility in vitro and the 4-fold increase in the bioavailability in vivo for the amorphous state as compared with nanocrystalline state, which stressed the key importance of the nanoparticle shape for its bioavailability.

Thus, the structural state should be taken into account in acquiring pharmacological kinetic profiles to characterize the activity of substances. These studies require the development of special methods and procedures. The morphology of drugs can be determined by X-ray diffraction technique; however, it should be borne in mind that the spectra of nanoparticles may contain reduced signals and crystal peaks of different intensity.

Many drugs are used in the form of colloidal suspensions for which the stability of the system is also important in addition to the particle size, solubility, and morphology. As the particles size decreases, it becomes more difficult to control the changes in the surface thermodynamics and the physical and chemical properties of the system. Stabilization of suspensions of nanoparticles involves the active use of different surfactants that can erect steric barriers. For the dissolution of nanoparticles, besides their diffusion, size, and surface area, one should take into account the thickness of the particle boundary layer, which has its own steady state.

For coarse crystals, the particle surface is assumed to be flat and linear with respect to the gradient of concentration in the boundary layer. For stabilization of a suspension, the absence of the secondary recrystallization is also important. The prospects of using aqueous drug suspensions are also limited by the substance concentration (the latter can be introduced as a single dose or in a larger volume in order to attain the therapeutic concentration). As an alternative to suspensions, the drugs are kept during long-term storage in the form of nanopowders. The conventional drying procedures are useless for nanoparticles due to their recrystallization. The most widely used methods are cryodrying and spray drying on freezing.[128] Both methods allow avoiding the aggregation of nanoparticles during the powder preparation. To prevent the possible binding of nanoparticles into aggregate bridges during vacuum cryodrying, the cryoprotectors such as polyvinyl alcohol, polyvinylpyrrolidone, hydroxypropyl methyl, and cellulose are used. The complications that arouse in the course of cryodrying were analyzed.[127]

As an alternative to the conventional cryodrying and cryospraying, a technology was proposed recently in which flakes of nanoparticles were subjected to special reversible filtration which allowed the material to be dried in a short time and at a low cost.[129]

Moreover, micronization up to nanosizes determines the development of drugs for the aerosol, intravenous, and intramuscular applications. For example, an aerosol of nanoparticles (3–200 nm) of a nonsteroidal anti-inflammatory drug, indometacin, was fabricated. The trials in mice have shown that this aerosol was more efficient than the conventional oral form.[130] Other applications of nanoparticles in medicine have been considered elsewhere.[131] The wider use of nanoparticles in medicine was partly limited by their relatively high cost and the insufficiently studied mechanisms of their interaction with living cells and tissues of the organism.

Nanoparticles can increase the selectivity of drug action, its dissolution rate, adsorption. and bioaccessibility. The selectivity at the delivery appears due to specific interaction with certain tissues. Nanoparticles are capable of decreasing the lifetime of small peptides and nucleic acids and affect the prolongation of the pharmacological action.

However, new side effects may be associated with the nanoparticles. Pharmacokinetic profiles of original drugs usually differed from those of their nanoencapsulated forms. When using nanoparticles, it is necessary to simultaneously monitor the positive and the negative effects of the reaction kinetics. The kinetic studies is the main criterion of taking into account the effects of size, charge, shape, and the surface and chemical properties that appeared when drugs consisting of nanoparticles encapsulated into liposomes or other carriers are used.[132] The major problem of the use of nanoparticles is ensuring their interaction with the target cell.

Different strategies of how to use nanoparticles for improving the accessibility of drugs are being developed.

However, besides hypothermia, none of the methods reached the step of preclinical trials.

As was mentioned above, the main direction of the use of drugs in medicine is their therapeutic effect. This problem is achieved by the drug delivery to the diseased organ. Liposomes and polymer-based drugs also lie in the basis of nanotechnologies. Moreover, new methods of drug delivery should be found, and the mechanisms that determine the drug interaction with cells and tissues should be explained. The use of nanoparticles and the peculiarities of their delivery to cancer cells were shown.[133] The size of nanoparticles was shown to play the leading role in their application. Rigid spherical particles measuring 100–200 nm were assumed to be suitable for long-term circulation being sufficiently large to avoid absorption by liver and simultaneously sufficiently small to avoid filtration by spleen.

The synthesis of aspheric and/or labile particles can substantially extend the circulation time of particles in vivo.[132]

The mentioned peculiarities are also reflected in the interaction of particles with liver and spleen, especially is those cases where the size variations along one dimension are operative and the particle shape is important.[132] The effect of particle morphology is still unclear.

In addition to the shape and the size of nanoparticles, their surface properties are important for their pharmacologic applications. The chemical properties of surfaces affect strongly the state of blood. Moreover, upon reaching the cell surface, a ligand binds selectively with the surface of the cell receptor and thus is incorporated into the system of the resulting nanoparticle. For an organelle as the target, the analogous phenomena were also observed.[132]

At present, the interests of nanotechnologies are shifted to the fields of biochemistry and medicine. Three main directions of the active use of metal nanoparticles in medicine can be specified.[134]

Firstly, it is the diagnosis of diseases in early stages and, in prospect, at the level of pathologies of individual cell. It was possible to identify proteins in blood serum, distinguish different bacterial strains, and differentiate healthy and cancerous cells.[135] The technology is based on 2 nm-core gold nanoparticles that could form complexes with different fluorophores. Under the effect of appropriate substances, these complexes are disrupted and fluorescence is observed.

Yet another example is nanoparticles with discrete absorption spectra. Following special treatment, they can be used for cellular labeling (for details, see Ref. 136). The targeted delivery of nanoparticles was achieved due to their conjugation with peptides or antibodies.

The second direction is the targeted drug delivery to injured tissues. Nanotechnologies assist in solving this problem, which, in turn, allows reduction of the drug doses, enhancement of their therapeutic effect, and increase in their safety.[137]

Drug delivery can become more efficient, provided nanoparticles are tagged with antibodies or other recognizing elements that enable highly selective binding of nanoparticles to antigens expressed on the surfaces of injured cells.

The exhaustive information on the use of nanotechnologies in oncology can be found in several reviews.[138–140]

It was shown[141] that metal-organic frameworks, a new class of hybrid materials composed of metal ions bound with one another by an organic ligand, can serve as drug carriers. Such systems are capable of biodegradation.

The third direction concerns the regenerative medicine. The latter is aimed at mobilizing the inherent properties of an organism for exerting control over such diseases as diabetes, osteoarthritis, and injuries of the cardiac muscle and the central nervous system. The fundamental principle of regenerative medicine is that the biocompatible materials and the signal drug molecules capable of initiating regenerative processes at the cell level can be delivered to the affected regions.

The regenerative medicine also embraces the trend in nanotechnology that is being actively developed, now aimed at imparting new physicochemical and therapeutic properties to the drugs used. Among the possible approaches to solving this problem is the synthesis and the investigations of properties of new polymorphic modifications of molecular crystals.[142]

9.4 CONCLUSION

The performed analysis of organic nanoparticles and materials on their basis allowed certain conclusions and generalizations to be drawn. It deserves special mention that all the methods of producing organic nanoparticles cited in this chapter require further development and improvement. First of all, the problems of stabilizing organic nanoparticles, enhancing the efficiency of the methods of their fabrication, and reducing their cost should be solved. In the industry, the only method used so far is the manufacturing of organic suspensions by the mechanical grinding, which is utilized by many companies. The other methods analyzed in this chapter have not yet reached the industrial level. In the field of chemistry of organic nanoparticles, the studies are now aimed at accumulation of experimental data.

In this chapter, we have demonstrated that stable crystalline organic nanoparticles of various sizes and structures different from those of the original substances can be pro duced from solutions or the gas phase containing the original substances in the molecular state. We believe that the possibility of their formations largely depends on the size of the single crystals that appear in the liquid or gas phases under certain conditions. For organic nanoparticles, the size effect is of the structural-molecular nature and manifested itself to the greatest extent in organic nanocrystals.

The following top-priority problems can be pointed out for studying and using organic nanoparticles:

- the extension of the range of organic compounds of different classes, capable of serving as the precursors of nanoparticles;
- the elucidation of experimental criteria and widening of the methods of preparation of organic nanoparticles and the elucidation of the relationship between the chemical activity and the size of organic nanoparticles;
- the analysis of the size, shape, and stability of nanoparticles, the trend to retain or change their properties, for example, spectral, in solutions, colloidal dispersions and in the crystalline state;
- the development of theoretical concepts on the size effect of organic nanoparticles on their spectral properties, based on the changes in the rigidity and elasticity of the crystal structures of organic compounds and the concomitant changes in the hierarchy of intermolecular interactions;
- the extension of the application field of organic nanoparticles in such fields as pharmaceutics, green chemistry, electronics, and optoelectronics.

The solution of mentioned problems will favor the origination and the development of organic nanochemistry.

REFERENCES

1. Ozin, G. A. *A C Arsenault Nanochemistry*; Royal Society of Chemistry: Cambridge, 2005.
2. Semiconductor Nanostructures, chapter 16, pp 539–578, In *Nanoscale Materials in Chemistry*, 2nd ed.; Klabunde, K.J., Richards, R.M., Eds.; Wiley: New York, 2009.
3. Sergeev, G.B. In Nanokhimiya: Uchebnoe Posobie Nanochemistry: The Manual, Moscow: Izd. KDU, 2006, p 336.
4. Asahi, T.; Sugiyama, T.; Masunara, H. *Acc. Chem. Res.* **2008**, *41*, 1790.
5. Zhao, Y. S.; Fu, H.; Peng, F.; Ma, Y.; Xiao, D.; Yao, J. *Acc. Chem. Res.* **2010**, *43*, 409.

6. Zhao, Y. S.; Fu, H.; Peng, F.; Ma, Y.; Xiao, D.; Yao, J. *Adv. Mater.* **2008**, *20*, 2859.
7. Shakhtshneider, T. P.; Boldyrev, V. V. *Reactivity of Molecular Solids*; Wiley: Chichester, 1999, p 271.
8. Ward, G. H. *Pharm. Res.* **1995**, *12*, 773.
9. Rasenack, N.; Steckel, H.; Muller, B. W. *Powder Technol.* **2004**, *143*, 291.
10. Steckel, N.; Rasenack, N.; Villax, P.; Muller, B. W. *Int. J. Pharm.* **2003**, *258*, 65.
11. Liversidge, E.M.; Liversidge, G.G. *Adv. Drug Del. Rev.* **2011**, *63*, 427.
12. US P. 5628981 (1997).
13. Arunkumar, N.; Deecaraman, M.; Rani, C. *Asian J. Pharm.* **2009**, *3*, 168.
14. Liversidge, E. M.; Liversidge, G. G.; Cooper, E. R. *Eur. J. Pharm. Sci.* **2003**, *18*, 113.
15. US P. 5518187 (1992).
16. Muller, R. H.; Jacobs, C.; Kayser, O., *Pharmaceutical Emulsions and Suspensions*. In *Nielloud*, F., Marti-Mestres, G., Eds.; Marcel Decker: New York, 2000.
17. Muller, R. H.; Jacobs, C.; Kayser, O. *Drug Pharm. Sci* **2003**, *126*, 135.
18. Scholer, N.; Krause, K.; Kayser, O.; Muller, R. H. *Antimicrob. Agents Chemother.* **2001**, *45*, 1771.
19. Akkar, A. *Poorly Soluble Drugs: Formulation by Nanocrystals and SolEmuls Technologies, Dissertation, Pharmazeutische Technologie*; Berlin: Freie Universitat, 2004.
20. Keck, C. M.; Muller, R. H. *Eur. J. Pharm. Biopharm.* **2006**, *62*, 3.
21. Muller, R. H.; Peters, K.; Becker, R.; Kruss, B. *Proc. Int. Symp. Control Rel. Bioact. Mater.* **1996**, *22*, 574.
22. Tamaki, Y.; Asahi, T.; Masuhara, H. *App. Surf. Sci.* **2000**, *168*, 85.
23. Sugiyama, T.; Asahi, T.; Yuyama, K.; Takeuchi, H.; Jeon, H.; Hosokawa, Y.; Masuhara, H. *Proc. SPIE* **2008**, *6891*, 37.
24. Nagare, S.; Senna, M. *Solid State Ionics* **2004**, *172*, 243.
25. Fukumura, H.; Masuhara, H. *Chem. Phys. Lett.* **1994**, *221*, 373.
26. Fujiwara, H.; Fukumura, H.; Masuhara, H. *J. Phys. Chem.* **1995**, *99*, 11844.
27. Hosokawa, Y.; Yashiro, M.; Asahi, T.; Masuhara, H. *J. Photochem. Photobiol. A* **2001**, *142*, 197.
28. Kostler, S.; Rudorfer, A.; Haase, A.; Satzinger, V.; Jakopic, G.; Ribitsch, V. *Adv. Mater.* **2009**, *21*, 2505.
29. Tachibana, T.; Nakamura, A. *Kolloid-Z. Z. Polym.* **1965**, *203*, 130.
30. Kasai, H.; Nalwa, H. S.; Oikawa, H.; Okada, S.; Matsuda, H.; Minami, N., et al. *Jpn. J. Appl. Phys.* **1992**, *31*, 1132.
31. Li, S.; He, L.; Xiong, F.; Li, Y.; Yang, G. *J. Phys. Chem. B* **2004**, *108*, 10887.
32. Kasai, H.; Kamatani, H.; Yoshikawa, Y.; Okada, S.; Oikawa, H.; Watanabe, A., et al. *Chem. Lett.* **1997**, *9*, 1181.
33. An, B.; Kwon, S.; Jung, S.; Park, S. Y. *J. Am. Chem. Soc.* **2002**, *124*, 14410.
34. Gesquiere, A. J.; Uwada, T.; Asahi, T.; Masuhara, H.; Barbara, P. F. *Nano Lett.* **2005**, *5*, 1321.
35. Fu, H. B.; Yao, J. N. *J. Am. Chem. Soc.* **2001**, *123*, 1434.
36. Mori, J.; Miyashita, Y.; Oliveira, D.; Kasai, H.; Oikawa, H.; Nakanishi, H. *J. Cryst. Growth* **2009**, *311*, 553.
37. Oh, S. W.; Zhang, D. R.; Kang, Y. S. *Mater. Sci. Eng. C* **2004**, *24*, 131.
38. Kamogawa, K.; Akatsuka, H.; Matsumoto, M.; Yokoyama, S.; Sakai, T.; Sakai, H.; Abe, M. *Colloids Surf. A* **2001**, *180*, 41.
39. Johnson-Buck, A.; Kim, G.; Wang, S.; Hah, H. J.; Kopelman, R. *Mol. Cryst. Liq. Cryst.* **2009**, *501*, 138.
40. Kwon, E.; Oikawa, H.; Kasai, H.; Nakanishi, H. *Cryst. Growth Des.* **2007**, *7*, 600.

41. D'Addio, Suzanne M.; Robert, K. *Adv. Drug Deliv. Rev.* **2011**, *63,* 417–426.
42. Kakran, M.; Sahoo, N. G.; Li, L.; Judeh, Z.; Wang, Y.; Chong, K., et al. *Int. J. Pharm.* **2010**, *383,* 285–292.
43. Nyambura, B. K.; Kellaway, I. W.; Taylor, K. M.G. *Int. J. Pharm.* **2009**, *372,* 140–146.
44. Liu, Y.; Cheng, C. Y.; Liu, Y.; Prud'homme, R. K.; Fox, R. O. *Chem. Eng. Sci.* **2008**, *63,* 2829–2842.
45. Bochenkov, V. E.; Sergeev, G. B. *Usp. Khim.* **2007**, *76,* 1084; *Russ. Chem. Rev.* **2007**, *76,* 1013.
46. Kang, L.; Wang, Z.; Cao, Z.; Ma, Y.; Fu, H.; Yao, J. *J. Am. Chem. Soc.* **2007**, *129,* 7305.
47. Yao, H.; Ou, Z.; Kimura, K. *Chem. Lett.* **2005**, *34,* 1108.
48. Yao, H.; Yamashita, M.; Kimura, K. *Langmuir* **2009**, *25,* 1131.
49. Ou, Z.; Yao, H.; Kimura, K. *Bull. Chem. Soc. Jpn.* **2007**, *80,* 295.
50. Margulis-Goshen, K.; Netivi, H. D.; Major, D. T.; Gradzielski, M.; Raviv, U.; Magdass, S. *J. Colloid Interface Sci.* **2010**, *342,* 283.
51. Ionov, D. S.; Sazhnikov, V. A.; Alfimov, M. V. *Ros. Nanotekhnol.* **2010**, *5,* 31.
52. Jung, J.; Perrut, M. *J. Supercrit. Fluids* **2001**, *20,* 179.
53. Bahrami, M.; Ranjbarian, S. *J. Supercrit. Fluids* **2007**, *40,* 263.
54. Perrut, M.; Clavier, J. Y. *Ind. Eng. Chem. Res.* **2003**, *42,* 6375.
55. Yu Zalepugin, D.; Til'kunova, N. A.; Chernyshova, I. V. *Sverkhkriti. Flyuidy. Teor. Prakt.* **2008**, *5* (1).
56. Young, T. J.; Mawson, S.; Johnston, K. P. *Biotechnol. Prog.* **2000**, *16,* 402.
57. Hezave, A. Z.; Esmaeilzadeh, F. *J. Supercrit. Fluids* **2010**, *52,* 84.
58. Pathak, P.; Meziani, M. J.; Desai, T.; Sun, Y. -P. *J. Am. Chem. Soc.* **2004**, *126,* 10842.
59. Jagannathan, R.; Irvin, G.; Blanton, T.; Jagannathan, S. *Adv. Funct. Mater.* **2006**, *16,* 747.
60. Chen, X.; Zhang, X.; Pan, J.; Zhang, W.; Yin, W. *Powder Technol.* **2005**, *152,* 127.
61. Reverchon, E.; Porta, G. D. *Powder Technol.* **1999**, *106,* 23.
62. Kordikowski, A.; Shekunov, T.; York, P. *Pharm. Res.* **2001**, *18,* 682.
63. Steckel, H.; Thies, J.; Muller, B. W. *Int. J. Pharm.* **1997**, *152,* 99.
64. De Marco, Iolanda; Reverchon, Ernesto *J. Supercrit. Fluids* **2011**, *58,* 295–302.
65. Reverchon, E.; De Marco, I. *Chem. Eng. J.* **2011**, *169,* 358–370.
66. Zhang, Z. B.; Shen, Z. G.; Wang, J. X.; Zhao, H.; Chen, J. F.; Yun, J. *Ind. Eng. Chem. Res.* **2009**, *48,* 8493.
67. Vemavarapu, C.; Mollan, M. J.; Lodaya, M.; Needham, T. E. *Int. J. Pharm.* **2005**, *292,* 1.
68. J. Jung, J.Y. Clavier, M. Perrut. In *Proceedings of the 6th International Symposium on Supercritical Fluids*; Versailles: France, 2003, p 1683.
69. Sergeev, G. B.; Batyuk, V. A. *Cryochemistry*; Mir?: Moscow, 1981.
70. Sergeev, G. B.; Smirnov, V. V. *Molekulyarnoe Galogenirovanie Olefinov*; *Molecular Halogenation of OlefinsMolecular Halogenation of Olefins*; Moscow State University: Moscow, 1985.
71. Gol'danskii, V. I.; Trakhtenberg, L. I. *Tunnel'nye Yavleniya v Khimicheskoi Fizike*; *Tunnel Effects in Chemical PhysicsTunnel Effects in Chemical Physics*; Nauka: Moscow, 1986.
72. Tret'yakov, Yu. D. *Soros. Obrazovat. Zh.* **1996**, *4,* 45.
73. Abdelwahed, W.; Degobert, G.; Stainmesse, S.; Fessi, H. *Adv. Drug Deliv. Rev.* **2006**, *58,* 1688.
74. Chen, G.; Wang, W. *Drying Technol.* **2007**, *25,* 29.
75. Sergeev, G. B.; Komarov, V. S. *Mol. Cryst. Liq. Cryst.* **2006**, *456,* 107; 14 A Yu Utekhina, G.B. Sergeev.
76. Russ. P. 2195264, 2001.
77. Sergeev, G. B.; Komarov, V. S.; Zvonov, A. V. *Vestn. Mosk. Univ., Ser. 2, Khim.* **1984**, *25,* 43.

78. Sergeev, G. B.; Komarov, V. S.; Zvonov, A. V. *Zh. Obshch. Khim.* **1986**, *56,* 1602.
79. Russ. P. 2366653, 2009.
80. Morozov, Yu. N.; Mikhalev, S. P.; Shabatin, V. P.; Kolotilov, P. N.; Sergeev, G. B. *Vestn. Mosk. Univ., Ser. 2, Khim.* **2010**, *4,* 288.
81. Pat 2295511 RF, 2005.
82. Sergeev, G. B.; Sergeev, B. M.; Morozov, Yu. N.; Chernyshev, V. V. *Acta Crystallogr. Sect. E* **2010**, *66,* 2623.
83. Polymorphism in Pharmaceutical Solids. In *Drugs and the Pharmaceutical Series*; Brittain, H. G., Ed.; Informa Healthere: New York, 2009.
84. Iski, E. V.; Jonston, B. F.; Florence, A. J.; Urquahart, A. J.; Syces, E. C.H. *ACS Nano* **2010**, *4,* 5061.
85. Chernyak, V.; Meier, T.; Tsiper, E.; Mukamel, S. *J. Phys. Chem. A* **1999**, *103,* 10294.
86. Spano, F. S.; Shaul, M. *Phys. Rev. A* **1989**, *40,* 5783.
87. Jiang, Q.; Shi, H. X.; Zhao, M. *J. Chem. Phys.* **1999**, *111,* 2176.
88. Fu, H.; Loo, B. H.; Xiao, D.; Xie, R.; Ji, X.; Yao, J.; Zhang, B.; Zhang, L. *Angew. Chem. Int. Ed.* **2002**, *41,* 962.
89. Xiao, D.; Xi, L.; Yang, W.; Fu, H.; Shuai, Z.; Fang, Y.; Yao, J. *J. Am. Chem. Soc.* **2003**, *125,* 6740.
90. Gonzalez-Rodriguez, D.; Janssen, P. G.A.; Martin-Rapun, R.; Cat, I.; Feyter, S.; Schenning, A. P.H.J.; Meijer, E. W. *J. Am. Chem. Soc.* **2010**, *132,* 4710.
91. An, B. K.; Kwon, S. K.; Park, S. Y. *Angew. Chem. Int. Ed.* **2007**, *46,* 1978.
92. Deans, R.; Kim, J.; Machacek, M. R.; Swager, T. M. *J. Am. Chem. Soc.* **2000**, *122,* 8565.
93. Luo, J.; Xie, Z.; Lam, J. W.; Cheng, L.; Chen, H.; Qiu, C., et al. *Chem. Commun.* **2001**, *1740*.
94. Palayangoda, S. S.; Cai, X.; Adhikari, R. M.; Neckers, D. C. *Org. Lett.* **2008**, *10,* 281.
95. Panthi, K.; Adhikari, R. M.; Kinstle, T. H. *J. Phys. Chem. A* **2010**, *114,* 4550.
96. Briseno, A. L.; Mannsfeld, S. C.B.; Lu, X.; Xiong, Y.; Jenekhe, S. A.; Bao, Z.; Xia, Y. *Nano Lett.* **2007**, *7,* 668.
97. Kimura, M.; Hara, Y.; Okeyama, T.; Inoue, S.; Shimoda, T. *IEICE Trans. Electron.* **2005**, *E88* (C(11)), 2043.
98. Fardy, M.; Yang, P. *Nature (London)* **2008**, *451,* 408.
99. Wang, Z.; Medforth, C. J.; Shelnutt, J. A. *J. Am. Chem. Soc.* **2004**, *126,* 16720.
100. Takazawa, K.; Kitahama, Y.; Kimura, Y.; Kido, G. *Nano Lett.* **2005**, *5,* 1293.
101. Zhao, Y. S.; Yang, W.; Xiao, D.; Sheng, X.; Yang, X.; Shuai, Z., et al. *Chem. Mater.* **2005**, *17,* 6430.
102. Wang, J.; Zhao, Y.; Zhang, J.; Zhang, J.; Yang, B.; Wang, Y., et al. *J. Phys. Chem. C* **2007**, *111,* 9177.
103. Fu, H.; Xiao, D.; Yao, J.; Yang, G. *Angew. Chem. Int. Ed.* **2003**, *42,* 2883.
104. Hu, J. S.; Guo, Y. G.; Liang, H. P.; Wan, L. J.; Jiang, L. *J. Am. Chem. Soc.* **2005**, *127,* 17090.
105. Zhang, X.; Zhang, X.; Shi, W.; Meng, X.; Lee, C.; Lee, S. *J. Phys. Chem. B* **2005**, *109,* 18777.
106. Zhao, Y. S.; Xiao, D.; Yang, W.; Peng, A.; Yao, J. *Chem. Mater.* **2006**, *18,* 2302.
107. Zhao, Y. S.; Fu, H.; Hu, F.; Peng, A.; Yao, J. *Adv. Mater.* **2007**, *19,* 3554.
108. Zhao, Y. S.; Xu, J.; Peng, A.; Fu, H.; Ma, Y.; Jiang, L., et al. *Angew. Chem. Int. Ed.* **2008**, *47,* 7301.
109. Zhao, Y. S.; Fu, H.; Hu, F.; Peng, A.; Yang, W.; Yao, J. *Adv. Mater.* **2008**, *20,* 79.
110. Wang, L.; Dong, L.; Bian, G. R.; Wang, L. Y.; Xia, T.; Chen, H. Q. *Anal. Bioanal. Chem.* **2005**, *382,* 1300.
111. Anastas, P.; Horvath, I. T. *Chem. Rev.* **2007**, *107,* 2169.

112. Bergstrom, C. A.S.; Wassvik, C. M.; Norinder, U.; Luthman, K.; Artursson, P. *J. Chem. Inf. Comp. Sci.* **2004,** *44,* 1477.
113. Mihranyan, A.; Stromme, M. *Surface Sci.* **2007,** *601,* 315.
114. Lindfors, L.; Skantze, P.; Rasmusson, M.; Zackrisson, A.; Ollson, U. *Langmuir* **2006,** *22,* 906.
115. Ambrus, R.; Kocbek, P.; Kristl, J.; Sibanc, R.; Rajko, R.; Szabo-Revesz, P. *Int. J. Pharm.* **2009,** *381,* 153.
116. Margulis-Goshen, K.; Kesselmanb, E.; Daninob, D.; Magdassi, S. *Int. J. Pharm.* **2010,** *393,* 231.
117. Dalvi, S. V.; Dave, R. N. *Int. J. Pharm.* **2010,** *387,* 172.
118. Wu, Libo; Zhang, Jian; Watanabe, Wiwik *Adv. Drug Deliv. Rev.* **2011,** *v. 63,* 456–469; issue 6, 30 may.
119. Patravale, V. B.; Date, A. A.; Kulkarni, R. M. *J. Pharm. Pharmacol.* **2004,** *56,* 827–840.
120. Yang, W.; Peters, J. I.; Williams, R. O., III. *Int. J. Pharm.* **2008,** *356,* 239–247.
121. Nutan, M. T.H.; Reddy, I. K. *Pharmaceutical Suspensions: From Formulation Development to Manufacturing*; Springer, 2009; pp 39–66.
122. Makhlof, A.; Miyazaki, Y.; Tozuka, Y.; Takeuchi, H. *Int. J. Pharm.* **2008,** *357,* 280–285.
123. Eerdenbrugh, B. V.; Vermant, J.; Martens, J. A.; Froyen, L.; Humbeeck, J. V.; Auguestijns, P.; Mooter, G. V.D. *J. Pharm. Sci.* **2009,** *98* (6), 2091–2103.
124. Garad, S.; Wang, J.; Joshi, Y.; Panicucci, R. *Pharmaceutical Suspensions: From Formulation Development to Manufacturing*; Springer, 2009, pp 39–66.
125. Murthy, S. K. *Int. J. Nanomed.* **2007,** *2* (2), 129–141.
126. Verma, S.; Huey, B. D.; Burgess, D. J. *Langmuir* **2009,** *25* (21), 12481–12487.
127. Yang, W.; Johnston, K. P.; Williams, R. O. *Eur. J. Pharm. Biopharm.* **2010,** *75,* 33–41.
128. Abdelwahed, W.; Degobert, G.; Stainmesse, S.; Fessi, H. *Adv. Drug Deliv. Rev.* **2006,** *58,* 1688–1713.
129. D'Addio, S. M.; Kafka, C.; Akbulut, M.; Beattie, P.; Saad, W.; Herrera, M., et al. *Mol. Pharm.* **2010,** *7,* 557–564.
130. Onishchuk, A. A.; Tolstikova, T. G.; Sorokina, I. V.; Baklanov, A. M.; Karasev, V. V.; Boldyrev, V. V.; Fomin, V. M. *Dokl. Akad. Nauk.* **2009,** *425,* 5.
131. Nazarov, G. V.; Galan, S. E.; Nazarova, E. V.; Karkishchenko, N. N.; Muradov, M. M.; Stepanov, V. A. *Khim.-Fam. Zh.* **2009,** *43,* 2.
132. Kumar, K. V.; Kaddour, I. A.; Gupta, V. K. *Ind. Eng. Chem. Res.* **2010,** *49,* 7257.
133. Petros*, Robby A.; DeSimone, Joseph M. *Nat. Rev. Drug Discov.* **August 2010,** *9,* 615–627.
134. *European Technology Platform on NanoMedicine, Vision Paper and Basis for a Strategic Research Agenda for NanoMedicine*; Luxembourg, 2005.
135. Bunz, U. H.F.; Rotello, V. M. *Angew. Chem. Int. Ed.* **2010,** *49,* 3268.
136. Delehanty, J. B.; Mattoussi, H.; Medintz, I. L. *Anal. Bioanal. Chem.* **2009,** *393.* , 1091.
137. Нанотехнологии в биологии и медицине; ЕВ Шляхто, Ed.; Изд-во СПбУ: Санкт-Петербург, 2009.
138. Ruoslahti, E.; Bhatia, S. N.; Sailor, M. J. *J. Cell Biol.* **2010,** *188,* 759.
139. Hartman, K. B.; Li, W. *Mol. Diagn. Ther.* **2008,** *12,* 1.
140. Lammers, T.; Hennink, W. E.; Storm, G. *Br. J. Cancer* **2008,** *99,* 392.
141. Huxford, R. C.; Rocca, J. D.; Lin, W. *Curr. Opin. Chem. Biol.* **2010,** *14,* 262.
142. *Polymorphism: in the Pharmaceutical Industry*; Hilfiker, R., Ed.; Willey: Weinheim, 2009.

Chapter 10

Size Effects in Nanochemistry

Chapter Outline

10.1 Models of Reactions of Metal Atoms in Matrices 276
10.2 Melting Point 278
10.3 Optical Spectra 281
10.4 Kinetic Peculiarities of Chemical Processes on the Surface of Nanoparticles 287
10.5 Thermodynamic Features of Nanoparticles 289
10.6 Magnetic Properties 293
10.7 Electrical/conducting Properties 294

The experimental results on the reactions of atoms, clusters, and nanoparticles formed by various elements of the periodic table make it possible to formulate the following definition: the size effects in chemistry represent a phenomenon that manifests itself in the qualitative changes in physicochemical properties and reactivity, depending on the number of atoms or molecules in a particle, and takes place in a range of less than 100 atomic/molecular diameters. The development of studies in nanochemistry will allow the above definition to be refined.

Nowadays, it is evident that it is the size effects that distinguish nanochemistry from chemical reactions occurring under ordinary conditions.

It is a practice to distinguish two types of size effects, namely, the intrinsic or internal effects and the external effects. The former effects are associated with specific changes in the bulk and surface properties of both individual particles and assemblies formed as a result of self-organization of particles. An external effect is a size-dependent answer to the action of an external field or some forces independent of the internal effect.

The experiments dealing with internal size effects are directed at solving the problems associated with electronic and structural properties of clusters. These properties are the chemical activity, the binding energy between atoms in a particle and between particles, and the crystallographic structure. The melting point and optical properties can also be considered as the functions of the particle size and geometry. The dependence of the spatial arrangement of electron levels is termed as a quantum-size effect.

10.1 MODELS OF REACTIONS OF METAL ATOMS IN MATRICES

The first description of argon-matrix-isolated metal clusters of the M_x stoichiometry was carried out by the Monte Carlo method. A comparison with experimental results has shown that in matrices, metal aggregates are formed in excessive amounts, which may exceed the statistical estimates by factors 10 and 1000.[1]

In a system of consecutive and parallel reactions, which involve interactions of metal atoms, dimers, trimers, and more complicated particles with ligands in low-temperature co-condensates, complex processes occur. Upon a collision with a cooled surface, a metal atom or a ligand molecule can retain their mobility in the upper layer of co-condensate for a certain time. In this mobile layer, the gradients of temperature and concentration can exist. It is assumed that such systems can be modeled by a great number of layers from 1 to n, where the first layer is the co-condensate surface. On freezing out, it becomes the second layer, in which metal atoms and ligand molecules lose their mobility.

In the first approximation, the kinetics of such diffusion-limited processes can be described by two constants k_M and k_L.[1] The following two models were considered: a quenching model and a steady-state model. For the metal (M)–ligand (L) scheme, a system of differential equations was written and numerically solved. In the quenching model, it is assumed that all reactions run in a time τ_q, after which they were immediately frozen out. This means that rate constants of individual stages are changed for an invariant average rate constant.

In the steady-state model, it is assumed that a sample is constantly deposited and frozen out for a sufficiently long time so that a steady state can be established in the mobile zone, when $dL/dM = dM/dt = 0$. In this case, the system of differential equations is modified by adding the terms, which take into account the rates of condensation and freezing out of metal and ligand.

When processing the experimental results with the aim of finding the ratio of different particles, we come up against a problem associated with the fact that the assumptions laid on the basis of these models do not allow to determine separately k_M and τ_q or k_L and τ_q. In the quenching model, it is the product of these values that enters into the equations, while in the steady-state model, k_M and k_L are divided by the freezing rate.

A comparison of the models with experimental results was carried out for low and high metal concentrations by the example of nickel interaction with nitrogen and carbon monoxide molecules. The following sequence of reactions that occur in the presence of an excessive ligand amount was considered most comprehensively:

$$M \xrightarrow[L]{k_L} ML \xrightarrow[L]{k_L} ML_2 \xrightarrow[L]{k_L} ML_3 \xrightarrow[L]{k_L} ML_4$$

Solution of the system of differential equations allowed the ratio [ML]/[ML$_2$]/[ML$_3$]/[ML$_4$] to be found and compared with experimental results. Table 10.1 shows the results for systems Ni–CO–Ar and Ni–N$_2$–Ar, which were found based on the quenching model.

Experimental data were obtained by analyzing the IR spectra for the system with a ratio CO/Ar = 1:50 and a metal content of no less than 1%. In calculations, the values $k_L \tau_q \approx 50$ and $[L] \approx 2 \times 10^{-2}$ were used. The results for the Ni–N$_2$–Ar system were taken from Ref. 2.

As seen from Table 10.1, the agreement with experimental data can be considered as adequate for the Ni–CO–Ar system and satisfactory for the Ni–N$_2$–Ar system.

The quenching model was applied for analyzing the Ni–CO–Ar system at high metal concentrations. Figure 10.1 compares experimental results[3] with calculations. The agreement of results was satisfactory, as noted by the authors.[1] Processing of experimental data in terms of the steady-state model yielded less satisfactory agreement with experiments.

The development of more sophisticated models for analyzing the chemical reactions, which involve metal particles of different sizes and occur at low temperatures, including those in argon matrices, is complicated by theoretical and experimental factors. The latter are associated with different ways of sample synthesis and different states of solids.

When using the matrix isolation method and various techniques of preparative chemistry, scientists are striving to eliminate the interfering chemical reactions, which can involve different-sized particles. The analysis of examples in other chapters shows that to date, the mainstream of the research on size effects in chemistry has shifted to the gas phase.

TABLE 10.1 Comparison of Experimental Data and Those Calculated in Terms of the Quenching Model

System/ratio	Ni–CO–Ar				Ni–N$_2$–Ar			
	ML	ML$_2$	ML$_3$	ML$_4$	ML	ML$_2$	ML$_3$	ML$_4$
Experiment	1	0.55	0.17	0.05	1	0.816	0.577	0.167
Calculation	1	0.53	0.18	0.06	1	0.8	0.43	0.24

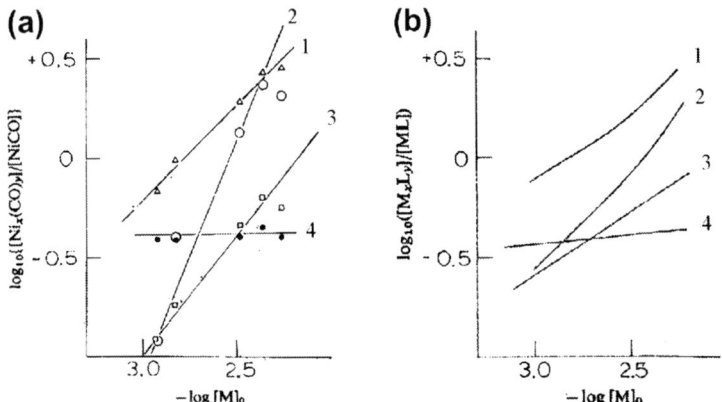

FIGURE 10.1 Distribution of products in Ni–CO–Ar system with a high metal content: (a) experimental results, (b) calculations: (1) Ni_2CO, (2) Ni_3CO, (3) $Ni_2(CO)_2$, and (4) $Ni(CO)_2$.[1]

10.2 MELTING POINT

The variations in the melting points of metals as a function of the particle size werer apparently among the first effects that attracted attention of many scientists. With a decrease in the size, the melting point can be reduced by several hundred degrees. Thus, for gold, the melting point decreased by 1000° with a transition from the compact metal that melts at 1340 K to 2 nm particles.[4] For the first time, the problem of the melting point versus particle size dependence came to light when studying Pb, Sn, and Bi particles by electron diffraction technique.[5]

The melting point marks the transition between solid and liquid phases. At this temperature, a solid-phase crystalline structure disappears giving way to a disordered liquid state. A strong decrease in the melting point with the size of metal particles can be reflected in their activity and selectivity. Indeed, as shown in certain examples above, recently a high reactivity of nanosize gold particles was discovered. Gold was never used earlier in electrocatalysis; however, quite a number of chemical reactions have been carried out on its nanoparticles. Gold nanoparticles were active in low-temperature combustion, hydrocarbon oxidation, hydrogenation of unsaturated compounds, and reduction of nitrogen oxides.[6]

The dependence of the melting point on the metal particle size was considered within the framework of two models. One of them employed the concepts of thermodynamics, while the other considered atomic vibrations.

From the standpoint of thermodynamics, a transition from the solid to the liquid state with an increase in the temperature starts with the appearance of an infinitely small liquid layer on the nanoparticle surface, while its core remains solid. Such a melting is caused by the surface tension, which reflects the liquid–solid interactions and the changes in the system's energy. The size dependence of the gold melting point was described for the sizes up to 2 nm.[4] For this purpose, two phenomenological models were used. The first model considered the equilibrium in a system formed by a solid particle, a liquid particle of equal mass, and their saturated vapors. The second model assumed the preliminary existence of a liquid layer around a solid particle and the equilibrium in such a system in the presence of a vapor phase. Both models agree with experimental results. A thermodynamic description of the melting point versus particle size dependence was given.[7,8] The peculiarities of thermodynamic approaches to the melting point versus the particle size dependence were considered in a review.[9]

A highly sensitive thin-film scanning calorimeter was used for studying the melting point versus size dependence for indium films 0.1–10-nm thick, deposited on the silicon nitride surface. In-depth studies of the initial step of film formation allowed a periodicity in the melting-point behavior to be observed, which was associated with the necessity of the formation of a coating consisting of a "magic" number of atoms.[10] By employing a calorimeter, the dependence of the melting-point decrease on the film thickness was observed and discussed in terms of classical thermodynamics.

The dependence of the melting point of metal nanoparticles on their size was also interpreted on the basis of the criteria put forward by Lindemann[11] and developed by other authors.[12] According to the concepts of Lindemann, a crystal melts when the root-mean-square (standard) deviation d of atoms in the crystal exceeds the average interatomic distance a, namely, $\delta/a \geq$ const. An increase in the temperature leads to an increase in the amplitude of vibrations. At a certain temperature, the latter becomes too wide and initiates the destruction of the crystal lattice, which results in the fusion of the solid. The surface atoms are more weakly bound, which under real conditions can result in their wider oscillations as compared with the atoms in the bulk at the same temperature. This effect can be described in terms of standard deviations of surface atoms δ_s and bulk atoms δ_v, namely $\alpha = \delta_s/\delta_v$, where parameter α takes the values from two to four. The share of surface atoms in a 3-nm spherical nanoparticle reaches approximately 50%, and their vibrations strongly affect the Lindemann criterion. This approach was used in describing the size dependence of the melting point of nanoparticles without invoking any thermodynamic notions.

A model, which considers the decrease in the nanoparticle temperature with a decrease in their size, was developed.[12] The following equation was proposed for the description of nanoparticle properties:

$$\frac{T_m(r)}{T_m(\infty)} = \exp\left[-(\alpha-1)\left(\frac{r}{3h}-1\right)^{-1}\right] \qquad (10.1)$$

where $T_m(r)$ and $T_m(\infty)$ are the melting points of a nanocrystal and a compact metal, respectively and h is the height of an atomic monolayer in a crystal.

Eq. (10.1) can be used for predicting the lowering of a crystal's melting point, provided that the parameter a, which is usually found from the corresponding experimental data, is known. Figure 10.2 shows the size versus melting point

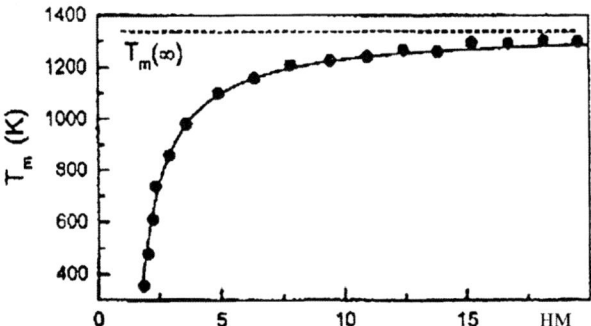

FIGURE 10.2 Dependence of gold melting point on the particle size; points indicate experimental data; solid lines are calculated using Eq. (10.1) for $\alpha=1.6$, $h=0.204$ nm.[4,12]

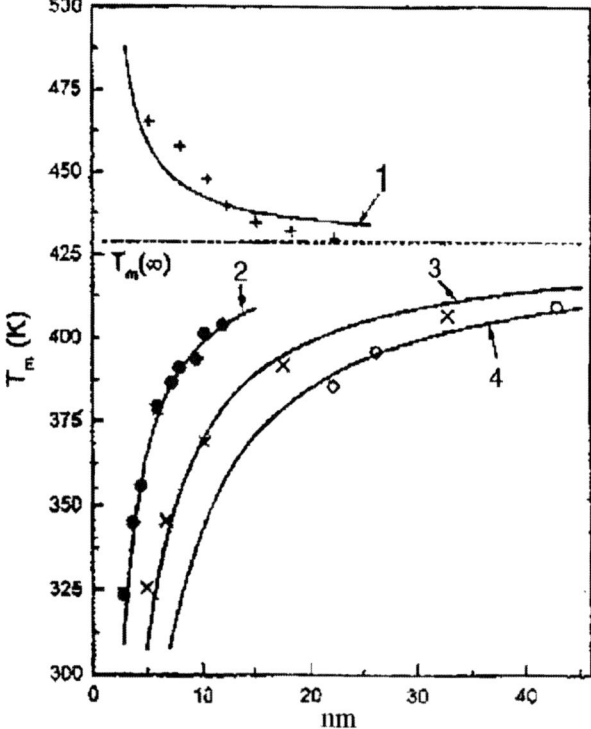

FIGURE 10.3 Dependence of indium melting point on the particle size; points indicate experimental data; solid lines are calculated using Eq. (10.1): (1) indium in aluminum matrix, $\alpha=0.57$; (2–4) indium in iron matrix, $\alpha=2.0$, 3.03, 4.04, respectively.[12]

dependence found for gold nanoparticles.[12] As seen from the figure, Eq. (10.1) adequately describes the experimental data.[4]

Certain nanocrystalline particles represent materials in which one metal is incorporated into another. For such particles, in contrast to compact materials, the melting point can both decrease and increase with the particle size. Figure 10.3 shows the results obtained for indium.[12] When indium nanocrystals are incorporated into iron, the melting point of particles decreases; however, their incorporation into aluminum increases the melting point. In Figure 10.3, both trends are shown as the size functions. The fact that Eq. (10.1) can also be applied to the cases when the melting point increases with a decrease in the particle size, i.e. when the parameter $\alpha < 1$, deserves mention. According to the equality $\alpha = \delta_s/\delta_v$, this occurs when the amplitudes of atomic oscillations on the surface exceed those in the bulk. Such a situation arises when the surface atoms strongly interact with the material of the matrix itself. Figure 10.3 confirms the adequate agreement with experimental results for this case too.

Thus, it was found that the melting point of free metal nanoparticles always decreases with a decrease in the particle size. For systems built of metallic matrices with incorporated nanoparticles of a foreign metal, the melting point can both decrease and increase with a decrease in the particle size. The corresponding experiments were carried out in high vacuum in the absence of stabilizing agents. For nanochemistry, it is of great importance to find the answer to the question of how the size of ligand-stabilized metal particles affects their melting point.

10.3 OPTICAL SPECTRA

The dependence of the energy properties of nanoparticles of metals and semiconductors on their size has been analyzed.[13]

For metals, the appearance of size-dependent metallic properties is an important feature. Thus, for mercury clusters, a gradual metal–insulator transition takes place between $N = 20$ and $N = 10^2$ atoms. An example of the external size effect is the collective electronic or lattice excitation known as the Mie resonance.

Quantum-size effects are associated with unusual spectra of electron energy levels arranged in a discrete fashion.

With the formation of a dimer, the energy level of an atom splits into two components. As the cluster grows, its levels split further and further and finally merge together to form a quasi-continuous absorption band of a solid. The bands are filled with electrons, the Fermi level appears, and the conduction is observed.

Simultaneously, the ionization potential of atoms and molecules transforms into the electron work function of the bulk metal. The splitting of one-electron energy levels δE approximately corresponds to the width of a cluster quasiband ΔE divided by the number of levels N, i.e. $\delta E = \Delta E/N$. For metals, the quantity ΔE and the Fermi-level energy E_F are usually of the same order of magnitude and equal to ca. 5 eV. Hence, in a cluster built of 100 atoms, the energy level splits to δE of ca. 50 meV.

The 3s level of an atom is partially involved in both the occupation of the conduction band of a solid and the extension of the ionization limit. Moreover, the 3p level is involved in a very broad unoccupied band located above E_F. The optical transitions of sodium atom D-lines between 3s and 3p correspond to the transitions between the relevant bands in a solid.

The transition energy is about 1.2 eV; however, the probability of prohibited transitions is high. A direct transition between 3s and 3p levels in a sodium atom requires energy of about 2.1 eV. Thus, despite being very illustrative, the diagrams of energy levels can lead to incorrect interpretation due to their insufficient periodicity.

In the general case, the interaction of clusters with photons does not result in electronic and/or vibrational excitation and can generate decomposition and even evaporation of clusters.

The peculiarities of optical spectra of clusters are determined by the oscillator strength. Moreover, the properties of spectra can be divided into those associated with single electron and collective excitations.

The collective excitation reflects the oscillating strength of delocalized conduction electrons. The resonance phenomena that determine the spectra of small and large clusters give rise to collective oscillations of electrons in a cluster. The excitation of collective oscillators of conduction electrons is conventionally considered as a surface plasmon.

For very fine clusters, a single electron acquires definite significance, while the role of collective excitation is still unclear. Moreover, the situation with the line broadening and the excitation-amplitude increase is also far from being entirely clear.

As a rule, the optical experiments with large clusters were carried out in matrices. The interpretation of spectra in such systems is additionally complicated by the interactions between particle–particle and particle–matrix and also by the wide size distribution of particles.

Figure 10.4 illustrates the changes in the optical properties at a transition from a sodium atom to solid bulk sodium. These properties vary in an intricate manner. A sodium atom has a well-separated doublet of D-line with $\lambda = 589.0$ and 589.6 nm (Figure 10.4a). These lines correspond to the transition from the ground state $2s_{1/2}$ to the first excited states $2p_{2/3}$ and $2p_{1/2}$. As the cluster grows, the individual lines give way to separate wide bands, although the optical spectrum still provides certain information on the individual features of its structure (Figure 10.4b). The optical spectra of clusters built of 8 sodium atoms already demonstrated a broad single line (Figure 10.4c). A similar line is typical of a cluster consisting of 10^5–10^6 atoms. The situation in Figure 10.4d corresponds to absorption by a large cluster in a sodium chloride salt matrix. If solid sodium forms a film (Figure 10.4e), its visible spectrum is structure free. Such a spectrum reflects the optical properties of free electrons in a metal.

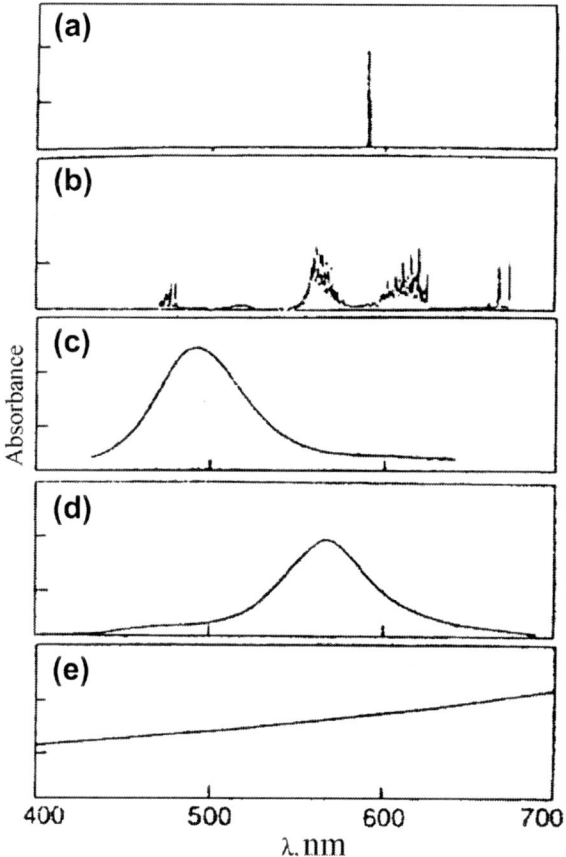

FIGURE 10.4 Optical properties of sodium atoms, clusters, and nanoparticles: (a) D-line of sodium atom, (b) Na_3 cluster, (c) Na_8 cluster, (d) sodium nanoparticle $2R < 10$ nm, (e) thin sodium film with thickness of 10 nm.[14]

For sizes intermediate to situations shown in Figure 10.4c and d and also Figure 10.4d and e, more complicated spectra are observed. The transitions similar to those considered for sodium were observed for silver, lithium, and copper.[14] Studying silver clusters in light-sensitive glasses has shown that the molecular structure disappears with an increase in the collective excitation as the cluster grows.

For particles with diameters greater than 10 nm, the permittivity $E(\omega)$ corresponds to the bulk-metal value and is size-independent. The properties of such clusters are considered in terms of classical electrodynamics. In the spectra of these clusters, the collective resonance, i.e. surface plasmons, predominates. The lagging effect of the electromagnetic field across a particle can shift and broaden the resonance with an increase in the particle size. Thus, for large clusters, the size dependence of optical properties is an external size effect.

For small clusters, the internal size effect is significant, and the permittivity depends on the particle size, i.e. $E = E(\omega, R)$. The properties mentioned above can be generalized as shown in Table 10.2.

The spectra of real systems are often complicated by the distribution of clusters over sizes and shapes. For clusters deposited on different surfaces, the nature of the carrier also affects the penetration depth of electromagnetic waves.

The energies of interactions between different metals, which involve equal numbers of atoms adsorbed on the same surface, were theoretically compared.[15] Interaction of square planar particles of silver, copper, nickel, and palladium with the ideal (0 0 1) MgO surface was studied with the aim of determining the bonding between the metal atoms in free samples in terms of the effect exerted by the MgO surface on the metal–metal bond. The following order of adsorption energy was found: $D(Ag_4/MgO) < D(Cu_4/MgO) \approx D(Pd_4/MgO) < D(Ni_4/MgO)$. The series found correlates with calculated binding energies of single atoms of the same metals. The binding energy was found to be 0.2–0.5 eV per atom. An intricate competition between the metal–metal and metal–oxygen bonds, which depended on the nature of the metal, was observed. It remained unclear whether these results can be extrapolated to larger particles of the same transition metals formed in the course of the interaction between the metal and the oxide surface.

The internal size effect can arise for different reasons and reflect certain changes in the cluster structure, e.g. its transition to the icosahedral structure. The appearance of a size effect can also be caused by the influence of the surface, which leads to an increase in electron localization and produces changes in the coordination number. The changes in the cluster properties concern the distances between the nearest "neighbors." For instance, this distance is 0.210 nm in Ag_2, 0.253 nm in Au_2, and 0.325 nm in the corresponding compact metals.[14]

The permittivity is defined as the average microscopic polarizability in clusters. It can change owing to the limited volume of averaging, the presence of a surface, a nonuniform charge density, local polarizability, etc. In aggregate, the properties mentioned above points to the changes in the atomic and electronic structures of a cluster and gives rise to size-dependent optical spectra.

Recently, new ligands, namely alkyl and aryl derivatives of carbodithioic acids (R–C(S)SH), were proposed for imparting different functional properties to

TABLE 10.2 The Dependence of Size Effect and Permittivity on the Particle Radius R

Cluster radius	$R \leq 10$ nm	$R > 10$ nm
Electrodynamic theory Mie	Independent of R	$f(R)$
Optical properties of the material	$\varepsilon = \varepsilon(R)$	Independent of R
Size effect	Internal	External

CdSe nanoparticles.[16] Their main advantages are the easiness of their exchange for phosphines, stability to photooxidation, and a possibility of grafting various molecules, e.g. an electroactive aniline tetramer.

The size dependence of optical properties of semiconductor nanocrystals opened up a possibility of using them as new building blocks in optoelectronics and bioengineering.[17–19] In turn, this stimulated the studies on synthesizing nanoscale semiconductors with controlled sizes. Thus, the preparation of CdS[20] and also the synthesis of CdTe were described.[21] Moreover, the methods for regulating the size and optical properties of semiconductor nanocrystals by either creating inorganic shells or doping were developed.[22–24] The optical properties of semiconductors could also be changed by compensating the surface defects with specially introduced organic molecules.[25] Recently, the kinetic control over optical properties of 3.3-nm CdTe nanocrystals was exerted by modifying their surfaces with 1-decanethiol.[26]

Figure 10.5 shows how the optical properties of CdSe particles depend on their size. As seen from Figure 10.5a, the absorption band of CdSe nanoparticles shifts to short wavelength region with the increase in their size. The dependence of the energy in the absorption band maximum on the radius of CdSe nanoparticles, which was calculated based on the data in Figure 10.5a and is shown in Figure 10.5b, is well approximated as $1/R^2$.

Recombination of light-induced charges gives rise to luminescence of nanoparticles, which is accompanied by a short wavelength shift with a decrease in the particle size (Figure 10.5c). The inverse problem, namely, the determination of the particle size from its spectrum has not yet been solved. This requires further development of the theory. However, the studies in which the plasmon-line width versus particle-size dependence were observed, using SEM, and electron spectroscopy techniques have already appeared.[27] Figure 10.6 shows such dependence for silver particles. As can be seen, the plasmon-line width is minimum for large particles and maximum for small particles.

The nanoparticle properties depend on an effect associated with the bond-length decrease. As a rule, in metal clusters, the interatomic distances are shorter as compared with compact metals. This effect, expressed as $\Delta R/R$, where R is the particle radius, is more pronounced for small particles. Thus, in rhodium clusters, $\Delta R/R$ is 17% for Rh_2, 10.9% for Rh_3, 7.4% for Rh_4, and 4.0% for Rh_{12}.[28] The particle size effect on the average bond length was calculated for nickel, and a decrease in the lattice parameter with a decrease in the particle size was predicted for palladium.[29]

While the STM examination provides information on the superficial structure of particles, their internal structure, particularly, the size dependence of the lattice parameter, was studied based on the results of high-resolution TEM. Figure 10.7 shows the dependence of the lattice parameter on the size of platinum particles.[30] As can be seen, the interatomic distance keeps shortening with a decrease in the particle size. For a 1-nm particle, this distance amounts to 90% of its value in compact platinum. On the other hand, for 3-nm particles,

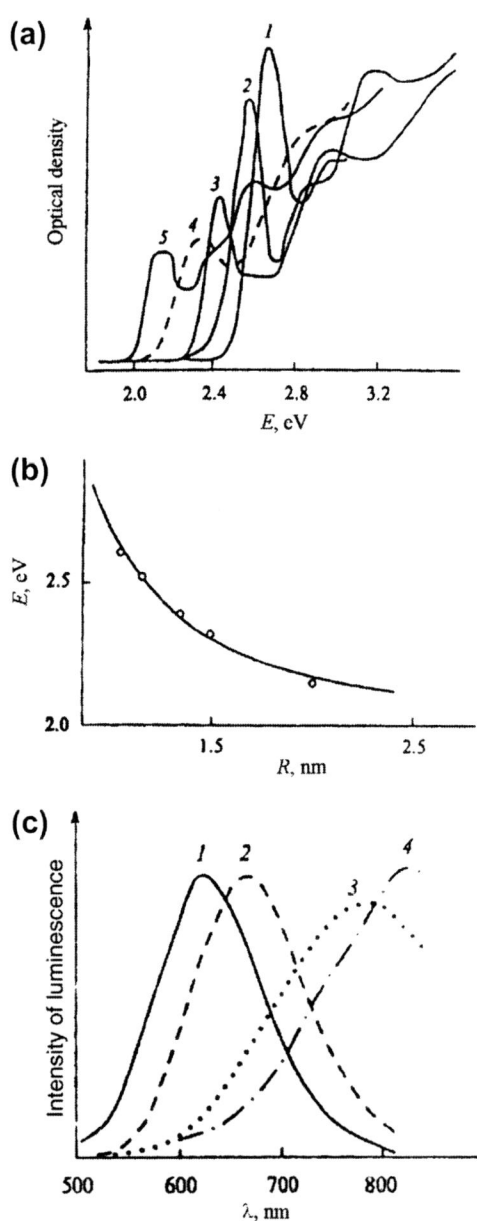

FIGURE 10.5 The effect of the size of semiconductor nanoparticles on their optical properties: (a) absorption spectra of CdSe nanoparticles of different size R (nm): (1) 2.1, (2) 2.2, (3) 2.7, (4) 3, (5) 4; (b) dependence of energy of maximum absorption of CdSe nanoparticles on their radii; (c) luminescence spectra of colloid CdSe nanoparticles in acetonitrile, R (nm): (1) 2.5, (2) 3.3, (3) 3.7, (4) 4.2.[50]

the lattice parameter approaches its value in the compact metal. Similar effects were observed for tantalum[31] and palladium[32] particles on thin aluminum films.

Strong size-induced changes in the lattice parameter should affect the chemical activity of nanoparticles. For reacting systems, the diffusion between particles and their interatomic distances are of no less importance than the particle size and morphology and should also be controlled.

10.4 KINETIC PECULIARITIES OF CHEMICAL PROCESSES ON THE SURFACE OF NANOPARTICLES

Surface reactions are of prime importance for the behavior of particles and their stabilization. For reagents adsorbed on the surface of nanoparticles, a

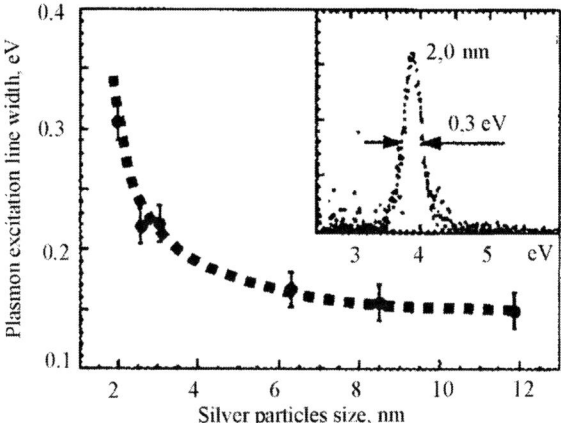

FIGURE 10.6 Dependence of the plasmon-line width of silver particles on their size.[27]

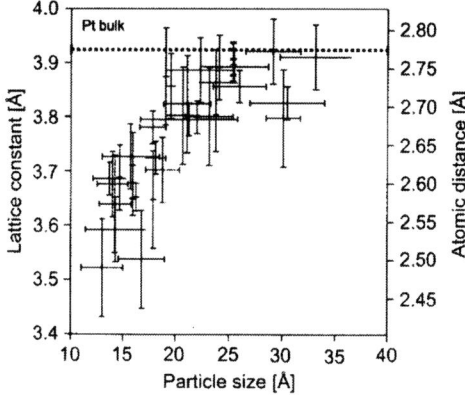

FIGURE 10.7 Lattice constant and interatomic distances for platinum particles on Al_2O_3–NiAl (110) as a function of size. Horizontal lines indicate the length and width of particles, vertical lines indicate the experimental error.[30]

chemical reaction can no longer be considered as a process in an infinite volume with a constant average density (concentration) of molecules. This is due to the fact that the nanoparticle size is small and commensurable with the size of the reacting species. In such systems, the kinetics of the bimolecular chemical reactions correspond to the kinetics in a limited volume and differ from the classical case.

Classical kinetics ignores fluctuations in reagent concentrations. Nanoparticles that contain small numbers of reacting molecules are characterized by comparatively wide variations in the number of reagents. This factor leads to a discord between the time variations in the reagent concentrations on different-sized particles and hence their different reactivity, which is dependent on the particle size.

For macroscopic samples, the general kinetic law is derived by averaging the kinetics over all nanoparticles. The description of processes in such systems employs a stochastic approach that, in place of concentrations, operates with the number of reagent molecules, which is a random value determined by statistical fluctuations in the number of the reacting species.[33]

The reactions that involve few molecules are usually diffusion controlled. It is assumed that they can be characterized by a rate constant similar for all reagent pairs. Based on this assumption, the kinetics of superficial bimolecular reactions was analyzed. For a reaction $A+A \rightarrow C$, the kinetics of reagent loss was described by the following equation:

$$\frac{\overline{N}(t)}{\overline{N}(0)} = \sum_{n=1}^{\infty} B_n \exp\left[-\frac{1}{2}n(n-1)kt\right]$$

where $N(t)$ and $N(0)$ are the current and initial concentrations of reagents on the surface of nanoparticles, which are averaged over a sample containing a macroscopic number of nanoparticles, and B_n is a complex function. In deriving this equation, an assumption on the equality of the rate constants k for all reagent pairs was employed. This assumption and, hence, the equation itself are applicable only for nanoparticles with sizes exceeding the size of reagent species at least by an order of magnitude.

For describing the kinetics of processes on the surface of nanoparticles, the Monte Carlo method was used.[33] This method allows solving the mathematical problems by simulating random processes and events. In fact, it employs the apparatus of the probability theory for solving the applied problems by means of a computer. The number of reagent species on the surface of each nanoparticle is small; hence, calculations are not too cumbersome. In calculations, the parameters of the process under study are selected in such a way as to obtain the best agreement of the calculated and experimental results.

The kinetics of adsorption, accommodation of adsorbents, and their migration over the surface to form clusters were analyzed.[34] The kinetics of nanosize catalysts[35] and the kinetics of oriented aggregation of nanocrystals[36] were also described.

Studies of the ligand exchange played an important role in understanding the processes of stabilization of metal particles by various ligands and in assessing the reactivity of resulting particles. These processes were studied by the example of gold nanoparticles.[37,38] Attention was focused on the dependence of the exchange processes on the ligand nature, the size of stabilized metal particles, and their charge. The gold particle core size was found to affect the electrochemical and spectroscopic properties of the stabilizing ligands.[39–41]

The kinetics of exchange reactions was studied by the example of gold nanoparticles.[37,42,43] It was found that the vertex and edge sites of the surface of gold nanoparticle exhibit higher kinetic reactivity as compared with the terrace-like core surface sites. NMR studies of ligand-exchange processes were carried out for different-sized gold nanoparticles stabilized by monolayers of phenylethanthiol, which was exchanged for para-substituted arylthiols p-X-PhSH, where X=NO_2, Br, CH_3, OCH_3, and OH. The second-order exchange rate constant for an $Au_{38}(ligand)_{24}$ particle was found to be very close to the corresponding constant of a larger $Au_{140}(ligand)_{53}$ particle, i.e. the chemical reactivity of different-sized particles was virtually identical in the initial segments of the kinetic curves.[43] Different numbers of defects on the terrace sites present on Au_{38} and Au_{140} particles resulted in different exchange rates for later stages. The reaction of 2.6-nm gold particles protected by the ligands with spin-labeled disulfides was studied by EPR technique.[44,45]

Based on the results discussed, we can conclude that there are substantial differences in the kinetic behavior of limited small-size systems and unlimited systems. This field of contemporary chemistry needs to be developed further.

10.5 THERMODYNAMIC FEATURES OF NANOPARTICLES

In nanoparticles, a substantial number of atoms pertain to the surface and their ratio increases with a decrease in the particle size. Correspondingly, the contribution of surface atoms to the nanocrystal energy increases.

The surface energy of a liquid is always lower than the surface energy of a corresponding crystal. A decrease in the nanoparticle size increases the share of surface energy and, hence, reduces the T_m of nanocrystals. As shown experimentally, this reduction can be very substantial.

The changes in the nanoparticle sizes determine other thermodynamic properties. The vacancy concentration increases with a decrease in the nanoparticle size (by a vacancy is meant a substitution for an atom in the lattice site). With a decrease in the particle size, the temperatures of polymorphous transitions and the lattice parameters decrease, while the compressibility and solubility increase.

Now, we illustrate the effect of size factors on the shift of the chemical equilibrium. According to chemical thermodynamics, the equilibrium in the transition from initial reagents A_i to products B_j can be written as follows:

$$\sum_i v_i A_i = \sum_j \mu_j B_j \qquad (10.2)$$

where v_i and μ_j are the corresponding stoichiometric coefficients.

The equilibrium constant at the fixed pressure and temperature is related to the changes in the isobaric potential by the following equation:

$$-kT \ln K_p = \Delta G^0 \qquad (10.3)$$

In the standard state, ΔG^0 is expressed by an equation

$$\Delta G^0 = \sum_j \mu_j G_{B_j}^0 - \sum_i v_i G_{A_i}^0 = G_f^0 - G_{in}^0 \qquad (10.4)$$

Changes in ΔG^0 are expressed as the changes in the standard enthalpy and entropy by the following equation:

$$\Delta G^0 = \Delta H^0 - T\Delta S^0 \qquad (10.5)$$

Reference sources provide ΔH^0 and ΔS^0 values only for compact substances.

The use of highly dispersed particles can further shift the equilibrium in a system. Theoretical studies in thermodynamics of small particles and experimental research showed that the particle size is an active thermodynamic variable, which together with other thermodynamic variables determines the system's state.

The size seems to play the role of the temperature. This fact can be used for reactions, the equilibrium of which is shifted to the initial reagents. In this case, if the total free energy G_{in}^0 of initial reagents is lower than G_f^0 and then $\Delta G^0 > 0$, the reaction proceeds.

The participation of nanoparticles can change the situation. The Gibbs potential of a dispersed reagent differs from its standard value in the compact phase. Dispersing of a reagent i, causes the following changes:[34]

$$\delta G_i = \frac{2}{3} \frac{A}{N_a} \frac{\sigma_i F_i}{\rho_i V_i} - kT(C_R - C_\infty) = \frac{A}{\rho_i N_A} \frac{2\sigma}{R_i} - kT(C_R - C_\infty) \qquad (10.6)$$

where s is the surface tension, F the surface area, V the volume of a dispersed particle, $C_R = C_\infty \exp(2\sigma/R\,\Delta V/kT)$ the relative number of vacancies per atom, R the particle radius, ΔV the volume change at a substitution of a vacancy for an atom in the lattice site, C_∞ the concentration of vacancies in the bulk, A the atomic mass, and N_A the Avogadro number.

In Eq. (10.6), the first term reflects the contribution of the surface energy and the second term corresponds to the contribution of the vacancies. If the final product is dispersed, the Gibbs potential also shifts according to Eq. (10.6).

A change in the isobaric potential $\Delta G = G_{in} - G_f$ for reaction (10.2) with the participation of dispersed reagents can be written as follows:

$$\Delta G = \Delta G^0 + \sum_j \mu_j \delta G_j - \sum_i v_i \delta G_i \qquad (10.7)$$

where summation is carried out over dispersed reagents. The reaction is possible if G_{in} exceeds G_f, i.e. $\Delta G < 0$, as illustrated by the following scheme:

```
     G_in
─────────────────────────
                              G_f
                         ─────────────
 δG_in=Σv_iδG_i          δG_f=Σμ_jδG_j
                              G^0_f
                         ─────────────
     G^0_in
─────────────────────────
```

By substituting Eq. (10.6) into Eq. (10.7) and, then, Eq. (10.7) into Eq. (10.3), we obtain an expression for the equilibrium constant K_{eq}:

$$K_{eq} = K_{eq}^\infty \exp\left\{\frac{2}{N_A}\left[\sum_i v_i \frac{\sigma_i A_i}{\rho_i R_i} - \sum_j \mu_j \frac{\sigma_j A_j}{\rho_j R_j}\right]\right\}\frac{1}{kT} - \sum_i v_i (C_R^i - C_\infty^i)$$

$$+ \sum_j \mu_j (C_R^j - C_R^j)$$

In this equation, K_{eq}^∞ is the constant of a massive sample determined by Eq. (10.3). The radius R_i characterizes the size effect. Estimates of δG_i for particles with $R < 10^3$ Å produced values in a range of 0.1–1.0 eV per atom. The quantities obtained mean that dispersion can initiate chemical reactions with such a barrier.

Interesting size-dependent thermodynamic peculiarities were revealed in studying the effect of pH on the behavior of cadmium chalcogenide nanocrystals coated with thiolates (deprotonated thiols).[46] The kinetic stability of small semiconductor nanocrystals in solutions was provided by their ligand coatings.

Such typical ligands as thiolates are Lewis bases, while the nanocrystal surface is positively charged, which is typical of Lewis acids. Their interaction represents a specific coordination in a nanocrystal–ligand complex. Dissociation of this complex with a decrease in solution pH can be considered as titration, which is characterized by a certain constant and free energy. The process is limited by slow diffusion of hydrogen atoms through the stabilizing ligand layer and can be likened to pseudosteady-state titration.[46]

Study of the effect of pH on CdS, CdSe, and CdTe nanocrystals stabilized with 3-mercapto-1-propanethiol has shown that at certain pH semiconductor crystals precipitate. Their relative stability decreases with an increase in hydrogen ion concentration, while the stability of complexes increases with a decrease in the crystal size. For CdSe crystals measuring 2.8 nm, a possibility of redispersing the nanocrystal precipitate was demonstrated,[46] although this process occurred with substantial hysteresis. Considering the example of CdSe, the equilibrium precipitation can be expressed as

$$(CdSe)_n - L_m + mH^+ \leftrightarrow ((CdSe)_n)^{m+} + mHL \quad K_{eq} \qquad (10.8)$$

where $((CdSe)_n)^{m+}$ and $(CdSe)_n - L_m$ designate ligand-free and ligand-coated nanocrystals, respectively. The positive charge on a ligand-free crystal should be compensated by a negative charge of the acid counter ions used for titration. K_{eq} is the equilibrium constant. Equilibrium (10.8) can be expressed by the following equations:

$$HL \leftrightarrow H^+ + L^- \quad K_a \qquad (10.9)$$

$$((CdSe)_n)^{m+} + mL^- \leftrightarrow (CdSe)^n - L_m \quad (K_s)^m \qquad (10.10)$$

Equation (10.10) describes the formation of a nanocrystal–ligand complex. Based on the experimental results, it was assumed that the constant K_s corresponds to the binding energy of all surface atoms of cadmium with deprotonated thiol ligands.

This assumption allows us to express K_s as

$$(K_s)^m = 1/[(K_a)^m K_{eq}] \qquad (10.11)$$

and

$$K_s = 1/[K_{eq}^{1/m} (K_a)] \qquad (10.12)$$

Equation (10.8) can be written as follows:

$$K_{eq} = \{[HL]^m [((CdSe)_n)^{m+}]\} / \{[(CdSe)_n - L_m]([H^+]_{eq})^m\} \qquad (10.13)$$

Combining Eqs. (10.12) and (10.13), we obtain

$$K_s = \{[(CdSe)_n - L_m]^{1/m} [H^+]_{eq}\} / \{[HL][((CdSe)_n)^{n+}]^{1/m} K_a\} \qquad (10.14)$$

In Eq. (10.14), all concentrations except $[H^+]_{eq}$ can be eliminated. This system is remarkable because the process is independent of both nanocrystal and ligand concentrations. Hence, Eq. (10.14) can be reduced to

$$K_s = [H^+]_{eq} / K_a \qquad (10.15)$$

Equation (7.15) makes it possible to assess the Gibbs free energy as the energy of the formation of a crystal–ligand bond ($\Delta_r G^0$)

$$\Delta_r G^0 = -RT \ln K_s = -RT \ln([H^+]_{eq} / K_a) \qquad (10.16)$$

Free energy values calculated from Eq. (10.16) depended on the size of the studied particles. The energy values calculated for semiconductors are similar to those obtained for thiol-stabilized gold particles.[47] No theoretical models have yet been calculated for this system; moreover, they may turn out to be more complicated than a relatively simple model based on the interaction of a negatively charged ligand with a positively charged core. First, the quantitative estimates of surface and interface energies of nanocrystals are necessary.[48] Moreover, thermodynamics should be corrected for size effects.[9]

Thermodynamic peculiarities of nanoparticles were analyzed in detail in reviews.[9,49]

In conclusion, we note once again that the main scope of nanochemistry is to study the size effects and reveal the relationship between the number of atoms taking part in a reaction and the resulting qualitative chemical changes. Size, shape, and organization of metal particles in the nanosize range directly determine the chemical activity of systems, the stability and properties of materials synthesized, and the advantages of their application in nanochemistry.

Specific heats are also affected by crystallite sizes. For example, palladium and copper nanocrystals (size determined by XRD line broadening) were pressed into pellets and specific heats determined using a differential scanning calorimeter, over a temperature range of 150–300 K. Both metals showed enhanced specific heats over normal polycrystalline pressed metal samples. For palladium, this enhancement was between 30 and 50%, while for copper about 10%.[51]

Several other reports have verified that nanocrystalline materials have higher specific heats, except at very low temperatures.[52] However, there is still little theoretical understanding. Small sizes, large number of surface atoms, and/or effects of grain boundaries must be involved, but early theories are not adequate to explain these results—even quantum theories by Einstein and Debye.[53,54]

10.6 MAGNETIC PROPERTIES

Nanoscale particles exhibit a treasure trove of interesting and useful magnetic properties. As an example, a ferromagnetic material, such as iron metal, can possess a long-range magnetic order when the single domains have magnetic vectors all lined up in the same direction. These single domains are about 14 nm for iron metal.[55] If one of these domains is so isolated that it has no neighbors, this nanoscale particle still has a strong internal magnetic field due to its many unpaired d-electrons. Thus, nanoscale magnetic particles are paramagnetic but with huge magnetic moments compared to paramagnetic molecules with one unpaired electron. They are "superparamagnetic."

Single domain magnetic particles reverse their magnetization vector via spin rotation since there are no domain walls (from neighboring single domains), rather than moving domain walls. Therefore, single domain particles have higher resistance to field reversal compared with multidomain systems, since it is more difficult to rotate the magnetization than to move a domain wall.

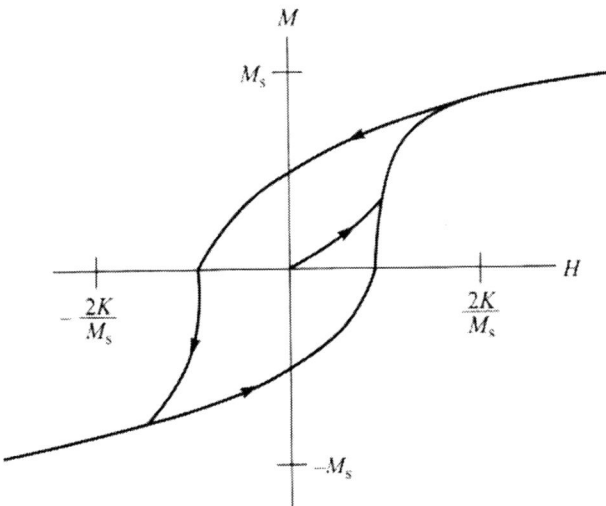

FIGURE 10.8 Hysteresis loops exhibited as a variable external magnetic field H (positive to right, negative to left) is applied. MS is the generated magnetic field strength of the sample. Coercivity is the intersection of the left magnetic line with the H axis, and is expressed as the energy needed to move from zone field to the intersection on the H axis (Coercivity = H_c).

"Coercivity" is the energy needed to reverse the field, Figure 10.8.

Resistance to magnetic field reversal in single-domain particles (coercivity) is due to anisotropy, which can be due to particle shape, stress, or crystallinity.

Returning to supermagnetism, if a particle gets very small, thermal energy can disrupt or overcome the resistance to magnetic-moment directional change. The particle can have a large moment and is free to respond to an applied field. An applied field would tend to align this large (or super) moment, but thermal energy would resist the alignment. Two key qualities remain for a superparamagnetic system: (1) absence of hysteresis, and (2) magnetic data taken at different temperatures superimpose of M vs. H/T°, so supermagnetism is temperature dependent. Thus, there is blocking temperature T_B, which is somewhat analogous to a Curie temperature for bulk materials.

The largest coercivities generally occur at the single domain size, while for small particles, H_c decreases due to thermal activation over the anisotropy barriers.

It can be appreciated that there are many practical uses for nanoscale magnetic particles for data storage, memory storage, refrigeration, and more. An excellent, more in-depth review is given by Sorensen.[55] If more details are needed, there are three excellent books available.[56–58]

10.7 ELECTRICAL/CONDUCTING PROPERTIES

As depicted in Figure 10.9, reduction in size of any substance, metal, semiconductor, or insulator, caused the eventual change from energy bands to orbitals.

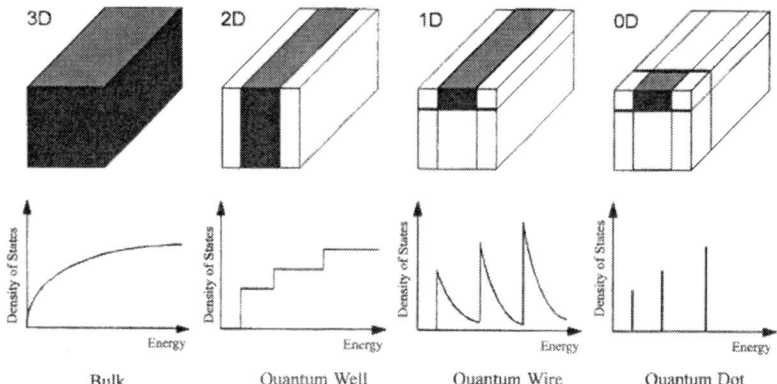

FIGURE 10.9 Formation of a zero-dimensional quantum dot by reduction of dimensions, and loss of band structure and evolvement of molecular/atomic orbitals.

Metals, which are conductors due to their band structure, possessing a conduction band only partially occupied by electrons, allow the electrons to move with little resistance in all directions for a crystalline sample. Resistance occurs when defects, grain boundaries, dislocations, and thermal vibrations (phonons) are present.

As metal particles become smaller and smaller, discrete energy levels finally dominate, and because of this, Ohm's Law no longer is valid. Thus, current–voltage response should change.

To probe this experimentally is quite difficult. A single nanoparticle needs to be placed between two very small electrodes, and the data on current vs. voltage collected. Nanoparticles are usually protected with an organic ligand, and this presents a problem to conductivity studies. Nevertheless, several studies have been reported.[59–61]

Indeed, nanoparticles of gold and palladium do show distinct "Coulomb blockades" (quantized behavior). A gold 55 cluster (1.4 nm) showed a Coulomb blockade at room temperature, while a larger Pd (17 nm) particle showed Ohm's law behavior at room temperature but quantized behavior at 4.2 K, the low temperature eliminating the most interfering phonons.

The nonconductive ligand coatings on these and other nanoparticles are easily tunneled through by electrons in an excited state. There is a great deal of possible practical uses of such conducting nanoparticle arrays, such as capacitors, semiconductors, batteries, and other devices.[62,63]

REFERENCES

1. Moskovits, M.; Huelse, J. E. *J. Chem. Soc. Far.* **1977**, *11*, 471–484.
2. Kuendig, E. P.; Moscovits, M.; Ozin, G. A. *Can. J. Chem.* **1972**, *50*, 3587.
3. Huelse, J. E.; Moskovits, M. *Surf. Sci.* **1976**, *57*, 125–142.

4. Buffat, Ph.; Borel, J. -P. *Phys. Rev. A* **1976**, *13*, 2287–2298.
5. Takagi, M. *J. Phys. Soc. Japan* **1954**, *9*, 3559.
6. Haruta, M. *Catal. Today* **1997**, *36*, 153–166.
7. Suzdalev, I. P.; Buravtsev, Yu. V.; Imshennik, V. K.; Novichikhin, S. V. *Chem. Phys. (Russ.)* **1993**, *12*, 555–566.
8. Suzdalev, I. P.; Imshennik, V. K.; Matveev, V. V. *Inorg. Mater. (Russ.)* **1995**, *31*, 807–810.
9. Suzdalev, I. P.; Suzdalev, P. I. *Russ. Chem. Rev.* **2001**, *70*, 177–210.
10. Efremov, M. Yu.; Schiettekatte, F.; Zhang, M.; Olson, E. A.; Kwan, A. T.; Berry, R. S.; Allen, L. H. *Phys. Rev. Lett.* **2000**, *58*, 3560–3563.
11. Lindemann, F. A. *Phys. Z* **1910**, *11*, 609.
12. Shi, F. G. *J. Mater. Res.* **1994**, *9*, 1307.
13. Sattler, K. In *Handbook of Thin Films Materials*; Nalwa, H. S., Ed.; Academic Press: New York, 2002; pp 61–97.
14. Kreibig, U.; Vollmer, M. *Optical Properties of Metal Clusters*; Springer: Berlin, 1995, 1–532.
15. Matveev, A. V.; Neyman, K. M.; Pacchioni, G.; Rosch, N. *Chem. Phys. Lett.* **1999**, *299*, 603–612.
16. Querner, C.; Reiss, P.; Bleuse, J.; Pron, A. *J. Am. Chem. Soc.* **2004**, *126*, 11574–11582.
17. Pardo-Yissar, V.; Katz, E.; Wasserman, J.; Willner, I. *J. Am. Chem. Soc.* **2003**, *125*, 622–623.
18. Wang, Y.; Tang, Z.; Correa-Duarte, A.; Liz-Marzan, L. M.; Kotov, N. A. *J. Am. Chem. Soc.* **2003**, *125*, 2830–2831.
19. Pinaud, F.; King, D.; Moore, H. -P.; Weiss, S. *J. Am. Chem. Soc.* **2004**, *126*, 6115–6123.
20. Pradhan, N.; Erfima, S. *J. Am. Chem. Soc.* **2003**, *125*, 2050–2051.
21. Zhang, H.; Zhou, Z.; Yang, B.; Gao, M. *J. Phys. Chem. B* **2003**, *107*, 8–13.
22. Reiss, P.; Bleuse, J.; Pron, A. *Nano Lett.* **2002**, *2*, 781.
23. Manna, L.; Scher, E. C.; Li, L. -S.; Alivisatos, A. P. *J. Am. Chem. Soc.* **2002**, *124*, 7136–7145.
24. Bailey, R. E.; Nie, S. *J. Am. Chem. Soc.* **2003**, *125*, 7100–7106.
25. Gaponik, N.; Talapin, D. V.; Rogach, A. L.; Eychmuller, A.; Weller, H. *Nano Lett.* **2002**, *2*, 803–806.
26. Akamatsu, K.; Tsuruoka, T.; Nawafune, H. *J. Am. Chem. Soc.* **2005**, *127*, 1634–1635.
27. Freund, H. -J. *Surf. Sci.* **2002**, *500*, 271–299.
28. Barreteau, C.; Spanjaard, D.; Desjonqueres, M. C. *Phys. Rev. B* **1998**, *58*, 9721–9731.
29. Silva, E. Z.; Antonelli, A. *Phys. Rev. B* **1996**, *54*, 17057–17060.
30. Freund, H. -J.; Klimenkov, M.; Nepijko, S.; Kuhlenbeck, H.; Baeumer, M.; Schloegl, R. *Surf. Sci.* **1997**, *391*, 27–36.
31. Nepijko, S. A.; Klimenkov, M.; Kuhlenbeck, H.; Zemlyanov, D.; Herein, D.; Schlogel, R.; Freund, H. *J. Surf. Sci.* **1998**, *412–413*, 192.
32. Nepijko, S. A.; Klimenkov, M.; Adelt, M.; Kuhlenbeck, H.; Schlogl, R.; Freund, H. *J. Langmuir* **1999**, *115*, 5309–5313.
33. Khairutdinov, R. F.; Serpone, N. *Prog. React. Kinet.* **1996**, *21*, 1–30.
34. Lidorenko, N. S.; Chizhik, S. P.; Gladkih, N. T.; Grigor'eva, L. K.; Kuklin, R. N. *Doklady AN SSSR* **1981**, *5*, 1114–1116.
35. Zhdanov, V. P.; Kasemo, B. *Surf. Sci. Rep.* **2000**, *39*, 25–104.
36. Penn, R. L. *J. Phys. Chem. B* **2004**, *108*, 12707–12712.
37. Song, Y.; Murray, R. W. *J. Am. Chem. Soc.* **2002**, *124*, 7096–7102.
38. Ionita, P.; Caragherogheopol, A.; Gilbert, B. C.; Chechik, V. *J. Am. Chem. Soc.* **2002**, *124*, 9048–9049.
39. Hicks, J. F.; Miles, D. T.; Murray, R. W. *J. Am. Chem. Soc.* **2002**, *124*, 13322–13328.
40. Lee, D.; Donkers, R. L.; Gangli, W.; Harper, A. S.; Murray, R. W. *J. Am. Chem. Soc.* **2004**, *126*, 6193–6199.

41. Jimenez, V. L.; Georganopoulou, D. G.; White, R. J.; Harper, A. S.; Mills, A. J.; Lee, D.; Murray, R. W. *Langmuir* **2004**, *20*, 6864–6870.
42. Donkers, R. L.; Song, Y.; Murray, R. W. *Langmuir* **2004**, *20*, 4703–4707.
43. Guo, R.; Song, Y.; Wang, G.; Murray, R. W. *J. Am. Chem. Soc.* **2005**, *127*, 2752–2757.
44. Ionita, P.; Caragheorgheopol, A.; Gilbert, B. C.; Chechik, V. *Langmuir* **2004**, *20*, 11536–11544.
45. Chechik, V. *J. Am. Chem. Soc.* **2004**, *126*, 7780–7781.
46. Aldana, J.; Lavelle, N.; Wang, Y.; Peng, X. *J. Am. Chem. Soc.* **2005**, *127*, 2496–2504.
47. Yu, W. W.; Qu, L.; Guo, W.; Peng, X. *Chem. Mater.* **2003**, *15*, 2854–2860.
48. Cahen, D.; Kahn, A. *Adv. Mater.* **2003**, *15*, 271–277.
49. Uvarov, N. F.; Boldyrev, V. V. *Russ. Chem. Rev.* **2001**, *70*, 265–284.
50. Alivisatos, A. P. *J. Phys. Chem.* **1996**, *100*, 13226–13239.
51. Rupp, J.; Birringer, R. *Phys. Rev. B* **1987**, *36*, 7888–7890.
52. Bai, H. Y.; Luo, J. L.; Jin, D.; Sun, J. R. *J. Appl. Phys.* **1996**, *79*, 361–364.
53. Einstein, A. *Ann. Phys.* **1906**, *22*, 180–190.
54. Debye, P. *Ann. Phys.* **1912**, *39*, 789–839.
55. Sorensen, C. M. Magnetism. In *Nanoscale Materials in Chemistry*, 1st Ed.; Klabunde, K. J., Ed.; Wiley Interscience: New York, 2001; chapter 6, pp 169–221.
56. Cullity, B. D. *Introduction to Magnetic Materials*; Addison-Wesley: Reading, MA, 1972.
57. Kittel, C. *Introduction to Solid State Physics*; Wiley: New York, 1971.
58. Jiles, D. *Introduction to Magnetism and Magnetic Materials*; Chapman and Hall: London, 1991.
59. Schmid, G. Metals. In *Nanoscale Materials in Chemistry*, 1st Ed.; Klabunde, Ed.; Wiley Interscience: New York, 2001; chapter 2, pp 15–59.
60. Bezryadin, A.; Dekker, C.; Schmid, G. *Appl. Phys. Lett.* **1997**, *71*, 1273.
61. Chi, L. F.; Hartig, M.; Drechsler, T.; Schwaack, T.; Seidel, C.; Fuchs, H.; Schmid, G. *Appl. Phys. Lett. A* **1998**, *A66*, 187.
62. Kalita, M.; Basel, M. T.; Janik, K.; Bossmann, S. H. Optical and electronic properties of metal and semiconductor nanostructures. In *Nanoscale Materials in Chemistry*, 2nd Ed.; Klabunde, K. J., Richards, R. M., Eds.; Wiley: New York, 2009; chapter 16, pp 539–578.
63. Dong, W.; Dunn, B. Nanomaterials in Energy Storage Systems. In *Nanoscale Materials in Chemistry*, 2nd Ed.; Klabunde, K. J.; Richards, R. M., Eds.; Wiley: New York, 2009; chapter 15, pp 519–535.

Chapter 11

Nanoparticles in Science and Technology

Chapter Outline
11.1 Catalysis on Nanoparticles	299	11.5 Applications of CNTs	326
11.2 Oxide Reactions	311	11.6 Nanochemistry in Biology and Medicine	329
11.3 Semiconductors, Sensors, and Electronic Devices	314	11.6.1 DNA-modified Nanoparticles	336
11.4 Photochemistry and Nanophotonics	323		

The use of nanoparticles for synthesizing new ceramic materials, ultradispersed powders, and consolidated and hybrid systems was considered in monographs and reviews were cited in the previous chapters and also in some recent reviews.[1-6] The detailed consideration of the diverse aspects of nanotechnology in the cited studies allows us to focus attention on those directions where the advantages of nanochemistry are most evident.

11.1 CATALYSIS ON NANOPARTICLES

The development of new catalysts based on metal nanoparticles continues to attract the keen attention of scientists. Methane combustion in air is stable at temperatures above 1300 °C. At these temperatures, noxious nitrogen oxides evolve and smog is formed. Hence, the quest for new catalysts of methane oxidation is a burning problem. Thus, a new catalytic material that ensures methane combustion at 400 °C was described.[7] For its synthesis, reversible microemulsions based on isooctane, water, and surfactants such as the adducts of polyethylene oxide with alcohols were used. Salts $Ba(OC_3H_7)_2$ and $Al(OC_3H_7)_3$ were dissolved in isooctane and mixed with the microemulsion at room temperature. The resulting solid crystalline nanosize barium hexaaluminate exhibited high catalytic activity toward methane combustion. Barium hexaaluminate particles retained their size and surface area at high temperatures. Moreover, they could be additionally modified by cerium, cobalt, manganese, and lanthanum. Modification by cerium oxide produced a composite that affords methane combustion at temperatures below 400 °C. Syntheses of new catalysts of methane oxidation and their characteristics were discussed in detail.[8]

In aqueous buffer microemulsions AOT–n-heptane, several reactions catalyzed by palladium nanoparticles were carried out. A reaction of N,N-dimethyl-n-phenylenediamine with $Co(NH_3)_5Cl^{2+}$ catalyzed by palladium nanoparticles was described.[9] In similar microemulsions at pH 5.6, the catalytic oxidation of N,N,N',N'-tetramethyl-n-phenylenediamine by $Co(NH_3)_5Cl^{2+}$ was studied.[10] The limiting stage of this reaction was associated with the adsorption of n-phenylenediamine on palladium particles with a radius of 2.5 nm, localized inside the microemulsion. This conclusion was based on the changes in the reaction's activation energy from 97 kJ/mol at 15 °C to 39 kJ/mol at 40 °C and also on the results of electrochemical measurements. The peculiarities of this reaction were examined in terms of a microreactor model.

The syntheses of platinum, palladium, rhodium, and iridium nanoparticles and their use in the catalytic reactions of hydrogenation of cyclooctene, 1-dodecene, and o-chloronitrobenzene were analyzed.[11] Metal nanoparticles were stabilized by an amphiphilic copolymer of 1-vinylpyrrolidone with acrylic acid (PVP–AA). To obtain nanoparticles, the metal salts were reduced with alcohols. The particle diameter was 0.74 nm for Ir, 1.93 nm for Rh, 2.2 nm for Pd, and 1.2–2.2 nm for Pt. The particle size was determined by electron microscopy.

The introduction of an Ni ion into a catalytic system PVP–AA–Pt increased the efficiency of the latter. Hydrogenation of chloronitrobenzene to chloroaniline at 330 K was carried out with 97.1% selectivity and 100% conversion. The introduction of Co^{2+} and Fe^{3+} ions reduced the hydrogenation selectivity up to 78.1 and 72.1%, respectively.

An analysis of the IR spectra showed that ions of nickel, iron, and cobalt were not reduced to zero-valence metals under the effect of hydrogen. This was associated with the fact that metal ions interacted with two C=O groups in a PVP–AA copolymer and with only a single C=O group in poly (1-vinylpyrrolidone).[12]

In catalysis, it is necessary to determine the size of the deposited metal particles and their distributions over sizes and surface areas. Such measurements for particles smaller than 2.5 nm induce diffuse scattering, which affects the results.[13]

When using chemisorption for assessing the properties of a catalyst, it is necessary to know the prehistory of the catalysts, which could have already taken part in reactions and might have contaminated surfaces, and also make allowance for the strong interaction of a metal with the carrier. A systematic study of a series of catalysts, containing 0.5% palladium particles of average diameter about 2 nm deposited on activated carbon, was carried out over a temperature range 573–973 K.[14] A combination of X-ray diffraction (XRD), small angle X-ray scattering (SAXS), and transmission electron microscopy (TEM) made it possible to estimate the effective size of palladium particles and to compare it with the results obtained for carbon monoxide chemisorption on supported palladium. The agreement of the results obtained by the two methods mentioned was observed for a stoichiometric ratio, Pd:CO=2.

A high activity was observed for nanoparticles built of metal cores surrounded by shells.[15–18] For electrochemical oxidation of carbon monoxide, gold nanoparticles measuring 2 and 5 nm and surrounded by decanethiol molecules were used as the catalyst.[16] Figure 11.1 illustrates this electrode reaction. The application of gold nanoparticles suggests cooperative strengthening of the catalytic activity.

The catalytic oxidation of carbon monoxide was used for air purification, conversion of automobile combustion gases, and in the technologies of new fuel cells based on oxidation of methanol and other hydrocarbons. Fabrication of optimal catalysts may involve changing the shape of cores, the structure and properties of the molecular shells, the nature of the core–shell bonding that affects the active sites, the defect packing, and the collective electronic properties of nanoparticles.

The high catalytic activity of gold nanoparticles deposited on oxides has been discussed.[19]

The role of F-centers was elucidated by considering the example of CO conversion to CO_2 and catalyzed by MgO-supported gold particles with an average size of 3.8–4.3 nm.[20] It was shown that the concentration of F-centers correlates with the catalytic activity, controlling the partial charge transfer from the center to a gold cluster. Similar peculiarities were observed for different oxides. Theoretical estimates of the transferred charge corresponded to 0.15 e for TiO_2 supports. It was also shown that gold adsorption enhances the activity of titanium oxide, promoting the migration of oxygen vacancies.[21] For gold particles deposited

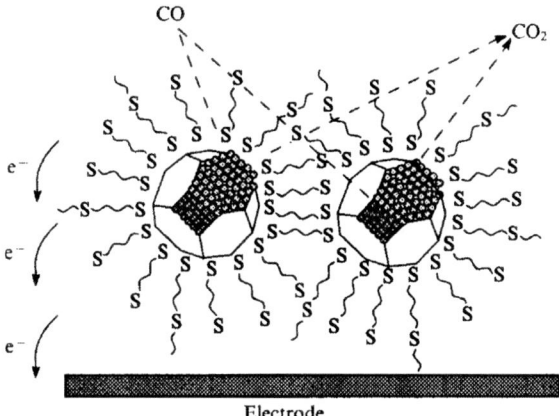

FIGURE 11.1 Catalytic oxidation of carbon monoxide on an assembly of thiol-coated gold particles on the electrode surface.[16]

on highly reduced ordered TiO_x films grown on Mo (112), it was shown that deeply reduced oxides favor the formation of a strong bonding between Au and reduced Ti atoms of the TiO_x support, yielding electron-rich Au.[22] These studies are consistent with the theoretical concepts based on the Au/TiO_2 system, which showed that the changes in the defect site number at the Au/TiO_2 interface determines the electronic and catalytic properties of gold particles.[23,24]

The synthesis of platinum nanoparticles by reduction with alcohols involved using poly(N-vinylformaldehyde), poly(N-vinylacetamide), poly(N-isopropylacrylamide), poly(N-vinylpyrrolidone), and chloroplatinic acid ($H_2PtCl_6 \cdot 6H_2O$). The synthesized particles had an average size of 2.0–2.5 nm.[25] The stability of the nanoparticles studied depended on the additions of KCl and Na_2SO_4 salts and on temperature variations. Hydrogenation of allyl alcohol was carried out in water and a salt solution (0.8 M Na_2SO_4). It was found that catalytic platinum particles are stable in a salt solution and exhibit an activity similar to that in water.

Studies of bimetallic catalytic systems have been developed. Bimetallic Pt–Au catalysts on graphite were prepared by selective deposition of gold on a carrier—a platinum film on graphite.[26] The latter monometallic catalyst was prepared by reducing H_2PtCl_6 in absolute ethanol. The subsequent reduction of $AuCl_4^-$ proceeded on the monometallic catalyst either in its original state or was modified by pretreatment with hydrogen. At pH 1, bimetallic particles were formed by the reaction:

$$3PtH + AuCl_4^- \rightarrow Pt_3Au + 4Cl^- + 3H^+$$

The average size of the particles was about 10 nm. Their properties and relevant reactions were studied by several techniques.

Gold being chemically inactive, its use as a catalyst escaped the attention of chemists until recently. However, gold ions Au^{n+} ($1 \leq n \leq 3$) incorporated into zeolites exhibited activity in the reaction:

$$H_2O + CO \rightarrow H_2 + CO_2$$

at 323 K.[27]

As a model of metal catalysts on carriers, lithographic systems consisting of Pt nanoparticles deposited on SiO_2 and Ag particles on Al_2O_3, which were synthesized by the electron-beam method, were used.[28] The average size of particles varied from 20 to 50 nm, and the interparticle distance varied from 100 to 200 nm. The thermal stability of fabricated systems was studied by electron microscopy and atomic force microscopy (AFM) techniques. The platinum-containing system was stable up to 973 K, while the silver-containing system was stable in vacuum and hydrogen up to 773 K and in oxygen up to 623 K. A model catalyst built of Pt nanoparticles on SiO_2, which represented a lithographic system obtained by an electron-beam method, was compared with platinum foil in the reactions of hydrogenation–dehydrogenation of cyclohexene at 100 °C.[29] As compared with platinum foil, a twofold increase in the reactivity was observed for a system containing platinum particles with 28-nm diameter; its selectivity increased by a factor of 3.

Iron-containing nanoparticles stabilized in polymeric matrices were used in the alkyl isomerization of dichlorobutanes. The catalytic activity was found to depend on the nature of the matrix and on the metal content.[30]

The photocatalytic reduction of bis(2-dipyridyl)disulfide (RSSR) to 2-mercaptopyridine (RSH) with water proceeded selectively on the surface of titanium dioxide. The process rate considerably increased when silver nanoparticles (0.24 mass%) measuring less than 1 nm were deposited on TiO_2.[31] Thiols are widely used in agrochemistry, petrochemistry, pharmaceutics, and for stabilization of metal nanoparticles.

Studies of RSSR adsorption, their absorption spectra, the effects of pH, the photolysis time, the amount of deposited silver, and the temperature allowed the authors[31] to put forward the following reaction mechanism:

$$RS-SR + Ag-TiO_2 \underset{}{\overset{k_a}{\rightleftharpoons}} 2RS-Ag-TiO_2 \quad (11.1)$$

$$2RS-Ag-TiO_2 \xrightarrow{I\varphi} 2RS-Ag-TiO_2\,(e^-\ldots h^+) \quad (11.2)$$

$$2RS-Ag-TiO_2\,(e^-\ldots h^+) \xrightarrow{k_{d1}} 2RS-Ag-TiO_2 \quad (11.3)$$

$$2RS-Ag-TiO_2\,(e^-\ldots h^+) \xrightarrow{k_{cs}} 2RS-Ag(e^-)-TiO_2\,(h^+) \quad (11.4)$$

$$2RS-Ag(e^-)-TiO_2\,(h^+) \xrightarrow{k_{d2}} 2RS-Ag-TiO_2 \quad (11.5)$$

$$2RS-Ag(e^-)-TiO_2\,(h^+)+H_2O \xrightarrow{k_0} 2RS-Ag(e^-)-TiO_2+2H^+ +\tfrac{1}{2}O_2 \quad (11.6)$$

$$2RS-Ag(e^-)-TiO_2 + 2H^+ \xrightarrow{k_r} 2RSH + Ag-TiO_2 \quad (11.7)$$

where K_a is the rate constant of equilibrium adsorption, I the intensity of light, φ the light absorption intensity, and k the rate constants of the corresponding stages.

In stage (11.1), the selective adsorption of RSSR on the surface of silver nanoparticles is accompanied by the rupture of the S–S bonds. In stage (11.2), excitation of nanoparticles in the bandgap of TiO_2 generates an electron pair (e^-)–hole (h^+). Most electron–hole pairs recombine in stages (11.3) and (11.5). In stage (11.4), charges are separated: silver nanoparticles take up from an electron, $AuCl_4^-$ while holes are transferred to the TiO_2 carrier. The positive potential of a hole is sufficient to oxidize water to form H^+ and an O_2 molecule (stage (11.6)). In stage (11.7), the catalytic system is regenerated and RSH is formed. In the opinion of the authors,[31] the deposited silver particles give rise to the following effects:

- acceleration of RSSR adsorption;
- spatial separation of positions corresponding to silver-particle reduction and supporting oxidation, i.e. the charge separation effect; and

- selective adsorption of the oxidizable compound (RSSR) and the reducer (H_2O) in the oxidation and reduction sites, which ensures the high selectivity of this process.

A comparison of the catalytic systems TiO_2 and $Ag-TiO_2$ has shown that both systems are characterized by equal activation energies of photocatalytic reduction (30.5±1.5 kJ/mol), which suggests that the limiting stage of this staged process is reaction (11.6) rather than photocatalysis.

The reduction of RSSR on TiO_2 in the absence of silver occurs as follows:

$$2RS \cdots TiO_2\,(e^-) + 2H^+ \xrightarrow{k_r} 2RSH + TiO_2$$

Recently, semiconductor nanoparticles have found extensive application in catalysis and photocatalysis. The relative catalytic activity of various nano-size particles made of semiconductor oxides and sulfides in polar (acetonitrile) organic solvents was studied.[32] This study was performed using the example of photooxidation of pentachlorophenol—a toxic compound used as a fungicide, a bactericide agent, and also for wood preservation. Figure 11.2 shows the dependence of the relative concentration of pentachlorophenol on the photolysis time in the presence of nanoparticles of different metals.

Mention should be made of an extremely strong dependence of the reaction rate on the particle size observed in the case of MoS_2, which in turn reflects the effect of the forbidden zone of a particle and is associated with the corresponding change in the reduction potential. In contrast to titanium dioxide, which is active in the UV range and is commonly used in electrocatalysis, molybdenum disulfide nanoparticles measuring 3 nm catalyze the reaction when illuminated with visible light. Nanoparticles of SnO_2 measuring 26 and 58 nm did not have any noticeable effect on the photooxidation of pentachlorophenol.

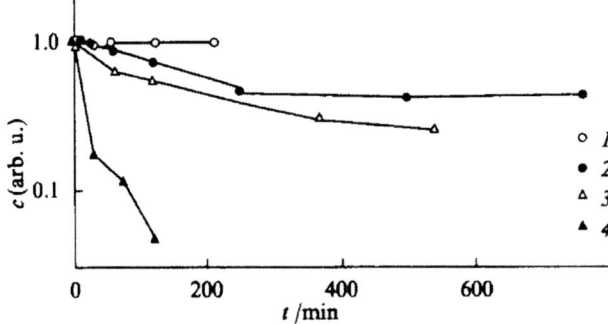

FIGURE 11.2 Changes in the relative concentration c of pentachlorophenol in water as a function of the time of irradiation with light of wavelength $400 < \lambda < 700$ nm in the (1) absence and (2–4) presence of catalysts: (2) CdS powder (0.1 mg/ml), (3) MoS_2 particles with diameter of 4.5 nm (0.036 mg/ml), and (4) MoS_2 particles with diameter of 3 nm (0.09 mg/ml).[32]

The preparation of molybdenum sulfide particles shaped as hollow nanospheres opened up new possibilities of using this substance in catalysis.[33] Such particles were synthesized by the ultrasound irradiation of a slurry containing $Mo(CO)_6$, S_8, and nanosize silicon particles in 1,2,3,5-tetramethylbenzene (isodecrene) under argon flow. Washing of MoS_2-coated silica with 10% HF served to remove silicon. The resulting hollow spheres of a diameter of ~50 nm exhibited high catalytic activity in the hydrodesulfurization of thiophene in the temperature range 325–375 °C. The increase in activity as compared with ordinary spherical MoS_2 particles was associated with the participation in the process of both internal and external surfaces of a hollow particle.

Molybdenum sulfides exhibited a high catalytic activity. As has been shown recently, MoS_3 nanoparticles with an average radius of 2.0–3.0 nm represent a new class of additives to hydrocarbon lubricants.[34] Addition of molybdenum sulfide provides a lower friction coefficient as compared with a conventional antifriction additive, molybdenum dithiocarbamate.

The actively developed synthesis of cluster anions, which contain 368 molybdenum atoms, opens up new prospects in nanochemistry and, in particular, in nanochemistry of molybdenum. Such compounds make it possible to realize reactions in the selected reaction centers of well-characterized nanosize samples.[35]

The development of a new direction in nanochemistry associated with chemical reactions carried out by means of AFM and scanning tunneling microscope (STM) probes can be predicted. Interesting examples of chemical modification of terminal functional groups in aggregates of organosilicon molecules by the catalytic impact of a palladium-covered AFM probe were considered.[36] The reactions studied are illustrated by the scheme below. Here, Z is the benzene-oxycarbonic group.

A minimum impact necessary for such chemical reactions is about 2.5 mN at a scanning rate up to 5 μm/s. Note that a typical activation energy of bimolecular reactions shown above amounts to nearly 50 kJ/mol. At the same time, according to estimates, the energy of surface deformation of an associate of organosilicon molecules is about 340 kJ/mol, which far exceeds the activation energy values shown above.

Catalytic scanning nanopens were proposed to be used in nanomodifying self-assembling monolayers.[37] Monolayers of $bis(\omega$-tertbutyldimethyl-siloxyundecyl) disulfide were deposited on gold, while 2-mercapto-5-benzimidazole sulfonic acid was deposited on the gold cantilever of a scanning microscope. An acidic pen induced local hydrolysis in the contact zone and fabricated samples of well-defined shape and size. In the catalytic region, dendritic wedges were formed with thiol functions incorporated into the catalyst domain.

The surface on which selective chemical changes occur can be used in sensitive optoelectronic devices, sensors, and units imitating biological devices. An approach to "chemical lithography," which involves the chemical modification of functional groups in the molecules of a sample, arranged along definite predetermined lines, was outlined.[38]

The catalytic reactions on metal nanoparticles under different conditions were comprehensively described in several reviews.[39–41] In the reactions that occur on the catalyst surface, the starting compounds are first adsorbed after which they migrate and react, and finally the products should be desorbed. The efficiency of the catalyst is determined by the coordination of all stages in the process, while elementary reactions can depend on the metal particle size in different ways. For a deeper understanding of the catalytic activity, it is necessary to know the number of adsorbed species and the number of metal atoms taking part in the catalytic process.

One of the approaches to the study of the catalytic activity, which has been actively developed recently, consists in the deposition of metal particles on thin films of different oxides. Studies of such a kind provide an approach to solving the problems associated with internal size effects and the nature of the carrier. For nickel particles (Ni_{11}, Ni_{20}, and Ni_{30}), the effect of their size on the dissociation of carbon monoxide was studied.[42] Nickel particles were synthesized by laser evaporation and supersonic expansion. Then, after mass-spectral separation, the particles were deposited on MgO films at a temperature of 90 K. The adsorption of CO molecules occurred at 240–260 K, while the associative desorption took place at 500–600 K. On Ni_{30} particles, the associative desorption proceeded more actively as compared with Ni_{11} and Ni_{20}. The experiments with labeled oxides $^{13}C^{16}O$ and $^{12}C^{18}O$ evidenced the presence of different adsorption and desorption sites in nickel particles, which were determined by the number of nickel atoms in a cluster and their different electronic properties.

The catalytic oxidation of carbon monoxide was studied on monodispersed platinum particles.[43] It was shown that each Pt_n particle ($8 \leq n \leq 20$) deposited on a magnesium oxide film was active at a certain temperature in the range 150–160 K. When a cluster grew by a single atom, e.g. at a transition from Pt_{14} to Pt_{15}, its reactivity increased by a factor of 3. The cluster activity was related to its electronic state, the changes in its morphology, and the interaction energy of boundary orbitals in oxygen molecules.

The catalytic oxidation of carbon monoxide ($CO + 0.5 O_2 \rightarrow CO_2$) on transition metals pertains to the main reactions used for controlling the air contamination. Earlier, it was believed that the catalytic oxidation of carbon monoxide is independent of the size of the palladium particles[44]; however, later it was demonstrated that palladium particles measuring less than 5 nm influence the kinetics of oxidation.[45] At the same time, for palladium clusters with preliminarily adsorbed oxygen, the kinetics of this reaction was independent of the size of the catalyst particles.[46] The aforementioned results obtained by the same group of scientists showed that the experimental procedure can affect the manifestation and reproducibility of size effects for one and the same particles and reactions.

The oxidation of carbon monoxide was used as a model reaction in comparing the activities of particles of different metals containing equal numbers of atoms.[47] Figure 11.3 shows the CO_2 yield as a function of the particle size, the metal type, and the temperature. As can be seen, all particles exhibited different reactivities. The fact that Au_8 particles were most active at low temperatures (140 and 200 K), while the particles with electron-closed shells Au_{13} were less active in the same temperature range, deserves mention. A strong size dependence was also typical of platinum particles. With a transition from Pt_8 to Pt_{20}, the amount of CO_2 formed increased by a factor of 9. Moreover, only large particles (Pt_n, $n > 14$) produced CO_2 at low temperatures (140 K). By comparing equal-size particles, e.g. containing 13 atoms, it is evident that palladium and rhodium are more active than gold and platinum.

Cyclotrimerization of acetylene to benzene on Pd_n particles ($1 \leq n \leq 30$) deposited on thin MgO films was studied.[48–50] It was found that even small clusters of Pd_n ($1 \leq n \leq 6$) produced benzene at 300 K. On large particles ($7 \leq n \leq 30$), benzene was synthesized at 430 K. The density-functional calculations made it possible to associate the high-temperature process with the surface diffusion of palladium

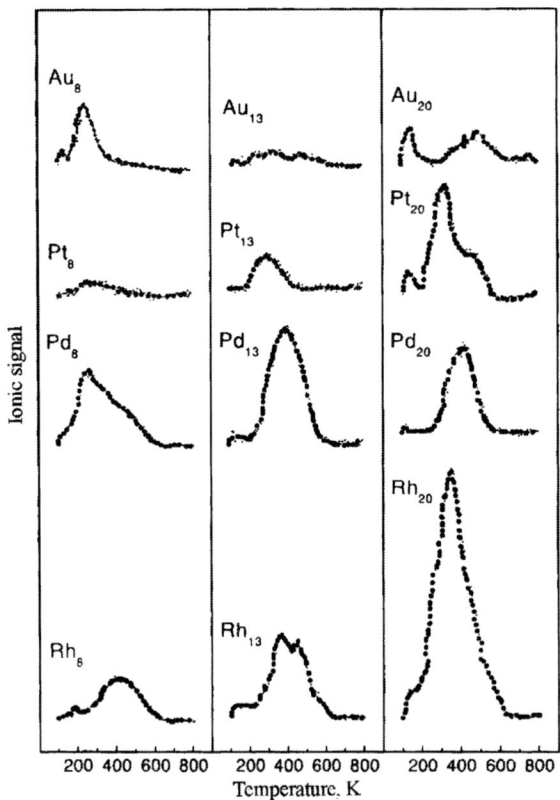

FIGURE 11.3 Dependence of the yield q of carbon dioxide during the oxidation of carbon monoxide on the temperature and the size of metal particles.[47]

atoms and the formation of three-dimensional clusters, which provided stronger bonding as compared with the two-dimensional structures. For supported palladium particles, a strong effect of the number of atoms on the efficiency and selectivity of acetylene conversion was observed. It was found that even a single palladium atom catalyzes cyclotrimerization of acetylene. On the palladium atoms and particles Pd_2 and Pd_3, benzene was formed at about 300 K. Moreover, small palladium clusters exhibited selectivity. Palladium atoms, dimers, and trimers could also produce substances other than benzene. Particles Pd_{4-6} polymerized acetylene and additionally produced C_4H_6, which was desorbed at 350 K. With a transition to larger particles, in addition to C_6H_6 and C_4H_6, a third product C_4H_8 (the structures of C_4H_6 and C_4H_8 are still unclear) was formed.

Of greatest interest are the catalytic properties of a single palladium atom. Theoretical studies have lead to a conclusion that palladium atom and MgO film form a cluster. The MgO surface and its point defects exhibit the electron-donor properties and increase the electron density on the palladium atom. The palladium atom diffuses over the surface and activates two acetylene molecules according to the following scheme:

$$Pd + 2C_2H_2 \rightarrow Pd(C_2H_2)_2 \rightarrow Pd(C_4H_4)$$

The $Pd(C_4H_4)$ complex then activates the third acetylene molecule. The optimal structures of the complexes were considered.[51] The activation was associated with an increase in the charge transferred from a palladium atom to the adsorbed molecule. Benzene formed was only weakly bound with the palladium atom and immediately desorbed. The reaction of the palladium atom with F-centers on the magnesium oxide surface was the limiting stage of this process.

An interesting example demonstrating how the size of metal nanoparticles and the nature of stabilizing ligands affect the catalytic activity and selectivity has been described.[52] Palladium nanoparticles measuring 3–4 nm, which were stabilized either by 1,10-phenanthroline or by 2-*n*-butylphenanthroline, served as the catalysts. Titanium dioxide used as the carrier was saturated with the catalyst (0.5 mol%). The reaction of triple-bond hydrogenation was carried out at room temperature and at a hydrogen pressure of 1 atm. As seen in Figure 11.4, when phenanthroline was used as the stabilizer, the formation of *cis*-2-hexene reached its maximum in 75–80 min with a selectivity of about 95%. Upon reaching the maximum conversion, *cis*-2-hexene started to transform into *trans*-2-hexene. The activity of palladium particles stabilized by 2-*n*-butylphenanthroline decreased and the selectivity increased. The maximum yield of *cis*-2-hexene was reached in 500 min with 100% selectivity; virtually no other products were formed. Only after 25–30 h, an insignificant amount of *trans*-2-hexene was found. Probably, the presence of the butyl residue in phenanthroline forms a catalyst surface that favors a reaction of semihydrogenation, which is difficult to realize under other conditions.

Subnanometer palladium clusters with a diameter of less than 0.7 nm were prepared and stabilized.[53] Clusters were synthesized within micelles of

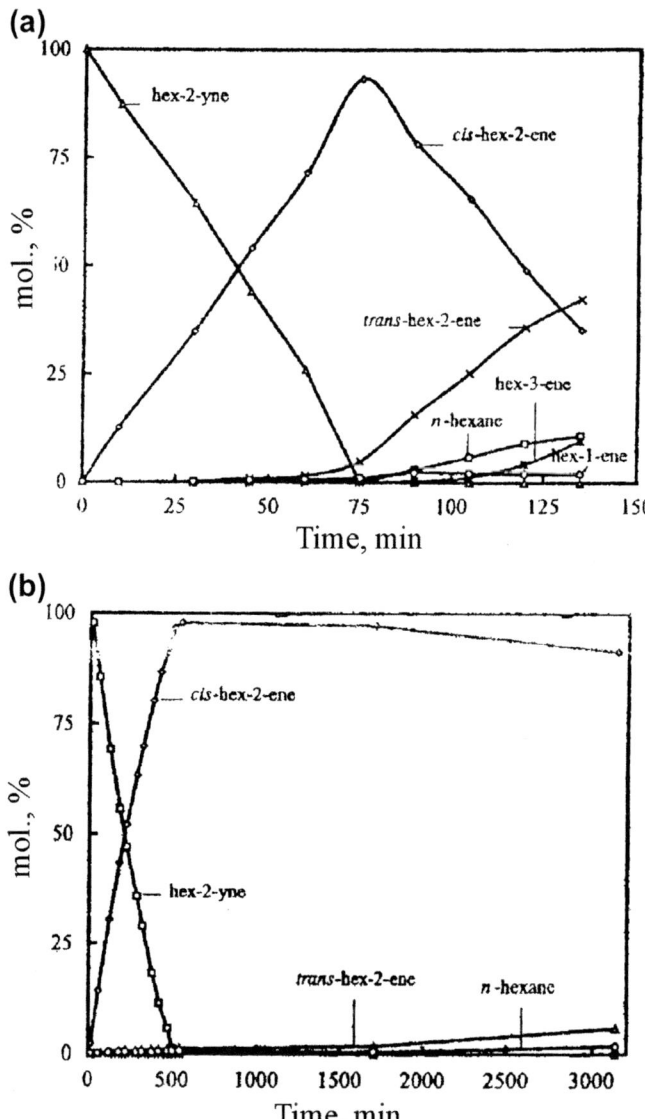

FIGURE 11.4 Selectivity and activity of palladium particles measuring 3–4 nm applied on titanium dioxide in the reaction of partial hydrogenation of triple bond[52]: (a) palladium particles stabilized with 1,10-phenanthroline and (b) palladium particle stabilized with 2-n-butylphenanthroline. (1) Hexine-2, (2) cis-hexene-2, (3) $trans$-hexene-2, (4) hexane, (5) hexene-3, and (6) hexene-1.

different morphologies formed by random copolymers via the ligand exchange from Pd(PPh$_3$)$_4$. Micelles containing Pd(0) were found to exhibit high catalytic activity in quinoline hydrogenation at room temperature and hydrogen pressure of 1 atm and also in the reaction of iodobenzene with ethylacrylate.

11.2 OXIDE REACTIONS

Oxides, like metals, find wide practical applications. The reactivity of metal oxides is lower as compared with the metals themselves; for this reason, the process of oxide formation is used for stabilizing metal nanoparticles. Recently, quite a number of reactions interesting from the standpoint of nanochemistry were realized.

A unique way of utilizing nanocrystalline zinc oxide was described.[54] Zinc oxide was synthesized by a modified sol–gel method according to the following reactions:

$$Zn(CH_2CH_3)_2 + 2(CH_3)_3COH \rightarrow Zn[OC(CH_3)_3]_2 + 2CH_3-CH_3$$
$$Zn[OC(CH_3)_3]_2 + 2H_2O \rightarrow Zn(OH)_2 + 2(CH_3)_3COH$$
$$Zn(OH)_2 \rightarrow ZnO + H_2O$$

The process of ZnO preparation involved three steps: synthesis, extraction, and activation of zinc oxide nanopowder. The last step consisted, in turn, of several consecutive steps of heat treatment. First, the powder was slowly heated to 90 °C and held at this temperature for 15 min. Next, the temperature was increased to 250 °C, the sample was exposed at this temperature for 15 min, and then slowly cooled to room temperature. The resulting zinc oxide represented crystalline nanoparticles measuring 3–5 nm with a specific surface area of about 120 m^2/g. Zinc nanooxide was used in the following reaction:

$$2ZnO + CCl_4 \rightarrow CO_2 + 2ZnCl_2$$

The process was carried out at 250 °C; CCl4 was added into the reaction vessel in portions at 7 min intervals. Carbon dioxide and unreacted CCl$_4$ were analyzed by gas chromatography. It was shown that nanocrystalline zinc oxide is more active as compared with commercial products. It was also shown that the adsorption of sulfur dioxide and destructive adsorption of diethyl-4-nitrophenylphosphate—a toxic organophosphorus compound—proceed on nanocrystalline zinc oxide with high effectiveness. In these processes, nanocrystalline zinc oxide displays a higher activity as compared with commercial samples.

A high activity of nanocrystalline metal oxides was utilized in reactions with compounds used as a chemical weapon. Nanocrystalline oxides of magnesium and calcium were shown to easily react with organophosphorus compounds.[55,56] Thus, 3,3-dimethyl-2-butylmethylphosphoxofluoride CH$_3$–(O)P(F)O–CH(CH$_3$)C(CH$_3$)$_3$, which represents a nerve-paralytic compound, reacted with nanocrystalline magnesium oxide according to the following scheme:

$$\underset{\underset{O-CH(CH_3)C(CH_3)_3}{|}}{\overset{\overset{O}{\|}}{CH_3-P-F}} \xrightarrow[-HF]{OH^-} \underset{\underset{O-CH(CH_3)C(CH_3)_3}{|}}{\overset{\overset{O}{\|}}{CH_3-P-O^-}} \xrightarrow{OH^-} HO\text{-}CH(CH_3)C(CH_3)_3 + CH_3PO_3^-$$

Magnesium oxide abstracts HF and, by using hydroxide groups available on the surface, completely converts the toxic compound into a nontoxic one. Liberated hydrogen fluoride reacts with MgO to form MgF_2 and H_2O.

Nanocrystalline oxides of alkali-earth metals were successfully utilized for deactivation of yperite and other toxic war agents. Autocatalytic dehydrohalogenation of 2,2'-dichlordiethylsulfide on nanocrystalline calcium oxide was studied.[54,56] The reaction proceeded according to the following scheme:

$$Cl-CH_2-CH_2-S-CH_2-CH_2-Cl \xrightleftharpoons{H_2O} HO-CH_2-CH_2-S-CH_2-CH_2-OH$$
$$\updownarrow$$
$$Cl-CH_2-CH_2-S-CH=CH_2 \rightleftharpoons CH_2=CH-S-CH=CH_2.$$

This reaction involves a competition between dehydrochlorination to give a divinyl compound and the substitution of a surface hydroxide for chlorine. The reaction products were analyzed by NMR technique. The products of 2,2-dichlorodiethylsulfide decomposition contained divinylsulfide (nearly 80%) and thioglycol and/or a sulfonium ion (20%) with a hydroxyl group, which probably provided its bonding with the alkali-earth metal surface. Along with yperite deactivation, the reactions of CaO with other phosphorus-containing compounds were studied. The kinetics of reactions of all compounds studied with CaO were characterized by fast initial stages and a slow subsequent diffusion-controlled stage.

In the examples shown above, nontoxic compounds were formed at room temperature, and the reaction rate was limited by the delivery of reagents. The presence of small amounts of water was beneficial for detoxication, which confirmed the promoting action of hydroxyl groups on the oxide surface. Catalytic dehydrohalogenation occurred when nanocrystalline calcium oxide was used together with water.

Yet another application of nanoparticles of transition-metal oxides deserves attention.[57] Particles of oxides of Co, Ni, Cu, and Fe measuring 1–5 nm were used as the electrode material in lithium cells (electrochemical capacitance 700 µAh/g). In this case, the reactions of Li_2O formation and decomposition and the corresponding reactions of reduction and oxidation of nanoparticles proceeded on electrodes made of CoO nanoparticles. The scheme of reversible reactions is illustrated by the example of CoO as follows:

$$2Li - 2e \rightleftharpoons 2Li^+ \qquad (11.8)$$

$$\frac{CoO + 2Li^+ + 2e^- \rightleftharpoons Li_2O + Co}{Co^0 + 2Li^+ \underset{(2)}{\overset{(1)}{\rightleftharpoons}} Li_2O + Co^0} \qquad (11.9)$$

Reaction (11.8) is probable and thermodynamically acceptable. Reaction (11.9) is unusual for electrochemistry. Li_2O was always considered as electrochemically inactive. The authors of the cited study[57] failed to electrochemically decompose Li_2O powders (mechanically ground powders of Li_2O and CoO were used). The possibility of reaction (11.9) was associated with the participation of nanoparticles and the increase in their electrochemical activity with a decrease in the particle size.

The use of the synchrotron X-ray diffraction technique allowed unusual structural states to be revealed in zirconium oxide nanopowders.[58] A ZrO_2 nanoparticle can comprise two or three different structures, viz. monoclinic, tetragonal, and cubic. Such particles, which were called "centaurs," point to the possibility of different polymorphous transitions in individual particle.

To date, attention is focused on the synthesis and physicochemical properties of the hybrid materials such as core–shell formation and on the particles comprising two and even three metals. New nanocrystalline materials of the core–shell type based on TiO_2 and MoO_3 were prepared and studied in detail.[59] Particles TiO_2–$(MoO_3)_x$ were synthesized by conucleation of metal oxides on a micelle surface. The reaction stoichiometry was represented as follows:[59]

$$(1-y)(NH_4)_2Ti(OH)_2(C_3H_4O_3)_{2(aq)} + y/8 \ Na_4Mo_8O_{26(aq)} + CTAC_{(aq)}$$

where CTAC is cetyltrimethylammoniumchloride and $y \leq 0.57$.

In the synthesized materials, the energy of light absorption correlated with the particle size. With a decrease in the size of TiO_2–MoO_3 particles from 8 to 4 nm, the absorption energy decreased from 2.9 to 2.6 eV. As a comparison, the energy of forbidden zones of compact TiO_2 and α-MoO_3 are 3.2 and 2.9 eV, respectively. The synthesized materials were more effective in the photocatalytic oxidation of acetaldehyde as compared with conventional titanium oxide manufactured by Degass (France).

Studies on preparation of nanorods with a diameter from 5 to 60 nm and a length of >10 μm based on $BaTiO_3$ and $SrTiO_3$,[60] TiO_2,[61] and SnO_2[62] were initiated. The preparation of oxide-based nanorods and nanotubes was surveyed in detail[63]. Vanadium dioxide was studied most actively. Polycrystalline vanadium dioxide nanorods were prepared.[64] Syntheses of metastable VO_2 nanobelts[65] and crystalline nanowires with an average diameter of 60 nm and a length of >10 μm[66] were reported. For synthesizing vanadium dioxide nanowires, an earlier developed method of high-temperature evaporation in an argon flow with subsequent condensation was used.[67]

To promote electron transport in metal nanoparticles, it was proposed to utilize the pH effect on the charge variation in stabilized coatings connected with the particle surface. As has been shown, the molecules of hard (solid) mercaptophenylacetylenes effectively bind gold and silver particles.[68] It was demonstrated that the electron transport in charged gold particles is described better in terms of classical concepts rather than in terms of quantum-mechanics.[69]

The use of metal nanoparticles in optical and electronic devices requires approaches that would solve the following problems:

- reliable location of electric contacts between individual nanoparticles;
- determination of characteristics of interparticle electromagnetic interactions in symmetrical well-organized aggregates of nanoparticles; and
- understanding of chemical properties of the nanoparticle surface and its effect on optical and electronic properties of these particles.

Certain approaches to the solutions of these problems have been considered.[70]

11.3 SEMICONDUCTORS, SENSORS, AND ELECTRONIC DEVICES

Semiconducting nanoparticles are extensively used in heterogeneous nanocatalysis and are of potential interest for laser equipment and for manufacturing of flat displays, light-emitting diodes, and sensors.

The development of heterostructures with spatially limited charge carriers, which are formed as a result of self-assembling of nanostructures on the surface of semiconductor systems was a great breakthrough in nanotechnology.[71] The spontaneous ordering of nanostructures makes it possible to realize the incorporation of narrow-gap semiconductors into high-energy gap matrices. The periodic ordered structures of such islet-type inclusions can appear as a result of heating the samples or at a long-term interruption of the growth of deposited semiconductors. The resulting heterostructures represent the sets of discrete levels separated by regions of forbidden states and exhibit the energy spectra similar to those of individual atoms. The described systems were used in the development of injection heterolasers.

Along with the physical methods of synthesizing new semiconductor systems, the methods of chemical synthesis continue to be developed and refined. The use of inversed micelles based on AOT for the production of ZnSe nanoparticles was initiated.[72] Particles of ZnSe measuring 5.7 nm were studied using X-ray diffraction, electron microscopy, light scattering, and luminescence techniques.

Japanese scientists proposed a new method for stabilizing semiconductors (using CdS, codeposition of CdS–ZnS) and metals (using gold).[73] They obtained inversed micelles and stabilized nanoparticles by in situ polymerization of (n-vinylbenzyl)dimethyl (cetyl)ammonium chloride initiated by either light or by the addition of azobisisobutyronitrile. The resulting polymer was dissolved in a polar medium to form transparent films with incorporated cadmium selenide nanoparticles measuring 4.7 to 6.3 nm. The authors characterized this method as the universal one, because it allowed them to obtain films and stabilize nanoparticles of metals and semiconductors.

According to the results of the studies,[74] the inner nuclei of vesicles based on α-phosphatidylcholine can be used as nanoreactors for growing monodispersed (with a deviation of ca. ±8%) nanocrystals of CdS, ZnS, and HgS of definite sizes.

A simple and inexpensive method of synthesizing stable and water-soluble nanocrystalline powders of zinc sulfide (particles measuring ca. 6 nm) in gram amounts was proposed.[75] Particles of ZnS covered with a cysteine layer were obtained by introducing sulfide into a preliminarily prepared solution of a zinc salt and cysteine. After the deposition, redissolution, and drying, the crystalline ZnS powders were obtained, which were stable for 30 months at 48 °C. Their stability was tested by carrying out photocatalytic decomposition of *p*-nitrophenol on freshly prepared and aged powders. The optical properties of the powders remained unchanged during their long-term storage. Highly concentrated aqueous solutions (up to 100 mM/l) of semiconductor particles were used as fluorescent biological labels.

Multilayer films containing CdSe nanoparticles were obtained.[76] Particles of CdSe measuring 1.7–2.0 nm were synthesized from dimethylformamide solutions of cadmium and selenium salts. The films containing CdSe particles were prepared by the consecutive formation of layers on the quartz or CaF_2 plates. The plate surface was first covered with layers of benzoic acid derivatives and polyvinylpyridine. A plate thus treated was placed into a formamide solution of CdSe to form a layer containing cadmium selenide particles. Multilayer (up to five layers) films were obtained by alternate additions of polyvinylpyridine and cadmium selenide. Such organized semiconducting films are of interest for the development of new types of optical fibers and nonlinear optical devices and can also be used as conducting films.

Semiconducting nanowires of definite diameters were fabricated using colloid solutions of nanoparticles of metal catalysts.[77–80] As an example, the results of controlled synthesis of GaP nanowires with diameters of 10, 20, and 30 nm and a length exceeding 10 μm can be shown.[80] The wires were fabricated by laser-generated sputtering of a solid target consisting of GaP and gold (catalyst). A solution with gold particles was spread onto a sample, which was put in a quartz tube. The tube was placed at the blown end of a furnace containing a solid GaP target. The furnace was heated to 700 °C, and the target was sputtered for 10 min by an excimer laser. The laser sputtering of a solid target was used for the simultaneous generation of nanosize metal-catalyst clusters and the reacting semiconductor atoms, which formed the semiconducting nanowire. The size of a catalyst nanoparticle determined the size of the resulting nanowire. As was shown in test experiments, no semiconducting wires were formed in the absence of gold particles. The correspondence between the sizes of nanoparticles and semiconducting nanowires allowed the authors[80] to conclude on the possibility of controlling the diameter of fabricated wires. A similar method of controlling the diameter of nanowires was considered.[79]

By considering the example of CdSe particles with sizes from 0.7 to 2 nm, the effect of the semiconductor nanoparticle size on the bandgap of this material was studied.[81,82] All the particles were formed using the methods of organometallic chemistry. Together with cadmium and selenium, the atoms of phosphorus and phenyl and propyl groups were present on their surface. Thus, one of the

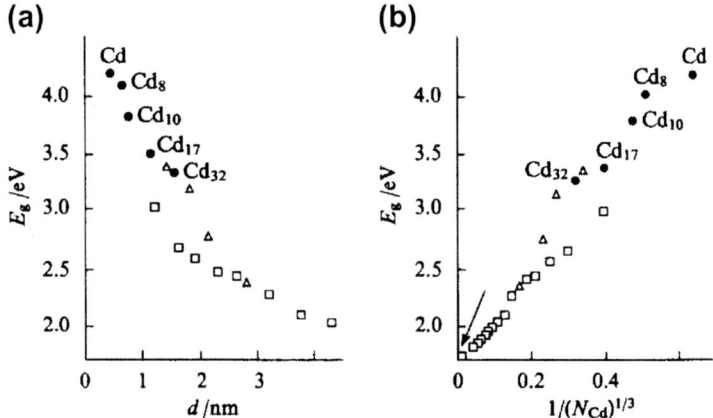

FIGURE 11.5 Bandgap width for CdSe nanoparticles as a function of their (a) diameter and (b) reciprocal radius; N is the number of cadmium atoms in a particle. Dark points correspond to experimental data, the rest of the symbols are the data of different authors for coarser particles.[82]

particles under study had a composition $Cd_{10}Se_4(SePh)_{12}(PPr_3)_4$, where Ph is phenyl and Pr is n-propyl. The atomic structure of the particles was determined by X-ray diffraction for individual crystals. Figure 11.5 shows the bandgap of nanoparticles synthesized and also for coarser particles of cadmium selenide as a function of their size. As can be seen, with an increase in the nanoparticle size the bandgap width E_g of a particle approaches the E_g value of the bulk CdSe samples.

New prospects in various branches of electronics were opened with the development of one- and two-dimensional bimetallic systems.[83] One of these branches is associated with synthesizing particles of the ME_2 type, particularly, $NbSe_2$.[84] Superconducting one-dimensional $NbSe_2$ nanotubes in a mixture with nanorods of 35–100 nm diameters and a length of several hundred nanometers were prepared by thermal decomposition of $NbSe_3$ in an argon flow at 700 °C. A synthesis in solution provided higher yields of products with higher purity and a narrow size distribution.[85–88] In a dodecylamine solution, two-dimensional plates and one-dimensional wires of $NbSe_2$ were synthesized.[20] Their precursors, $NbCl_3$ and Se, were mixed in dodecylamine in an argon flow and then heated at 280 °C for 4 h to yield a black suspension. Slow cooling (~58 C/min) of this suspension to room temperature produced two-dimensional lamellar structures 10–70 nm thick. Rapid quenching of a solution obtained by extraction of this suspension in hexane yielded $NbSe_2$ nanowires of a diameter of 2–25 nm and a length up to 10 μm.

Optoelectronics employs devices based on two-dimensional semiconductor heterostructures. Combining semiconductors with different bandgaps makes it possible to control fundamental parameters such as the bandgap width, the effective masses and mobilities of charge carriers, the refractive index, and the electron energy spectrum.[71,89]

The dynamic properties of charge carriers at the liquid–semiconductor interface are important for photocatalysis, solar energy conversion, and photoelectrochemistry. The peculiar properties of charge carriers, i.e. electrons and holes, which include their absorption and recombination in certain nanosize semiconductor systems, were studied by laser techniques.[90] Suspensions of cadmium sulfide nanoparticles were used for initiating the acrylonitrile polymerization. The resulting polyacrylonitrile (PAN) and the products of its partial hydrolysis were employed as the templates, which could regulate the shape of CdS nanoparticles and composite nanowires CdS/PAN with a diameter less than 6 nm and a length of 200 nm to 1 μm.[91]

A nanosize electronic circuit breaker built of gold nanoparticles and compounds containing redox groups was described.[92] As the redox group, bipyridyl groups that entered the composition of N,N-di-(10-mercaptodecyl)-4,4′-bipyridinium dibromide were used, the synthesis of which has been described.[93] When gold nanoparticles were bound with a film of bipyridinium bromide, the reduction of the bipyridine group on a gold electrode proceeded easily and reversibly:

$$\text{bipy}^{2+} + e^- \rightleftharpoons \text{bipy}^{+\bullet}$$

where bipy^{2+} is a bipyridine group and $\text{bipy}^{+\bullet}$ is a radical.

The operation of an electronic circuit breaker measuring 10 nm, which is based on the changes in the chemical states of molecules, is illustrated in Figure 11.6, reproduced from Ref. 94 and based on results of Ref. 93. The authors of the latter study reported that their system corresponded to no more than 60 organic molecules and its operation required no more than 30 electrons. When a molecule contains bipyridine in its reduced state $\text{bipy}^{+\bullet}$, a considerable tunneling current flows in the circuit nanocluster–molecule–electrode. If a certain threshold voltage is applied to the gold electrode, the tunneling current drastically decreases. The threshold voltage corresponds to the oxidation of $\text{bipy}^{+\bullet}$ to bipy^{2+}. Thus, an electrochemical switcher, the state of which is determined by the potential required for reducing a molecule by a single electron, was created. To date, these devices operate rather slowly and have an insufficient amplification factor. However, they can find application in cases where amplification is of little importance, e.g. as chemical sensors for detecting single molecules or unit chemical reactions. Moreover, based on such systems, the memory devices for computers can be developed.[94]

New prospects in designing sensors and optoelectronic devices based on nanoparticles were opened up.[95] It was shown that pore-free crystals of organoplatinum compounds can reversibly take up and evolve sulfur dioxide without breaking the crystal.[96] In other words, the crystals breathe. Figure 11.7 illustrates the corresponding transformations.

When interacting with SO_2, a crystal turns orange in a minute. The change in its color is accompanied by the transition of square-planar platinum complexes into square-pyramidal complexes that contain SO_2 as the fifth ligand. In doing

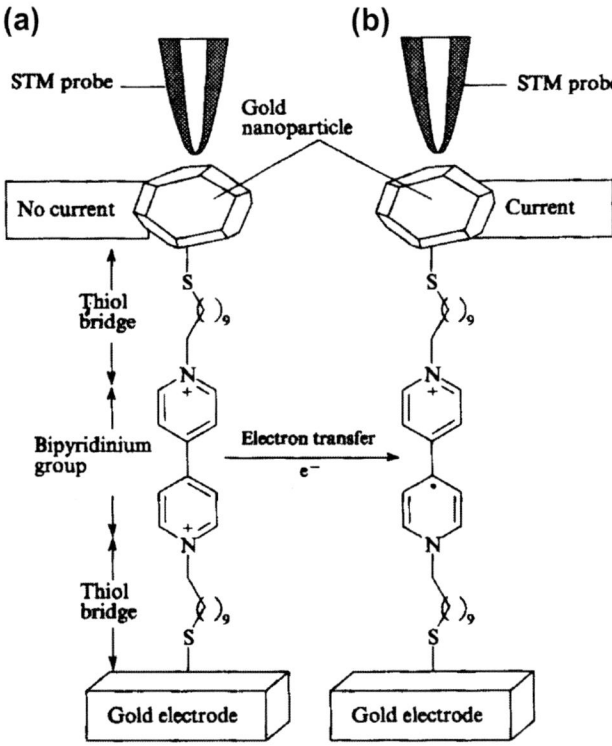

FIGURE 11.6 Illustration of operation of an electronic switcher: (a) bipyridine group is in the oxidized state and no current flows and (b) as a result of addition of one electron, the bipyridine group is reduced and current flows.[94]

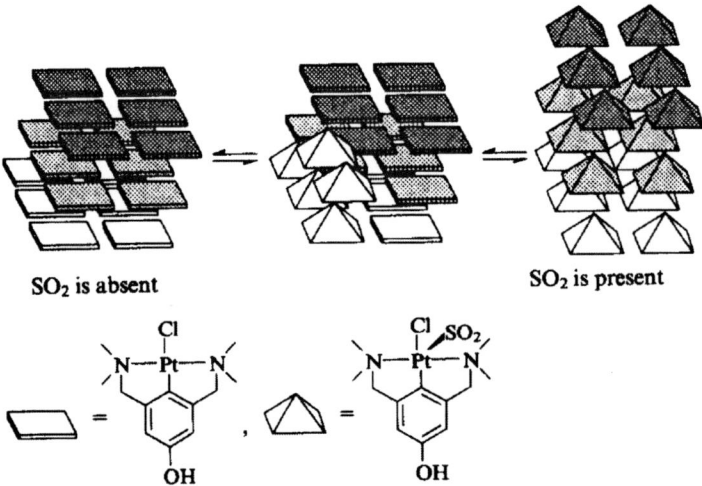

FIGURE 11.7 Crystals that can reversibly adsorb SO_2.[95,96]

so, the crystal increases in volume by 25%, while retaining the ordered lattice structure. Yet, a more interesting result is obtained if a crystal with increased volume is exposed to air. In this case, the crystal "breathes out" SO_2 and relaxes to its original, colorless, and SO_2-free state. This process can be repeated many times without breaking the crystal. In the opinion of the authors,[95] such crystals can be used as optical switchers and sulfur dioxide sensors. It is possible that analogous compounds would reversibly interact with chlorine, carbon dioxide, and other gases. There is good reason to believe that the substances that bind gases in solutions will also bind them in the solid state.

Polypyrrole nanowires 200 nm thick and 13 µm long were proposed to be used as pH sensors.[97] A method for synthesizing polypyrrole nanowires sensitive to biomaterials, based on using alumina templates, was described.[98] Polypyrrole nanowire biosensors of controlled composition and size (200 nm), which could contain built-in electrical contacts, incorporated quantum dots, and biosensitive molecules were suggested for detecting biomolecules, particularly, DNA.[99]

Nanoparticles of a definite stoichiometric composition and a size of 1–4 nm exhibited unusual properties with temperature variations. Palladium clusters measuring 17 nm behaved as compact palladium at room temperature but demonstrated the presence of a Coulomb barrier at 4.2 K. At the same time, ligand-stabilized gold particles $Au_{55}(PPh_3)_{12}Cl_6$ (PPh designates triphenylphosphate) measuring 1.4 nm had a Coulomb barrier at room temperature. As was shown,[100] a Coulomb energy barrier for the electron transport through a cluster can be changed by organizing clusters into the ligand-stabilized three-dimensional structures. The length and the chemical nature of these ligands, which are called "spacers," change the intercluster distance and the Coulomb barrier. The aforementioned peculiarities in the behavior of metal nanoparticles have already found application in practice. Based on the temperature dependence of the Coulomb barrier, a sensor for temperature variations in a range of 1–30 K was developed (Nanoway, Finland).[101]

At present, considerable attention is drawn to the development of sensitive units of sensor materials based on semiconductor oxides and heterostructures. The effect of ethanol and nitrogen dioxide vapors on the properties of SnO_2 nanocrystals, supported by single-crystal silicon and tin dioxide nanocrystals doped with Ni, Pd, and Cu, was studied.[102] Doped SnO_2 nanocrystals were synthesized by the pyrolysis of aerosols. Their average size was 6–8 nm, according to the XRD data. The effect of the gas-molecule adsorption on the volt–ampere characteristics of heterostructures, which manifested itself in the changes in the potential barrier height at the heterojunction and in the changes in tunneling processes at the interface, was demonstrated.

A new generation of chemical sensors can be synthesized by employing nonuniform nanosystems and unusual electronic and physicochemical properties of metal nanoparticles entering into such systems. In these systems, the compact metal cores of nanoparticles are surrounded by outer amorphous shells so that the electron transfer under the effect of adsorbed gas molecules proceeds via the interacting nanoparticles bound into assemblies. The sensor properties of non-uniform–nonequilibrium–nanocomposite films were discussed.[103] Oxide–oxide (SnO_2–TiO_2), metal–oxide (Cu–SiO_2), and metal–polymer (Cu–poly-n-xylylene) systems were studied. The sensor activity of samples toward hydrogen, humid air, and ammonium was determined by the conductivity changes. From our viewpoint, to obtain new effective sensor materials, in addition to employing nonequilibrium synthetic conditions, studies with oxide particles of different sizes should be carried out.

Many fields of nanotechnology are based on physical and chemical interactions, involving nanoparticles of particular size and shape. In order to use the nanomaterials as sensors and catalysts, one has to understand the peculiarities of both the synthesis and interaction mechanism during the sensing/catalytic act.

In recent years, the interest of researchers and engineers to gas-sensitive materials has grown substantially due to the progress in nanotechnology. This interest is primarily connected to the promising electronic properties of nanomaterials, their size dependence, and the possibility of controlling the material structure by using new experimental techniques. More and more materials and devices are produced every year with the use of nanotechnology.[104]

Nowadays, most of the commercial metal-oxide gas sensors are manufactured using the screen printing method on small and thin ceramic substrates. The advantage of this technique consists in the fact that the thick films of metal-oxide semiconductors are deposited in batch processing, thus leading to small deviations of characteristics for different sensor elements.

Despite the well-established fabrication technology, this method has a number of drawbacks and needs to be improved. First of all, the power consumption of a screen-printed sensor can be as high as 1 W,[105] which prevents us from using it in the battery-driven devices. Another technological problem is the proper mounting of the overall hot ceramic plate to ensure the good thermal isolation between the sensor element and the housing.

These problems promote the development of substrate technology and the extensive research on the preparation of sensitive layers.

Nanotechnology has provided new tools for controlling the microstructure of sensitive layers, allowing one to obtain materials with narrow grain size distributions. The studies of thin SnO_2 films have shown the strong increase of sensitivity with the reduction of the oxide particle size to the nanometer scale.[106] A systematic analysis of the dependence of SnO_2 sensitivity on grain size was reported.[107]

Using the low-temperature physical vapor deposition and the setup (Figure 2.5), the gas-sensitive nanostructured thin films of lead were obtained. A careful control over the deposition parameters such as the deposition rate, evaporation rate, and substrate temperatures allows one to obtain condensates with the required structure and also opens up the possibility of chemically modifying the surface and grain boundaries.[108,109] As examples, the microstructures of two samples containing nearly the same amounts of lead but deposited at different rates are

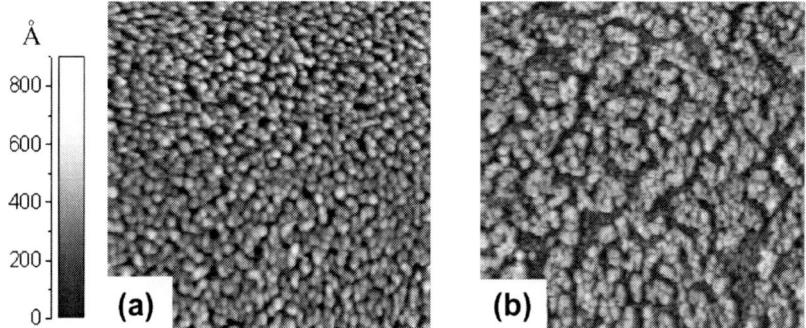

FIGURE 11.8 AFM images of Pb condensates, deposited at 80 K after annealing to room temperature and exposure to the air. Scan area is 5 m × 5 m. (a) 34.3 ML deposited at 0.05 ML/s, no conductance onset during the deposition was observed. (b) 28.1 ML deposited at 0.20 ML/s, the conductance onset was observed.

depicted in Figure 11.8. The microstructure influences the electrical properties; thus, only sample (b) was found to be sensitive to humidity. After the deposition and controlled annealing, the particles can be oxidized (totally or partly) to form a highly porous sensitive layer.

Sputtering oxidation method was successfully used for preparing NO_2-sensitive nanostructured SnO_2 films.[110] The high sensitivity to 5 ppm of NO_2 at 130 °C was demonstrated by the work function measurements. Other methods, based on a modified sol–gel[111] technique as well as on the original approach that uses the mechanochemical milling,[112] were also employed for improving the gas sensor performance of nanostructured metal-oxide materials. Besides enhancing the characteristics of existing gas-sensitive elements, nanotechnology also promotes the development of new types of materials for microelectronics and sensorics.[104] The typical examples are the thin polymer and molecular films[113,114] and composite materials.[115] The cross-sensitivity of such materials to humidity was also investigated.[116]

Yet another example is the functional oxide nanobelts—a quasi-one-dimensional nanomaterial with well-defined chemical composition, crystallographic structure, and surface.[117,118] Such materials exhibit several important properties, in particular, gas sensitivity.[119] The synthetic method is a simple process, in which the solid starting material is evaporated at an elevated temperature and the resultant vapor phase condenses under definite conditions (temperature, pressure, atmosphere, substrate, etc.).

The process is carried out in a horizontal tube furnace, which consists of an alumina tube, a rotary pump system, and a gas supply and control system. The right-hand end of the tube is connected to a rotary pump, and the carrier gas enters from the left end. The pressure in the system is kept around 2×10^{-3} Torr. Alumina substrates are placed downstream the boat for collecting the growth products.

A very promising approach is to increase the surface area of a semiconductor by preparing porous materials. Thus, both the SnO_2 and TiO_2 mesoporous powders, which were fabricated by self-assembling of a surfactant followed by treatment with phosphoric acid, and the conventional tin oxide powders with surfaces modified by mesoporous SnO_2 exhibited a higher sensor performance as compared with the corresponding metal-oxide powder materials with lower specific surface areas.[120,121]

Porous silicon fabricated by electrochemical etching also exhibited a high gas sensitivity.[122] It was found that different electroless coatings influence the sensitivity toward a particular gas. By using the Fast Fourier Transform analysis of the sensor signal, it is possible to separate the drift and exclude false sensor signals.

One of the most popular nanomaterials nowadays, carbon nanotubes (CNTs), is also used to detect gases. One of the first CNT-based gas sensors was found to be able to detect low concentrations of NO_2 and NH_3 gases at room temperature.[123] These results initiated the further experimental and theoretical[124] investigations in this field. For instance, it was found that CNTs can be added to the SnO_2 substrate to form an NO_2-sensitive material.[125] The doping increases the sensitivity as compared with pure SnO_2 sensors. Another application involves using CNTs as the cathodes in ionization-based gas sensors.[126] In this case, the gas is ionized in an electric field and is identified by a unique breakdown voltage. The individual CNTs create very high electric fields near their ultrafine tips; hence, the combined effect of several billions of CNTs should increase the overall field and accelerate the gas-breakdown process. This allows the ionization of gases to be carried out at voltages of up to 65% lower than that in conventional ionization sensors.

Thus, nanotechnology opens up new possibilities for the production of gas-sensitive materials of new types and also allows one to control the structure of conventional semiconductor metal-oxide materials on the nanoscale level. New nanomaterials that appear today, such as the highly ordered arrays of metal/semiconductor core–shell nanoparticles,[127] may soon be used in gas-sensor applications.

Experimentally, it was found that doping of SnO_2 with Cu enhances the sensitivity and selectivity toward H_2S. The same effect was observed for heterostructures $CuO–SnO_2$[128] and $SnO_2–CuO–SnO_2$.[129] This phenomenon was explained by a decrease in the barrier height in p-CuO-n-SnO_2 owing to the chemical transformation of highly resistive CuO into well-conducting Cu_2S.

Humidity, which is usually present in almost all target environments for gas-sensor applications, is known to strongly influence conductivity and sensitivity. In the presence of water vapor, the sensitivity to CO increased,[130] while it decreased with hydrocarbons.[131]

The sensitivity of SnO_2 samples to CO was studied at an elevated temperature in atmospheres with different humidity.[132]

Diffuse-reflectance infrared Fourier transform spectroscopy is a very productive tool for in situ investigations of such systems.[133] Using this technique, it was shown that the hydrated proton plays the donor role, rather than the rooted hydroxyl group.[132]

Among the main trends in the gas-sensor technology, the creation of sensor arrays or "electronic noses" should be mentioned.[134] Such multisensor systems can be fabricated on a single substrate, which can involve gas sensors of different types and, necessarily, the signal-processing systems.[135] There are many examples of successive production of their macroscopic prototypes, which are able to discriminate the mixtures of gases,[136–139] volatile organic compounds,[140] and odors.[141] An important part in such systems is the pattern-recognition method, which is usually realized using artificial neural networks.[142] The typical metal-oxide sensor mechanism is discussed in review.[143]

Nanotechnology with its tools for the production of clean, structurally pure, and perfectly ordered materials is very promising for the preparation of new gas sensors with desired characteristics. Along with this, it helps the researchers for a deeper understanding of the sensing mechanism at the atomic level, which will undoubtedly promote rapid progress in this field.

At the same time, the creation of new materials for the gas-sensor applications is necessary for solving a number of basic problems of nanochemistry. Among them, the following problems are considered to be most important:

- monitoring the size, structure, and stability of metals, semiconductors, and hybrid nanomaterials by the controlled conditions of synthesis, in particular, with the use of low-temperature synthesis;
- understanding the kinetics and thermodynamics of self-organization processes that take place on the surface of nanoparticles and extending the working temperature interval; and
- determination of the effect of the shape of nanoparticles, particularly, nanowires and nanotubes of different materials on their chemiresistive properties.

The solution of these problems in combination with the search for new nanosystems will allow the preparation of new sensor and catalytic materials.

11.4 PHOTOCHEMISTRY AND NANOPHOTONICS

Recently, studies devoted to kinetics and dynamics of reactions involving metal nanoparticles appeared. A photochemical reaction with participation of *n*-dodecanethiol-covered silver nanoparticles dispersed in cyclohexane was studied in the picosecond time interval.[144] Figure 11.9 illustrates the photoreaction mechanism.

According to electron microscopy studies, nanoparticles $Ag_xSC_{12}H_{25}$ had an average size of 6.2 nm and resembled polyhedrons. During short-term photolysis, nanoparticles first decomposed to finer particles (<2 nm). An increase in the photolysis time to 9 min led, in contrast, to coarsening of particles to

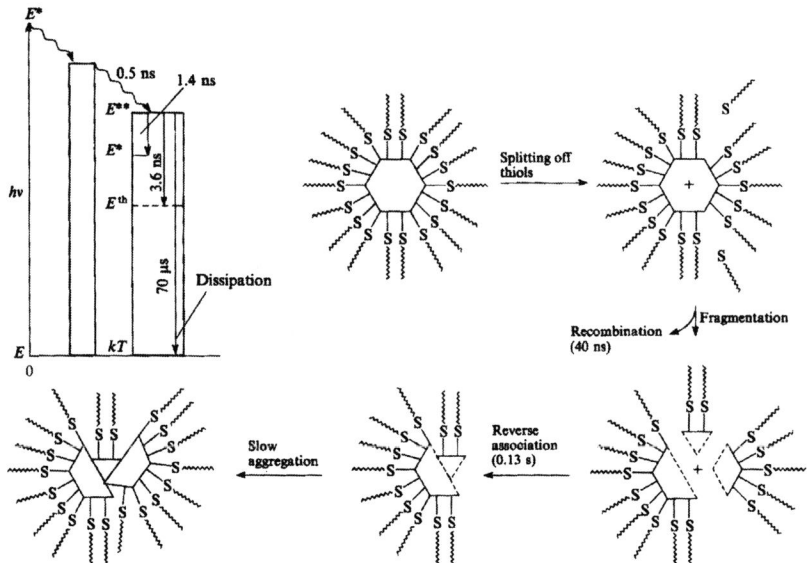

FIGURE 11.9 Illustration of a photoreaction involving $Ag_xSC_{12}H_{25}$ nanoparticles dispersed in cyclohexane.[144]

20 nm. Photochemical transformations were detected by the temporal changes in the intensity and width of the silver plasmon absorption line (designated in Figure 11.9 by vertical heavy lines). As was found, the absorption depended nonlinearly on the light intensity. The kinetics of photolysis of particles was studied in a nanosecond range, and the corresponding kinetic constants were found. The changes in the dielectric properties of the environment occurred in 0.5 ns. Under the effect of optic excitation, a part of alkanethiol molecules were split out in a time (3.6 ns) determined by the energy E_{th} liberated as heat. The ratio of particles with split-out thiol groups depended on the excitation pulse energy. Within a period of 40 ns (recombination time), the thiol-coated particles also underwent fragmentation. The subsequent photolysis partially prevents the disintegration of nanoparticles.

In contrast to silver-based nanoparticles that undergo light-induced photochemical reactions, for gold-based nanoparticles, shape changes are more likely under similar conditions. For the latter particles, fragmentation is drastically suppressed in more polar solvents or with the shortening of the alkanethiol molecular chain.[145]

Silver nanoparticles were used in photochemical transformations of phenazine and acridine.[146] When illuminated, both molecules, which were adsorbed on the surface of silver nanoparticles, decomposed with the break of the N–C bond in a single-photon process. The reaction rate and the degree of photodecomposition depended on the light wavelength. When photolyzed in a short wavelength range, phenazine decomposed to give graphite. The rate of phenazine decomposition was higher as compared with that of acridine, which apparently was associated with different orientation of their molecules on the surface of silver nanoparticles. The authors of this study,[146] have not yet found an explanation for the decomposition of these molecules on the silver nanoparticle surface. New bands at 543 and 619 cm^{-1}, which were close to the bands observed at the adsorption of quinazoline on the silver surface, were revealed in the acridine spectra.[147] It was noted that the rate and the depth of the described photoreactions were unrelated to one another. It was assumed that to be decomposed, the molecules of phenazine and acridine should either be adsorbed on the definite surface sites or should possess definite geometrical configurations, which would provide the participation of only a small part of molecules.

The optical properties of gold nanoparticles measuring 2.5, 9, and 15 nm, which were synthesized by γ-radiolysis of $KAuCl_4$ salt in water and stabilized in polyvinyl alcohol, were studied.[148] Solutions containing 2.5-nm gold particles did not attenuate transmitted light, while those with larger particles drastically reduced the intensity of a laser pulse with a wavelength of 530 nm. The observed effects were attributed to the appearance of a large number of light-scattering centers formed as a result of laser pulse-generated evaporation of gold nanoparticles. It took several nanoseconds for the light-scattering centers to be formed; the greater part of these centers relaxed to the initial state; at the second pulse, a part of them degraded into small particles.

Time-resolution laser measurements were employed for studying electron–phonon interaction in metal particles, which included gold particles measuring 2 to 120 nm and bimetallic particles containing Au, Ag, Pt, and Pd. Most of the detailed studies were carried out for particles built of either Au cores with Ag shells or Au core with Pb shells. All particles were synthesized by radiolysis.[149] For gold, in contrast to, e.g. silver, the time scale of electron–phonon interaction was shown to depend on the particle size.[150]

Nanophotonics is a new field that studies the optical properties of objects with sizes much smaller than the wavelength of light.

In connection with this, we mention only those studies, which considered the possibilities of synthesizing and utilizing nanosize quantum dots, quantum wells, and quantum wires in inorganic semiconductors.[26] The efforts of scientists were largely concentrated on studying linear optical effects, whereas the development in the field of nonlinear optical nanoscale phenomena has just begun.[151] The problem and prospects of nanophotonics have been analyzed.[152]

One of the directions aimed at the production of cheap and renewable energy is based on the development of organic solar cells. The use of hybrid materials offers great possibilities in this field. Solar cells were obtained by quaternary self-organization of various porphyrins (donors) and fullerenes C_{60} and C_{70} (acceptors) by their clusterization with gold nanoparticles on nanostructured tin dioxide films.[153–156] The prepared composition proved to have 45 times higher power conversion efficiency as compared with a reference system containing both single components of porphyrin and fullerene. A combination of porphyrin and fullerene is considered as an ideal donor–acceptor couple, making the acceleration of photoinduced electron transfer and the slowing of charge recombination possible, which results in generation of long-lived charge-separated states with a high quantum yield.[157–159]

11.5 APPLICATIONS OF CNTs

CNTs can be used as both massive articles and miniature devices. In the first case, vast amounts of nanotubes are used as fillers in composites, current sources, adsorbents, and accumulators of gases. In the second case, nanotubes are used for manufacturing various electronic devices, field electron emitters, superstable probes for microscopes, and sensors.

As various application fields are rapidly developed, it is difficult to embrace everything; this is why we show those examples, which, from our viewpoint, are most closely related to nanochemistry.

Polymethylmethacrylate filled with oriented nanotubes served as a material for fabricating strong fibers.[160] Polymeric membranes filled with nanotubes were created for detecting large neutral molecules.[161] Great expectations in electronics are associated with the use of nanotubes. As was repeatedly stressed, the transition from the currently exploited micrometer sizes (0.1–1.0 μm) to nanometer sizes (1–10 nm) entails the changes in the material properties. In nanochemistry, a single atom or molecule is the limit of miniaturization; however, it is difficult to connect leads to such species. Hence, in electronics, great attention is drawn to CNTs, which, depending on their structure, can exhibit conduction of either metallic or semiconductor types.[162]

Methods used for synthesizing single-walled nanotubes and the potentialities of the development of devices on their basis were considered.[163] In doing so, attention was drawn to the fact that in contrast to the traditional way, which is based on an approach from the top-downward or from large to small, the elaborate promising methods are based on an approach from the bottom-upward, or from small to large, from simple to complex.

The effect of the chemical nature of adsorbed molecules on the characteristics of electronic devices was discussed.[164] For particles of nanoscale sizes, adsorption of foreign molecules can change their electronic properties. Yet another problem of using nanotubes as electronic devices is connected with the noises caused by the electric current flow.[165] The molecules adsorbed on the external surface of a carbon nanotube were supposed to be the source of these noises. In turn, this effect can be used for the development of highly sensitive sensors.[166]

The chemical fabrication of electronic circuits involving multiwalled nanotubes was described.[167] Carbon tubes were obtained by catalytic pyrolysis of methane.

The chemical methods of assembling structures from nanotubes have just begun to develop. Recently, much attention was paid to the preparation of Y-shaped tubes, which was accomplished by the pyrolysis of a mixture of gaseous nickelocene and thiophene at 1000 °C.[168] The yield of tubes with a joint angle of ca. 90° and a diameter of 40 nm approached 70%. The volt–ampere characteristics of the synthesized tubes were asymmetric with respect to the bias current. This points to the rectifying properties of the synthesized tubes. Among the other chemical methods, these results deserve special attention.[169]

Great efforts were concentrated on the research and possible applications of emission properties of CNTs. Nanotubes are promising as an emission material because of their high length-to-diameter ratio, small curvature radius of the tip, and the enhanced conductivity, heat conduction, and chemical stability. The studies dealing with field emission of tubes still demonstrate many unsolved problems and inconsistences. To use nanotubes in the manufacture of displays, it is necessary to reduce their cost and develop technologies for fabrication of extended surfaces with uniform and reproducible characteristics. Work in this direction is being carried out in many countries. The prospects of the development and application of emission properties of nanotube-based materials were discussed.[170]

Nanotubes can emit electrons at a comparatively low applied voltage. This allowed a microwave generator to be developed based on CNTs.[171] From the viewpoint of its developers, such systems can be used for miniaturization of mobile-telephone communication stations and elongation of their service life. In designing telephones, one can also use the properties of nanotubes to change the electrical characteristics under the effect of mechanical stresses.[171]

CNTs are assumed to be competitive with metal hydrides as the reliable high-capacity systems for hydrogen storage.[172] Hydrogen has advantages over hydrocarbon fuel, because its combustion produces only water vapors that do not contaminate air. Accumulation of hydrogen in CNTs was discussed.[173]

The sensitivity of single-walled nanotubes to gases and their ability to change their electric resistance and thermal emf during the gas adsorption were observed.[174] Such properties of tubes make it possible to use them as sensor materials. Based on single-walled tubes, sensors for nitrogen dioxide, ammonium, and oxygen were proposed.[123,165] Sensors based on nanotubes are characterized by short response times and high sensitivity. As compared with conventional solid-state oxide sensors, the sensitivity of transducers based on nanotubes increases by several orders of magnitude. Thus, the response of such a transducer to the presence of 0.02% NO_2 consisted in the increase in conductivity by three orders of magnitude in 10 s, while the introduction of 1% NH_3 decreased the conductivity by two orders of magnitude in 2 min. To return the system to the initial state, it had to be heated to 400 °C, while its relaxation at room temperature proceeded very slowly. Density-functional calculations have shown that the energy of the NO_2-molecule bonding with a single-walled tube is approximately 0.9 eV,[123] their interaction resembles chemisorption, and the charge transfer increases the number of holes. The interaction with ammonium represents physical adsorption. Being a Lewis base, ammonium donates electrons to nanotubes, thus decreasing the number of holes.

The use of nanotubes as gas sensors was analyzed.[175] Measuring several square micrometers, the nanotube-based sensor materials are miniature, relatively cheap, and can be used at room temperatures. Similar to the sensors based on metal nanoparticles, the use of CNT-based sensors is associated with the problem of selectivity that arises when analyzing complex mixtures of gases.

For determination of the composition of liquids, the shifts of the IR spectrum lines observed upon immersion of single-walled nanotubes into liquids were proposed for using as the indicator. The spectral shift depends on the surface tension of liquids, which can also be used in analysis.[176]

Nanotubes have already found application as probes for microscopes. STM probes are traditionally fabricated from tungsten or platinum; cantilevers for AFM are made of silicon or Si_3N_4. It was proposed to use nanotubes as probes.[177] By attaching a single-walled tube to a silicon cantilever, the service life of the latter could be increased. The use in cantilevers of chemically inert, hydrophobic, thin, and flexible CNTs allowed narrow cavities to be analyzed and soft biological samples to be studied.

The use of hydrocarbon pyrolysis allowed a single nanotube to be grown immediately on the microscope probe.[178] A current most promising method combines the chemical deposition of a tube from the gas phase onto a special cartridge, which is accomplished by catalytic pyrolysis, and the electric field-promoted transfer of the tube from the cartridge to the probe's tip.[179]

Single-walled CNTs were proposed for using as nanoelectrodes in electrochemical applications.

The fabrication of tubes with carboxyl groups at their ends was described.[180,181] Tubes were obtained by oxidation in air at 700 °C, which allowed a probe microscope with a high chemical sensitivity and an ability to analyze substances at the atomic and molecular levels to be developed. Probes with amine, hydrocarbon, and biologically active groups were obtained. In principle, various functional groups can be attached to the carboxyl group, which makes possible the fabrication of probes with different target functions. It was found that a probe with carboxyl groups exhibits acidic properties, a probe with amine groups has alkaline properties, and hydrocarbon groups render hydrophobic properties. In contrast to the probes fabricated of silicon oxide with functional groups attached to their side surface, in nanotube-based probes, functional groups are attached only to their ends. Moreover, an open end with a diameter of 1.4 nm contains about 20 atoms, which allows fabricating probes with single functional groups of known structure. The tips of such probes can be used for carrying out selective chemical reactions. Cantilevers modified with CNTs allowed the AFM resolution to be enhanced and can be used in studying aqueous solutions, which is of special interest when analyzing biological samples.

It seems quite natural to use nanotubes as supports for catalysts and sorbents. This idea was brought forth virtually immediately after the discovery of nanotubes.[182] Catalysts can be obtained by filling both the internal and external surfaces of tubes. A study was carried out with multiwalled CNTs. Ruthenium-based catalysts were prepared in the course of arc discharge synthesis of nanotubes, which yielded multiwalled tubes with a specific area of 27 m^2/g. This catalyst was used in the hydrogenation of cinnamic aldehyde.[183] The yield of hydrocinnamic alcohol was increased to 80%, while the selectivity increased from 30 to 92%. The effectiveness of different carriers in catalytic hydroformylation of propylene was compared.[184]

The unique properties of CNTs can probably be realized in full measure within the framework of molecular nanotechnology for the fabrication of functional devices and structures by their assembly from atoms and molecules. Such self-reproducible assembling should proceed in accordance with the laws of chemistry, which, for the case of interaction of several separate atoms and molecules, operate in a different way as compared with the common chemical reactions that involve vast amounts of molecules. Different types of probe microscopes play a definite role in solving this problem. Thus, the development of a scanning probe microscope of the tweezers type opened up great possibilities.[185] The tweezers' function is performed by two gold leaflets separated by a glass layer and furnished with tips formed by nanotubes with a diameter of 100 nm. The electric field allows one to move small particles by jointing and opening the nanotubes. By decreasing the tube diameter, one can handle particles measuring 1–2 nm, i.e. single molecules and tens of atoms.

For self-assembling of CNTs, it was proposed to use their solubilization in a solution of tetraoctylammonium bromide in tetrahydrofuran (THF), followed by subjecting them to a DC electric field.[186] Under the effect of the field, the tubes self-assembled into stretched bundles anchored to the positive electrode, thus forming a film. At a voltage ~40 V, all tubes from the suspension were deposited on the electrode; at voltages higher than 100 V, the bundles were aligned normally to the electrodes. The demonstrated method of aligning single-walled CNTs in the electric field opens up new possibilities of achieving electric contacts in nano and microdevices.

Studies devoted to filling tubes with different substances have shown that CNTs can be considered as a sort of chemical reactor. Results obtained, to date, already allow us to the reason that chemical reactions in the internal voids of tubes differ from those occurring under ordinary conditions. Such phenomena as crystal fusion and liquid crystallization also differed from those observed under ordinary conditions.[187]

The synthetic methods are being constantly developed and modernized. Syntheses of tubes were accomplished under hydrothermal conditions by bombarding a mixture of nanosize graphite and iron with hydrogen atoms.[188] Nanochemistry plays the key role in the production of tubes uniform in size and structure; their synthesis is among the current burning problems.

11.6 NANOCHEMISTRY IN BIOLOGY AND MEDICINE

Nowadays, a new direction in nanochemistry, which is aimed at synthesizing and applying the systems built of metal nanoparticles (mainly, gold, and silver) and various biological molecules (DNA, peptides, and oligonucleotides), is actively developed.

To be used in biology or medicine, various nanoparticles that usually possess hydrophobic stabilizing coatings should be transferred to aqueous solutions. The most straightforward and well-developed method is to exchange hydrophobic ligands for hydrophilic ones.[189–191] Ordinary ligand exchange was often used for the stabilization of nanoparticles in water. Additional advantages are offered by using stabilizing molecules containing several groups capable of interacting with a nanoparticle.[192,193]

The hydrophilic part of a stabilizer molecule should either be charged, e.g. negatively charged carboxy groups, or have polymeric "brushes" of polyethylene glycol or dextran.[194] For stabilizing nanoparticles in water, surface silanization processes were used, when the hydrophobic ligand shell was replaced by a layer of silane molecules.[195,196]

Further strategies in stabilizing nanoparticles are based on using amphiphilic polymeric molecules.[197] The hydrophobic part of an amphiphilic molecule coordinates with the hydrophobic stabilization shell around a nanoparticle, whereas its hydrophilic part points outward into the solution, thus facilitating water solubility. This procedure was used for transferring Au, $CoPt_3$, Fe_2O_3, and CdSe/ZnS nanoparticles to water.[198] The advantage of this procedure consists in the absence of ligand exchange, which allows applying it to virtually any particle. A certain drawback is associated with an increase in the particle volume owing to the presence of several stabilizing layers. At the same time, the presence of considerable amounts of amphiphilic polymers facilitates the incorporation of various functional groups immediately into the nanoparticle shell, thus providing its binding with biologically active molecules.

Several methods were used for linking biomolecules to nanoparticles. The simplest method involves adsorption of biomolecules on the surfaces of nanoparticles or their stabilizing shells.[199] However, the method of chemical bonding of biomolecules to nanoparticles proved to be most advantageous. The resulting systems were used as sensors for detecting molecules[200,201] or labeling cells.[202,203]

Nowadays, attention is focused on DNA–Au conjugates that contain different numbers of DNA molecules per particle. Nanoparticles with exact numbers of DNA can be separated using gel electrophoresis.[204] Other methods were developed for sorting nanoparticles with controlled number of biomolecules.[205,206] The number of DNA molecules that can be attached to a single gold particle is determined by the size of the latter.[207] The methods of labeling cells were surveyed by several authors.[208,209]

Nanocrystals and especially quantum dots can serve as convenient cell labels. They were shown to be sufficiently biocompatible and reduce the tendency to photobleaching.[210] Moreover, upon the cell division, the nanoparticles are passed to both daughter cells; and, therefore, the label is not lost.[199,211] For example, nanocrystals were introduced into specific cells of Xenopus embryos to follow their development.[212] Furthermore, cell-labeling method can be extended by using magnetic nanoparticles in drug delivery.[213] Drugs are immobilized on the surface of magnetic nanocrystals, and the resulting assemblies are directed by a magnetic field to the injured tissue.[214] Each nanoparticle type has its own intrinsic properties. Semiconductors can serve as fluorescent labels, while metals and their oxides form magnetic labels. For many biological and medical applications, hybrid nanoparticles with different properties are a good choice. Thus, hybrid nanoparticles combining fluorescent and magnetic properties could be obtained by the direct growth of fluorescent CdS particles on magnetic FePt nanoparticles.[215] The obvious advantage of these particles is a negligible gap between their domains, which allows investigating the coupling between different domains of hybrid particles. No energy transfer or tunneling is possible when nanoparticles are farther than a few nanometers apart. Another important problem is the preparation of hybrid nanoparticles containing more than two components. Along with the preparation of hybrid materials, much effort is concentrated on the creation of assemblies of hybrid materials with exactly defined structures. Some success in this direction has already been achieved.[216]

Building organized assemblies involving biomolecules entails a number of disadvantages. First, biological molecules used for binding nanoparticles also act as spacers that create a gap between the nanoparticles. Moreover, the linkage of biomolecules to the nanoparticles is not stiff but rather flexible. Such flexibility suggests that nanoparticles are not bound in rigid assemblies, and the distance between nanoparticles in an assembly can fluctuate. Additional difficulties are due to the thermal instability of biomolecules and a possibility of their destruction in solutions.

The optical properties of aggregates of gold particles bound by DNA fragments, which included 27–72 nucleotide pairs, were considered.[217] Alkanethiol-modified gold particles with diameters of ca. 15 nm were used. In studying how the length of oligonucleotide chains affects the properties of the resulting aggregates, it was shown that the chain length defines the position of the metal plasmon peak. The sizes of nanoparticle aggregates were determined by a kinetic method, because their growth rate depended on the length of oligonucleotide chains and on interparticle distances. Among the factors determining the growth rate of nanoaggregates, the rate of binding DNA-linkers with complementary DNA on the surface of gold nanoparticles (k_1) and the rate of aggregate growth (k_2) are the most important. Linkers are the untwisted segments of the double-chain DNA that binds densely packed DNA stabilized by proteins and polyamines. Optical changes were also observed when nanostructures formed by the two longest nucleotides linkers were annealed at temperatures below the melting point. Thus, DNA linkers can be used for kinetic control over the aggregate growth.

Combinations of nanocrystals with biomolecules were proposed for using in new detection schemes.[218] A synthesis of CdS/ZnS core–shell semiconductor quantum dots with Au nanoparticles was described. The structures obtained had one quantum dot in the center and a discrete number of Au nanocrystals (one to seven) attached to it through a DNA.[219] The use of DNA as the scaffolding material allowed controlling the Au–quantum-dot distance and the number of gold nanocrystals around the central quantum dot. Such systems can be used as traditional biological labels as well as for studying the interaction between nanoparticles and quantum dots.

A positively charged complex of DNA and CdSe nanorods that exhibited high luminescence was synthesized. The complexes could self-assemble to form either a filamentary netline or spherical nanostructures. Filamentary nanorods of CdSe and DNA exhibited strong linear polarized photoluminescence owing to the unidirectional orientation of nanorods along the fibers.[220] The processes of self-assembling of DNA–CdSe and some other nanostructures were analyzed in detail.[221]

Gold nanoparticles of 2-nm diameters covered with monolayers of trimethyl ammonium were found to recognize and stabilize peptide α-helices.[222]

A new method of incorporating biomaterials into living cells was proposed.[223] The method is based on electrosputtering (electrostatic spraying) of metal particles that carry large electric charges and have high rates of sputtering. When metal particles covered with genes were dispersed by means of elaborate equipment, they disintegrate to liquid droplets under the effect of the external nonuniform field. The resulting gene fragments had a charge of the same sign as the metal particles and were present at a high concentration. Fibroblast cells of monkeys were used as the living cells; a suspension of fluorescent protein-labeled plasmids and also plasmids with gold particles measuring 5–10 nm (plasmids are extrachromosomal formations representing closed rings of double-chain DNA) served as the biomaterial. By using UV-fluorescence microscope, it was shown that a plasmid suspension with gold could penetrate into the cells. Of particular importance is the fact that a plasmid suspension that did not contain gold also penetrated into the cells and was incorporated into its DNA. From the viewpoint of the authors,[223] this deserves most attention, because they managed to do without gold, which is traditionally used for penetration into cells. It was noted that electrosputtering opens up new possibilities for transfection and gene therapy.

Nanoparticles containing admixtures of gadolinium ions were proposed as a new contrasting material to be used in magnetic resonance studies in medicine. Such particles with a diameter of 120 nm are sufficiently small to penetrate blood vessels. They were used in obtaining images of heart and gastrointestinal tract of rats.[224]

The possibility of using different types of quantum dots for labeling biological cells, tissues, bacteria, and viruses was actively explored. As compared with organic dyes, these labels exhibited higher photostability and sensitivity. However, the toxicity of quantum dots was reported by the example of CdSe.[210] Notwithstanding, the applications of animal quantum dots as biolabels for diagnosing diseases were actively developed.[186,220,221] A one-step synthesis of water-dispersible multifunctional core–shell quantum dots CdS:Mn/ZnS measuring 3.1 nm, which was performed in water–oil emulsions, was proposed.[225] The fluorescence, radioopacity, and photostability of such quantum dots were studied earlier.[226] These properties were attributed to the effective surface passivation by zinc-sulfide crystals that surround the crystalline core CdS:Mn. Cadmium ions provided radioopacity of quantum dots, while magnesium ions were responsible for the magnetic properties. The effectiveness of CdS:Mn/ZnS quantum dots as biolabels was demonstrated in studying brain tissues of rats.

Various thiols were actively used for stabilizing nanoparticles of different metals, especially, gold. Recently, attention was focused on self-organization processes that affect the chemical and catalytic properties of metal nanoparticles.[227–229] The exchange between the protective ligand coating of a gold nanoparticle and *para*-substituted benzyl hydroxyalkyl nitroxide was studied by EPR technique.[230]

The synthesis of gold nanoparticles coated with dipeptides containing thiol groups $SH-(CH_2)_{11}CO-His-Phe-OH$ was described.[231] By using this peptide together with $HS-(CH_2)_7-CONH-(CH_2OCH_2O)_3-CH^3$, the protected gold clusters were synthesized and then employed in the hydrolysis of esters.[227] By the example of 2,4-dinitrophenylbutanoate, a 300-fold increase in the rate of hydrolysis was observed in the presence of nanoparticles.

Self-assembling of biological molecules into various nanoarchitecture assemblies can assist the conversion of information contained in them into physicochemical signals. More detailed studies of self-assembling were conducted for large molecules such as proteins,[232] oligopeptides,[233] and nucleic acids.[234] For small biomolecules, self-assembling of adenosine-5′-triphosphate and a dichlorine-substituted thiocarbocyanine dye into excitation-delocalized nanowires was described. According to TEM studies, the latter were ~10 nm wide and several micrometers long.[235]

Spherical polystyrene particles can also be used for labeling biomolecules. These particles were treated with surfactants, which represented a mixture of polyethylene and polyethylene glycols swelled in toluene. The latter solvent was removed by heating to 98 °C. This process was stabilized by a surfactant that contained biomolecules and fluorescent molecules attached to its active groups.[236]

Enzymes are powerful biocatalysts in water but display much weaker activity in organic solvents owing to poor solubility. Nanopores of nanostructured amphiphilic networks prepared by the nanophase separation of hydrophilic and hydrophobic phases can serve as nanoreactors, which allow using enzymes in organic solvents.[237] An enzyme is entrapped in the hydrophilic domain, while the substrate is introduced into the hydrophobic phase. The potentials of this method were demonstrated for the oxidative coupling of *N,N*-dimethyl-*p*-phenylene diamine and phenol with *tert*-butyl hydroperoxide in hexane to yield an indophenol dye, which was catalyzed by the network-entrapped horseradish peroxidase.

The chemical activities of solid organic drug substances and their biological activities strongly depend on their polymorphic modification, supramolecular organization and structural ordering, and particle size and shape. These specific features determine the biopharmaceutical and therapeutic activities of the drug forms. Production of structure and size-modified drug substances was realized via cryoformation of the solid-phase substance by a metastable state obtained by low-temperature condensation of their vapors on the cooled surface.[238] Obtaining structural modifications of drug substances was realized using static and flow vacuum setups. Organic vapor condensation was carried out on polished copper cube or glass tube cooled by liquid nitrogen. The samples of gabapentine (1-aminomethyl)cyclohexaneacetic acid, moxomidine 4-chloro-N-(4,5-dihydro-1H-imidazol-2-yl)-6-methoxy-2-methyl-5-pyrimidinamine, and carvedilol 1-(9H-carbazol-4-yloxy)-3-[[2-(2-methoxyphenoxy)ethyl]amino]-2-propanol were studied. Physical and chemical properties of cryomodified substances were studied by UV and FTIR spectroscopy, microscopy, and by X-ray diffraction. The composition of the former compounds and condensation products were controlled by chromatographic methods. By using gabapentin, the formation of three different forms was established. One of them is new where the size of particles obtained at 77 K was equal to 0.5 µm. The size of the particle increases with an increase in the condensation temperature. Minimization of particle size during cryoformation was also found (for moxomidine and carvedilo). The modified drug substances possess biological activities different from the former substances.

Nanostructures built of biological molecules represent supramolecular assemblies, which can easily undergo various chemical modifications. Peptide nanotubes similar to CNTs as regards their morphology were synthesized. The diphenylaniline-based tubes were proposed for use in electrochemistry.[218]

Mixed protective monolayers were proposed for application on metal nanoparticles, e.g. gold, to carry out specific reactions with biological molecules such as proteins.[239]

Studies directed at the elaboration of inorganic, biologically compatible, materials are carried out. One of such substances is hydroxyapatite $Ca_{10}(PO_4)_6(OH)_2$. The synthesis of hydroxyapatite crystals of different shapes and sizes was accomplished.[240–242] In one of the studies,[240] hydroxyapatite particles were obtained by the rapid mixing of solutions of $Ca(OH)_2$ and H_3PO_4. For analysis, the particles were sampled from the solution at different times upon the addition of acid. In 10^4 s, all particles represented nanoplates with hydroxyapatite lattice. The analysis of experimental results revealed the appearance of a new form, which was named two-dimensional crystalline hydroxyapatite. The synthesized substance is a highly effective drug, which stimulates osteogenesis at the implantation of bone tissue in living organisms.

The possibilities of application of biological molecules for identification of widely used inorganic materials were analyzed. This was accomplished using the principles of selective binding known in molecular biology. The use of selective binding of peptides with various semiconductors for the development of nanocrystalline assemblies was proposed.[243] It was shown that the use of nanocrystalline semiconductors allows one to extract certain peptides, the latter being linked with high specific bonds with the surface of these semiconductors. As substrates, five different single-crystal faces of the following semiconductors: GaAs (100), GaAs (111) (with gallium atoms on the surface), GaAs (111) (with arsenic atoms on the surface), InP (100), and Si (100) were chosen. It was found that among the wide diversity of randomly chosen peptides, each substrate was picked and selectively bonded with a definite sequence of amino acids. The resulting nanocrystalline assemblies were examined by means of antibodies labeled with 20 nm gold particles and also by means of transmission and fluorescence microscopes, photoelectron spectroscopy, AFM, and STM.

Figure 11.10 demonstrates two different approaches to the formation of assemblies of biomolecules on the surface of inorganic materials.[244] The development of these approaches is a challenging problem in material science of the twenty-first century. The role of the surface in the self-organization of biomolecules, particularly peptides, was discussed.[245] Problems of biomimetism, the interaction of biological molecules with inorganic surfaces, and bioengineering were surveyed.[246]

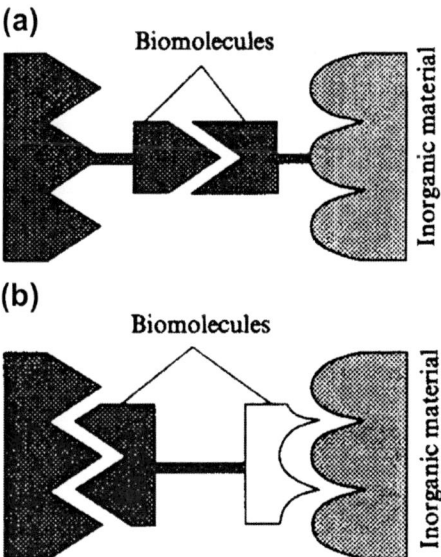

FIGURE 11.10 Two approaches to assembling inorganic materials into more complex structures using biological molecules: (a) the use of complementarity of two biomolecules and (b) the use of the interaction of a biomolecule with an inorganic material.[244]

The application of various nanoparticles in biology and medicine poses certain problems. Most metal and semiconductor nanoparticles are synthesized and stabilized in organic solvents. To use them, e.g. in biochemistry, it is necessary to transfer them into aqueous solutions, whereas their further use in medicine requires the attachment of drugs to them. To realize such transformations, which are additionally complicated by self-organization of particles, attempts were undertaken to use the key–lock principle, which was developed for enzyme reactions, on the molecular level.

In recent years, several approaches to synthesizing these materials were developed. One approach is based on assembling the receptor–ligand-mediated groupings of building blocks to form new multifunctional structures.[204] Another approach consists in the arrangement of ligand-modified blocks on a surface that is patterned with receptor molecules.[247,248] The latter approach was used for the fabrication of conducting DNA-based metallized molecular wires.[249] A method for labeling of specific receptors in a cell with ligand-modified building blocks was developed.[250,251] Such labels most often represent organic fluorophores.

The prospects and problems of synthesizing two- and three-dimensional nanostructures based on biological principles were discussed.[252] From the viewpoint of the authors, in this century, the strategy of synthesizing complex assemblies from simpler components will be actively developed. Biological structures can be used as surface detectors for organizing the linkage of large organic and inorganic blocks. In fact, this will allow one to synthesize new materials by employing the principles developed in the course of evolution for assembling complex functional systems.

11.6.1 DNA-modified Nanoparticles

Although DNA-nanoparticle species have been mentioned earlier, some recent developments need to be briefly added, and two excellent reviews are mentioned.[253,254]

Actually, the DNA–Au nanoparticles' (NPs') field started in 1996 with two papers published simultaneously.[255,256] The gold nanoparticle provides an excellent scaffold on which biomolecules can be anchored. These DNA molecules need to be prepared with a thiol-end group, and once prepared, these species are stable, soluble in water, and most importantly, exhibit intense optical properties that are dependent on particle size, shape, and state of aggregation.[257]

Combining the recognition properties of DNA with the optical properties of gold allows these new materials to exhibit unique properties and is leading to many useful bio-related devices.[254] The first of these devices is illustrated in Figure 11.11.

It has been learned that the amount of DNA attached to a gold nanoparticle can vary greatly depending on the salt concentration and on particle size.[258]

Assembly of DNA–Au NPs into highly ordered crystals have produced both FCC and BCC structures[259,260] in solution, which were analyzed by small angle

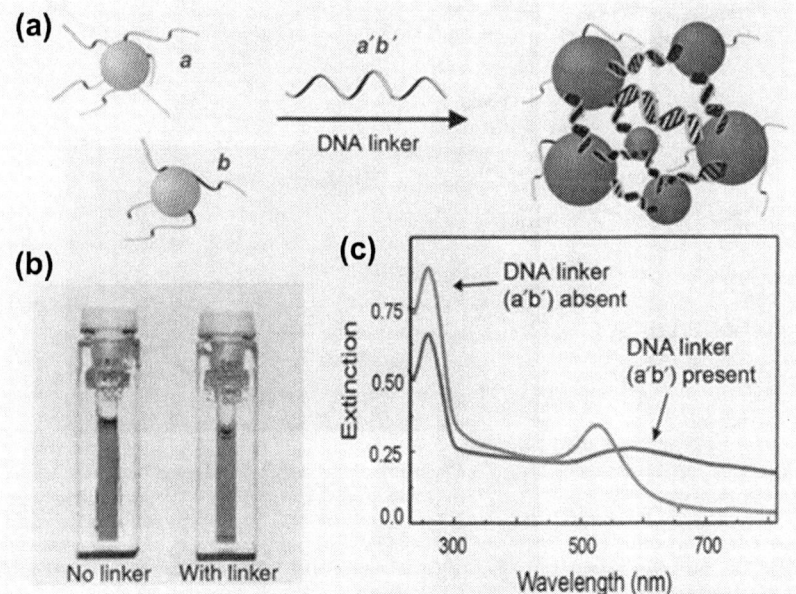

FIGURE 11.11 (a) DNA–Au NPs assembled using a complimentary DNA linker (b) color change from red to blue upon addition of linker (c) electronic extinction spectrum before and after.[254] (For color version of this figure, the reader is referred to the online version of this book.)

X-ray scattering. Applications of these DNA–Au NPs take advantage of their intense optical properties, strong binding properties, short melting transitions, and high DNA loadings.[253]

These unique properties have allowed them to be used in the design of several diagnostic systems: detection of DNA, protein detection, metal ion detection, and small organic molecule detection. Current work is directed to therapeutics, and progress will depend a lot on how these materials behave in vivo and how they can be eliminated from the human body.[254]

REFERENCES

1. Roco, M. C.; Williams, S.; Alivisatos, P., Eds. *Nanotechnology Research Directions: IWGN Workshop Report-vision for Nanotechnology in the Next Decade*; Kluwer: New York, 1999, pp 1–360.
2. Andrievsky, R. A. *Russ. Chem. J.* **2002,** *46,* 50–56.
3. Eberhardt, W. *Surf. Sci.* **2002,** *500,* 242–270.
4. Physical chemistry of ultradispersed systems Parts 1,2 (Russ.). 2001. Ekaterinburg, UrO RAN. Ref Type: Conference Proceeding.
5. *Ros. Khim. Zhurn* **2002,** *46,* 3–98.
6. *Springer Handbook of Nanotechnology*; Bhushan, B., Ed.; Springer-Verlag: Berlin, 2004; pp 1–1222.

7   Zarur, A. J.; Ying, J. Y. *Nature (London)* **2000**, *403*, 65–67.
8   McCarty, J. G. *Nature (London)* **2000**, *403*, 35–36.
9   Spiro, M.; De Jesus, D. M. *Langmuir* **2000**, *16*, 2464–2468.
10 De Jesus, D. M.; Spiro, M. *Langmuir* **2000**, *16*, 4896–4900.
11 Tu, W.; Liu, H.; Liew, K. Y. *J. Colloid Interface Sci.* **2000**, *229*, 453–461.
12 Liu, J.; Shin, Y.; Nie, Z.; Chang, J. H.; Wang, L. -Q.; Fryxell, G. E.; Samuels, W. D.; Exarhos, G. J. *J. Phys. Chem. A* **2000**, *104*, 8328–8339.
13 Fagherazzi, G.; Canton, P.; Riello, P.; Pinna, F.; Pernicone, N. *Catal. Lett.* **2000**, *64*, 119–124.
14 Fagherazzi, G.; Canton, P.; Riello, P.; Pernicone, N.; Pinna, F.; Battagliarin, M. *Langmuir* **2000**, *16*, 4539–4546.
15 Wang, L.; Rocci-Lane, M.; Brazis, P.; Kannewurf, C. R.; Kim, Y. -I.; Lee, W.; Choy, J. -H.; Kanatzidis, M. G. *J. Am. Chem. Soc.* **2000**, *122*, 6629.
16 Maye, M. M.; Lou, Y.; Zhong, C. -J. *Langmuir* **2000**, *16*, 7520–7523.
17 Maye, M. M.; Zheng, M.; Leibowitz, F. L.; Ly, N. K.; Zhong, C. -J. *Langmuir* **2000**, *16*, 490–497.
18 Maye, M. M.; Zhong, C. -J. *J. Mater. Chem.* **2000**, *10*, 1895–1901.
19 Mills, G.; Gordon, M. S.; Metiu, H. *Chem. Phys. Lett.* **2002**, *359*, 493–499.
20 Yan, Z.; Chinta, S.; Mohamed, A. A.; Fackler, J. P.; Goodman, D. W. *J. Am. Chem. Soc.* **2005**, *127*, 1604–1605.
21 Rodriguez, J. A.; Liu, G.; Jirsak, T.; Hrbek, J.; Chang, Z.; Dvorak, J.; Maiti, A. *J. Am. Chem. Soc.* **2002**, *124*, 5242–5250.
22 Chen, M. S.; Goodman, D. W. *Science* **2004**, *306*, 252–255.
23 Lopez, N.; Janssens, T. V.W.; Clausen, B. S.; Xu, Y.; Mavrikakis, M.; Bligaard, T.; Norskov, J. K. *J. Catal.* **2004**, *223*, 232–235.
24 Lopez, N.; Norskov, J. K.; Janssens, T. V.W.; Carlsson, A.; Puig-Molina, A.; Clausen, B. S.; Grunwaldt, J. -D. *J. Catal.* **2004**, *225*, 86–94.
25 Chen, C. -W.; Tano, D.; Akashi, M. *J. Colloid Interface Sci.* **2000**, *225*, 349–358.
26 Del Angel, P.; Dominguez, J. M.; Del Angel, G.; Montoya, J. A.; Lamy-Pitara, E.; Labruquere, S.; Barbier, J. *Langmuir* **2000**, *16*, 7210–7223.
27 Mohamed, M. M.; Salama, T. M.; Ichikawa, M. *J. Colloid Interface Sci.* **2000**, *224*, 366–371.
28 Rupprechter, G.; Eppler, A. S.; Avoyan, A.; Somorjai, G. A. *Stud. Surf. Sci. Catal. A* **2000**, *130*, 215.
29 Eppler, A. S. *Top. Catal.* **2000**, 33–41.
30 Zagorskaya, O. V.; Zufmfan, V. Yu.; Rostovshikova, T. N.; Smirnov, V. V.; Gubin, S. P. *Russ. Chem. Bull. (Russ.)* **2000**, 854–857.
31 Tada, H.; Teranishi, K.; Inubushi, Y.; Ito, S. *Langmuir* **2000**, *16*, 3304–3309.
32 Wilcoxon, J. P. *J. Phys. Chem. B* **2000**, *104*, 7334–7343.
33 Dhas, N. A.; Suslick, K. S. *J. Am. Chem. Soc.* **2005**, *127*, 2368–2369.
34 Parenago, O. P.; Bakunin, V. N.; Kuz'mina, G. N.; Suslov, A. Yu.; Vedeneeva, L. N. *Doklady Chem.* **2002**, *383*, 86–88.
35 Mueller, A.; Roy, S. *Russ. Chem. Rev.* **2003**, *71*, 981–992.
36 Blackedge, C.; Engebreston, D. A.; McDonald, J. D. *Langmuir* **2000**, *16*, 8317–8323.
37 Liu, B.; Zeng, H. C. *Langmuir* **2004**, *20*, 4196–4204.
38 Hong, S.; Mirkin, C. *Science* **2000**, *288*, 1808.
39 Bukhtiyarov, V. I.; Slin'ko, M. G. *Russ. Chem. Rev.* **2001**, *70*, 147–160.
40 Sergeev, G. B. *Russ. Chem. Rev.* **2001**, *70*, 809–826.
41 Bonnemann, H.; Richards, R. M. *Eur. J. Inorg. Chem.* **2001**, *2001*, 2455–2480.
42 Heiz, U.; Vanolli, F.; Sanchez, A.; Schneider, W. D. *J. Am. Chem. Soc.* **1998**, *120*, 9668–9671.

43. Heiz, U.; Sanchez, A.; Abbet, S.; Schneider, W. D. *J. Am. Chem. Soc.* **1999**, *121*, 3214–3217.
44. Rumpf, F.; Poppa, H.; Boudart, M. *Langmuir* **1988**, *4*, 722–728.
45. Becker, C.; Henry, C. R. *Surf. Sci.* **1996**, *352–354*, 457–462.
46. Piccolo, L.; Becker, C.; Henry, C. R. *Appl. Surf. Sci.* **2000**, *164*, 156–162.
47. Heiz, U.; Sanchez, A.; Abbet, S.; Schneider, W. D. *Chem. Phys.* **2000**, *262*, 189–200.
48. Abbet, S.; Sanchez, A.; Heiz, U.; Schneider, W. D.; Ferrari, A. M.; Pacchioni, G.; Rosch, N. *Surf. Sci.* **2000**, *454*, 984–989.
49. Abbet, S.; Heiz, U.; Ferrari, A. M.; Giordano, L.; Di Valentin, C.; Pacchioni, G. *Thin Solid Films* **2001**, *400*, 37–42.
50. Heiz, U.; Schneider, W. D. *J. Phys. D: Appl. Phys.* **2000**, *33*, R85–R102.
51. Ferrari, A. M.; Giordano, L.; Rosch, N.; Heiz, U.; Abbet, S.; Sanchez, A.; Pacchioni, G. *J. Phys. Chem. B* **2000**, *104*, 10612–10617.
52. Schmid, G.; Maihack, V.; Lantermann, F.; Peshel, S. *J. Chem. Soc., Dalton Trans.* **1996**, *5*, 589–595.
53. Okamoto, K.; Akiyama, R.; Yoshida, H.; Yoshida, T.; Kobayashi, S. *J. Am. Chem. Soc.* **2005**, *127*, 2125–2135.
54. Carnes, C. L.; Klabunde, K. J. *Langmuir* **2000**, *16*, 3764–3772.
55. Wagner, G. W.; Bartram, P. W.; Koper, O.; Klabunde, K. J. *J. Phys. Chem. B* **1999**, *103*, 3225–3228.
56. Wagner, G. W.; Koper, O. B.; Lucas, E.; Decker, S.; Klabunde, K. J. *J. Phys. Chem. B* **2000**, *104*, 5118–5123.
57. Poisot, P.; Laurelle, S.; Gruqeon, S.; Dupont, L.; Tarascon, J. -M. *Nature* **2000**, *407*, 496–499.
58. Shevchenko, V. Ya.; Khasanov, O. L.; Yur'ev, G. S.; Pokholkov, Yu. P. *Doklady Phys. Chem.* **2001**, *377*, 121–124.
59. Elder, S. H.; Cot, F. M.; Su, Y.; Heald, S. M.; Tyryshkin, A. M.; Bowman, M. K.; Gao, Y.; Joly, A. G.; Balmer, M. L.; Kolwaite, A. C.; Magrini, K. A.; Blake, D. M. *J. Am. Chem. Soc.* **2000**, *122*, 5138–5146.
60. Urban, J. J.; Yun, W. S.; Gu, Q.; Park, H. *J. Am. Chem. Soc.* **2002**, *124*, 1186–1187.
61. Cozzoli, P. D.; Kornowski, A.; Weller, H. *J. Am. Chem. Soc.* **2003**, *125*, 14539–14548.
62. Cheng, B.; Russel, J. M.; Shi, W.; Zhang, L.; Samulski, E. T. *J. Am. Chem. Soc.* **2004**, *126*, 5972–5973.
63. Patzke, G. R.; Krumeich, F.; Nesper, R. *Angew. Chem. Int. Ed.* **2002**, *41*, 2446–2461.
64. Gui, Z.; Fan, R.; Mo, W.; Chen, X.; Yang, L.; Zhang, S.; Hu, Y.; Wang, Z.; Fan, W. *Chem. Mater.* **2002**, *14*, 5053–5056.
65. Liu, J.; Li, Q.; Wang, T.; Yu, D.; Li, Y. *Angew. Chem. Int. Ed.* **2004**, *43*, 5048–5052.
66. Guiton, B. S.; Gu, Q.; Prieto, A. L.; Gudiksen, M. S.; Park, H. *J. Am. Chem. Soc.* **2005**, *127*, 498–499.
67. Dai, Z. R.; Pan, Z. W.; Wang, Z. L. *Adv. Funct. Mater.* **2003**, *13*, 9–24.
68. Novak, J. P.; Feldheim, D. L. *J. Am. Chem. Soc.* **2000**, *122*, 3979–3980.
69. Templeton, A. C.; Wuelfing, W. P.; Murray, R. W. *Acc. Chem. Res.* **2000**, *33*, 27–36.
70. McConnell, W. P.; Novak, J. P.; Brousseau, L. C., III; Fuierer, R. R.; Tenent, R. C.; Feldheim, D. L. *J. Phys. Chem. B* **2000**, *104*, 8925–8930.
71. Alferov, Zh. I. *Semiconductors* **1998**, *32*, 1–14.
72. Quinlan, F. T.; Kuther, J.; Tremel, W.; Knoll, W.; Risbud, S.; Stroeve, P. *Langmuir* **2000**, *16*, 4049–4051.
73. Hirai, T.; Watanabe, T.; Komasawa, I. *J. Phys. Chem. B* **2000**, *104*, 8962–8966.
74. Korgel, B. A.; Monbouquette, H. C. *Langmuir* **2000**, *16*, 3588–3594.
75. Kho, R.; Torres-Martinez, C. L.; Mehra, R. K. *J. Colloid Interface Sci.* **2000**, *227*, 561–566.

76 Hao, E.; Lian, T. *Langmuir* **2000**, *16,* 7879–7881.
77 Duan, X.; Lieber, C. M. *Adv. Mater.* **2000**, *12,* 298–302.
78 Duan, X.; Lieber, C. M. *J. Am. Chem. Soc.* **2000**, *122,* 188–189.
79 Holmes, J. D.; Johnson, K. P.; Doty, R. C.; Korgel, B. A. *Science* **2000**, *287,* 1471.
80 Gudiksen, M. S.; Lieber, C. M. *J. Am. Chem. Soc.* **2000**, *122,* 8801–8802.
81 Soloviev, V. N.; Eichhoefer, A.; Fenske, D.; Banin, U. *J. Am. Chem. Soc.* **2001**, *123,* 2354–2364.
82 Soloviev, V. N.; Eichhoefer, A.; Fenske, D.; Banin, U. *J. Am. Chem. Soc.* **2000**, *122,* 2673–2674.
83 *Nanoparticles: From Theory to Application*; Schmid, G., Ed.; Wiley-VCH: Weinheim, 2005; pp 1–444.
84 Nath, M.; Kar, S.; Raychaudhuti, A. K.; Rao, C. N.R. *Chem. Phys. Lett.* **2003**, *368,* 690–695.
85 Jiang, X.; Mayers, B.; Wang, Y.; Cattle, B.; Xia, Y. *Chem. Phys. Lett.* **2004**, *385,* 472–476.
86 Li, L. S.; Pradhan, N.; Wang, Y.; Peng, X. *Nano Lett.* **2004**, *4,* 2261–2264.
87 Yu, W. W.; Falkner, J. C.; Shih, B. S.; Colvin, V. L. *Chem. Mater.* **2004**, *16,* 3318–3322.
88 Babayan, Y.; Barton, J. E.; Greyson, E. C.; Odom, T. W. *Adv. Mater.* **2004**, *16,* 1341–1345.
89 Pinna, N.; Garnweitner, G.; Beato, P.; Niederberger, M.; Antonietti, M. *Small* **2005**, *1,* 112–121.
90 Zhang, J. Z. *J. Phys. Chem. B* **2000**, *104,* 7239–7253.
91 Chen, M.; Xie, Y.; Chen, H.; Qiao, Z.; Zhu, Y.; Qian, Y. *J. Colloid Interface Sci.* **2000**, *229,* 217–221.
92 Gittins, D. I.; Bethell, D.; Schiffrin, D. J.; Nichols, R. J. *Nature (London)* **2000**, *408,* 67–69.
93 Gittins, D. I.; Bethell, D.; Nichols, R. J.; Schiffrin, D. J. *J. Mater. Chem.* **2000**, *10,* 79–83.
94 Feldheim, D. *Nature (London)* **2000**, *408,* 45–46.
95 Albrecht, M.; Lutz, M.; Spek, A. L.; van Koten, G. *Nature (London)* **2000**, *406,* 970–974.
96 Steed, J. W. *Nature (London)* **2000**, *406,* 943–944.
97 Yun, M.; Myung, N.; Vasquez, R. P.; Lee, C.; Menke, E.; Penner, R. M. *Nano Lett.* **2004**, *4,* 419–422.
98 Hernandez, R. M.; Richter, L.; Semancik, S.; Stranick, S.; Mallouk, T. E. *Chem. Mater.* **2004**, *16,* 3431–3438.
99 Ramanathan, K.; Bangar, M. A.; Yun, M.; Chen, W.; Myung, N. V.; Mulchandani, A. *J. Am. Chem. Soc.* **2005**, *127,* 496–497.
100 Schmid, G. *J. Chem. Soc., Dalton Trans.* **1998**, *7,* 1077–1082.
101 http://www.nanoway.fi, 2003.
102 Vasil'ev, R. B.; Gas'kov, A. M.; Rumyantseva, M. N.; Ryzhikov, A. S.; Ryabova, L. I.; Akimov, B. A. *Semiconductors* **2000**, *34,* 955–959.
103 Galyamov, B. Sh.; Zav'yalov, S. A.; Kupriyanov, L. Yu *Russ. J. Phys. Chem.* **2000**, *74,* 387–392.
104 Fu, L.; Cao, L.; Liu, Y.; Zhu, D. *Adv. Colloid Interface Sci.* **2004**, *111,* 133–157.
105 Simon, I.; Barsan, N.; Bauer, M.; Weimar, U. *Sens. Act. B* **2001**, *73,* 1–26.
106 Tan, O. K.; Zhu, W.; Yan, Q.; Kong, L. B. *Sens. Act. B* **2000**, *65,* 361–365.
107 Kennedy, M. K.; Kruis, F. E.; Fissan, H.; Mentha, B. R.; Stappert, S.; Dumpich, G. *J. Appl. Phys.* **2003**, *93,* 551–560.
108 Bochenkov, V. E.; Zagorskii, V. V.; Sergeev, G. B. *Sens. Act. B* **2004**, *103,* 375–379.
109 Bochenkov, V. E.; Karageorgiev, P.; Brehmer, L.; Sergeev, G. B. *Thin Solid Films* **2004**, *458,* 304–308.
110 Karthigeyan, A.; Gupta, R. P.; Scharnagl, K.; Burgmair, M.; Zimmer, M.; Sharma, S. K.; Eisele, I. *Sens. Act. B* **2001**, *78,* 69–72.

111 Taurino, A. M.; Epifani, M.; Toccoli, T.; Iannotta, S.; Siciliano, P. *Thin Solid Films* **2003**, *436*, 52–63.
112 Cukrov, L. M.; McCormick, P. G.; Galatsis, K.; Wlodarski, W. *Sens. Act. B* **2001**, *77*, 491–495.
113 Pedrosa, J. M.; Dooling, C. M.; Richardson, T. H.; Hyde, R. K.; Hunter, C. A.; Martin, M. T.; Camacho, L. *J. Mater. Chem.* **2002**, *12*, 2659–2664.
114 Richardson, T. H.; Dooling, C. M.; Worsfold, O.; Jones, L. T.; Kato, K.; Shinbo, K.; Kaneko, F.; Tregonning, R.; Vysotsky, M. O.; Hunter, C. A. *Colloids Surf. A* **2002**, *198–200*, 843–857.
115 Godovsky, D. Y. *Adv. Polymer Sci.* **2000**, *153*, 163–205.
116 Bochenkov, V. E.; Stephan, N.; Brehmer, L.; Zagorskii, V. V.; Sergeev, G. B. *Colloids Surf. A* **2002**, *198–200*, 911–915.
117 Wang, Z. L. *Annu. Rev. Phys. Chem.* **2004**, *55*, 159–196.
118 Comini, E.; Guidi, V.; Malagu, C.; Martinelli, G.; Pan, Z.; Sberveglieri, G.; Wang, Z. *J. Phys. Chem. B* **2004**, *108*, 1882–1887.
119 Comini, E.; Faglia, G.; Sberveglieri, G.; Pan, Z.; Wang, Z. L. *Appl. Phys. Lett.* **2002**, *81*, 1869–1871.
120 Shimizu, Y.; Hyodo, T.; Egashira, M. *J. Eur. Ceram. Soc.* **2004**, *24*, 1389–1398.
121 Egashira, M.; Shimizu, Y.; Hyodo, T. *Mater. Res. Soc. Symp. Proc.* **2005**, *828*, A1.1.1–A1.1.10.
122 Lewis, S.; Cole, J.; Hesketh, P. *Mater. Res. Soc. Symp. Proc.* **2005**, *828*, A1.7.1–A1.7.6.
123 Kong, J.; Franklin, N.; Zhou, C.; Chapline, M.; Peng, S.; Cho, K.; Dai, H. *Science* **2000**, *287*, 622–625.
124 Peng, S.; Cho, K.; Qi, P.; Dai, H. *Chem. Phys. Lett.* **2004**, *387*, 271–276.
125 Wei, B. -Y.; Hsu, M. -C.; Su, P. -G.; Lin, H. -M.; Wu, R. -J.; Lai, H. -J. *Sens. Act. B* **2004**, *101*, 81–89.
126 Bogue, R. W. *Sens. Rev.* **2004**, *24*, 253–260.
127 Lei, Y.; Chim, W. -K. *J. Am. Chem. Soc.* **2005**, *127*, 1487–1492.
128 Chowdhuri, A.; Sharma, P.; Gupta, V.; Sreenivas, K.; Rao, K. V. *J. Appl. Phys.* **2002**, *92*, 2172–2180.
129 Yuanda, W.; Maosong, T.; Xiuli, H.; Yushu, Z.; Guorui, D. *Sens. Act. B* **2001**, *79*, 187–191.
130 Hahn, S. H.; Barsan, N.; Weimar, U.; Ejakov, S.; Visser, J. H.; Soltis, R. E. *Thin Solid Films* **2003**, *436*, 17–24.
131 Schmid, W.; Barsan, N.; Weimar, U. *Sens. Act. B* **2003**, *89*, 232–236.
132 Barsan, N.; Weimar, U. *J. Phys.: Condens. Matter* **2003**, *15*, R813–R839.
133 Harbeck, S.; Szatvanyi, A.; Barsan, N.; Weimar, U. *Thin Solid Films* **2003**, *436*, 76–83.
134 Gong, J.; Chen, Q.; Fei, W.; Seal, S. *Sens. Act. B* **2004**, *102*, 117–125.
135 Hagleitner, C.; Hierlemann, A.; Lange, D.; Kummer, A.; Kerness, N.; Brand, O.; Baltes, H. *Nature* **2001**, *414*, 293–296.
136 Lofdahl, M.; Eriksson, M.; Lundstrom, I. *Sens. Act. B* **2000**, *70*, 77–82.
137 Traversa, E.; Sadaoka, Y.; Carotta, M. C.; Martinelli, C. *Sens. Act. B* **2000**, *65*, 181–185.
138 Carotta, M. C.; Martinelli, G.; Crema, L.; Gallana, M.; Merli, M.; Ghiotti, G.; Traversa, E. *Sens. Act. B* **2000**, *68*, 1–8.
139 Hahn, S. H.; Barsan, N.; Weimar, U. *Sens. Act. B* **2001**, *78*, 64–68.
140 Lee, D. -S.; Kim, Y. T.; Huh, J. -S.; Lee, D. -D. *Thin Solid Films* **2002**, *416*, 271–278.
141 Maekawa, T.; Suzuki, K.; Takada, K.; Kobayashi, T.; Egashira, M. *Sens. Act. B* **2001**, *80*, 51–58.
142 Qin, S. J.; Wu, Z. *J. Sens. Act. B* **2001**, *80*, 85–88.
143 Vasiliev, R. B.; Ryabova, L. I.; Rumyantseva, M. N.; Gaskov, A. M. *Russian Chem. Rev.* **2004**, *73*, 939–956.

144　Ah, C. S.; Han, H. S.; Kim, K.; Jang, D. -J. *J. Phys. Chem. B* **2000**, *104*, 8153–8159.
145　Ah, C. S.; Han, H. S.; Kim, K.; Jang, D. -J. *Pure Appl. Chem.* **2000**, *72*, 91.
146　Jeong, D. H.; Suh, J. S.; Moskovits, M. *J. Phys. Chem. B* **2000**, *104*, 7462–7467.
147　Jeong, D. H.; Jang, N. H.; Suh, J. S.; Moskovits, M. *J. Phys. Chem. B* **2000**, *104*, 3594–3600.
148　Francois, L.; Mostafavi, M.; Belloni, J.; Delouis, J. -F.; Delaire, J.; Feneyrou, P. *J. Phys. Chem. B* **2000**, *104*, 6133–6137.
149　Hodak, J. H.; Henglein, A.; Hartland, G. V. *J. Phys. Chem. B* **2000**, *104*, 9954–9965.
150　Hodak, J. H.; Henglein, A.; Hartland, G. V. *J. Chem. Phys.* **2000**, *112*, 5942–5947.
151　Shen, Y.; Jakubczyk, D.; Xu, F.; Swiatkiewicz, J.; Prasad, P. N.; Reinhardt, B. A. *Appl. Phys. Lett.* **2000**, *76*, 1–3.
152　Shen, Y.; Friend, C. S.; Jiang, Y.; Jakubczyk, D.; Swiatkiewicz, J.; Prasad, P. N. *J. Phys. Chem. B* **2000**, *104*, 7577–7587.
153　Hasobe, T.; Imahori, H.; Kamat, P. V.; Ahn, T. K.; Kim, S. K.; Kim, D.; Fujimoto, A.; Hirakawa, T.; Fukuzumi, S. *J. Am. Chem. Soc.* **2005**, *127*, 1216–1228.
154　Imahori, H.; Kashiwagi, Y.; Endo, Y.; Hanada, T.; Nishimura, Y.; Yamazaki, I.; Araki, Y.; Ito, O.; Fukuzumi, S. *Langmuir* **2004**, *20*, 73–81.
155　Hasobe, T.; Imahori, H.; Kamat, P. V.; Fukuzumi, S. *J. Am. Chem. Soc.* **2003**, *125*, 14962–14963.
156　Yamada, H.; Imahori, H.; Nishimura, Y.; Yamazaki, I.; Ahn, T. K.; Kim, S. K.; Kim, D.; Fukuzumi, S. *J. Am. Chem. Soc.* **2003**, *125*, 9129–9139.
157　Li, K.; Schuster, D. I.; Guldi, D. M.; Herranz, M. A.; Echegoyen, L. *J. Am. Chem. Soc.* **2004**, *126*, 3388–3389.
158　D'Souza, F.; Smith, P. M.; Zandler, M. E.; McCarty, A. L.; Itou, M.; Araki, Y.; Ito, O. *J. Am. Chem. Soc.* **2004**, *126*, 7898–7907.
159　Imahori, H. *J. Phys. Chem. B* **2004**, *108*, 6130–6141.
160　Haggenmueller, R.; Gommans, H. H.; Rinzler, A. G.; Fischer, J. E.; Winey, K. I. *Chem. Phys. Lett.* **2000**, *330*, 219–225.
161　Sun, L.; Croocks, R. M. *Langmuir* **1999**, *15*, 738–741.
162　Wilder, J. W.G.; Venema, L. C.; Rinzler, A. G.; Smalley, R. E.; Dekker, C. *Nature (London)* **1998**, *391*, 59–62.
163　Dai, H.; Kong, J.; Zhou, C.; Franklin, N.; Tombler, T.; Cassel, A.; Fan, S.; Chapline, M. *J. Phys. Chem. B* **1999**, *103*, 11246–11255.
164　Lefebvre, J.; Antonov, R. D.; Radosavljevic, M.; Lynch, J. F.; Llagumo, M.; Johnson, A. T. *Carbon* **2000**, *38*, 1745–1749.
165　Collins, P. G.; Fuhrer, M. S.; Zettl, A. *Appl. Phys. Lett.* **2000**, *76*, 894–896; 865.
166　Service, R. *Science* **1999**, *285*, 2053–2057.
167　Soh, H. T.; Quate, C. F.; Morpurgo, A. F.; Marcus, C. M.; Kong, J.; Dai, H. *Appl. Phys. Lett.* **1999**, *75*, 627–629.
168　Satishkumar, B. C.; Thomas, P. J.; Govindaray, A.; Rao, C. N.R. *Appl. Phys. Lett.* **2000**, *77*, 2530–2532.
169　Venema, L. *Nature (London)* **1998**, *407*, 959.
170　Monteiro, O. R.; Mammana, V. P.; Salvadori, M. C.; Ager, J. W., III; Dimitrievic, S. *Appl. Phys. A* **2000**, *71*, 121.
171　Tombler, T. W.; Zhou, C.; Alexeyev, L.; Kong, J.; Dai, H.; Liu, L.; Jayanthi, C. S.; Tang, M.; Wu, S. -Y. *Nature (London)* **2000**, *405*, 769–772.
172　Chambers, A.; Park, C.; Baker, R. T.K.; Rodriguez, N. M. *J. Phys. Chem. B* **1998**, *102*, 4253–4256.
173　Tarasov, B. P.; Goldshleger, N. F.; Moravsky, A. P. *Russ. Chem. Rev.* **2001**, *70*, 131–146.
174　Sumanasekera, G. U.; Adu, C. K.W.; Fang, S.; Eklund, P. C. *Phys. Rev. Lett.* **2000**, *85*, 1096–1099.

175 Modi, A.; Koratkar, N.; Lass, E.; Bingqing, W.; Ajayan, P. M. *Nature* **2003**, *424*, 174.
176 Wood, J. R.; Zhao, Q.; Frogley, M. D.; Meurs, E. R.; Prins, A. D.; Peijs, T.; Dustan, D. J.; Wagner, H. D. *Phys. Rev. B* **2000**, *62*, 7571–7575.
177 Dai, H.; Hafner, J. H.; Rinzler, A. G.; Colbert, D. T.; Smalley, R. E. *Nature (London)* **1996**, *384*, 147.
178 Hafner, J. H.; Cheung, C. L.; Lieber, C. M. *J. Am. Chem. Soc.* **1999**, *121*, 9750–9751.
179 Stevens, R.; Nguyen, C.; Cassel, A.; Delzeit, L.; Meyyappan, M.; Han, J. *Appl. Phys. Lett.* **2000**, *77*, 3453–3455.
180 Wong, S. S.; Woolley, A. T.; Joselevich, E.; Lieber, C. M. *Chem. Phys. Lett.* **1999**, *306*, 219–225.
181 Wong, S. S.; Woolley, A. T.; Joselevich, E.; Cheung, C. L.; Lieber, C. M. *J. Am. Chem. Soc.* **1998**, *120*, 8557–8558.
182 Rodriguez, N. M.; Kim, M. -S.; Baker, R. T.K. *J. Phys. Chem.* **1994**, *98*, 13108–13111.
183 Planeix, J. M.; Coustel, N.; Coq, B.; Brotons, V.; Kumbhar, P. S.; Dutartre, R.; Geneste, P.; Bernier, P.; Ajayan, P. M. *J. Am. Chem. Soc.* **1994**, *116*, 7935–7936.
184 Zhang, Y.; Zhang, H. -B.; Lin, G. -D.; Chen, P.; Yuan, Y. -Z.; Tsai, K. R. *Appl. Catal. A* **1999**, *187*, 213–224.
185 Kim, P.; Lieber, C. M. *Science* **1999**, *286*, 2148.
186 Ballou, B.; Lagerholm, B. C.; Ernst, L. A.; Bruchez, M. P.; Waggoner, A. S. *Bioconjug. Chem.* **2004**, *15*, 79–86.
187 Carraro, C. *Phys. Rev. B* **2000**, *61*, R16351–R16354.
188 Calderon Moreno, J. M.; Swamy, S. S.; Fujino, T.; Yoshimura, M. *Chem. Phys. Lett.* **2000**, *329*, 317–322.
189 Wuister, S. F.; Swart, I.; van Driel, F.; Hickey, S. G.; de Mello Donega, C. *Nano Lett.* **2003**, *3*, 503–507.
190 Chen, Y.; Ji, T.; Rosenzweig, Z. *Nano Lett.* **2003**, *3*, 581–584.
191 Mayya, K. S.; Caruso, F. *Langmuir* **2003**, *19*, 6987–6993.
192 Pinaud, F.; King, D.; Moore, H. -P.; Weiss, S. *J. Am. Chem. Soc.* **2004**, *126*, 6115–6123.
193 Kim, S.; Bawendi, M. G. *J. Am. Chem. Soc.* **2003**, *125*, 14652–14653.
194 Wilhelm, C.; Billotey, C.; Roger, J.; Pons, J. N.; Bacri, J. -C.; Gazeau, F. *Biomater.* **2003**, *24*, 1001–1011.
195 Kobayashi, Y.; Horie, M.; Konno, M.; Rodriguez-Gonzalez, B.; Liz-Marzan, L. M. *J. Phys. Chem. B* **2003**, *107*, 7420–7425.
196 Parak, W. J.; Gerion, D.; Zanchet, D.; Woerz, A. S.; Pellegrino, T.; Micheel, C.; Williams, S. C.; Seitz, M.; Bruehl, R. E.; Bryant, Z.; Bustamante, C.; Bertozzi, C. R.; Alivisatos, A. P. *Chem. Mater.* **2002**, *14*, 2113–2119.
197 Petruska, M. A.; Bartko, A. P.; Klimov, V. I. *J. Am. Chem. Soc.* **2004**, *126*, 714–715.
198 Pellegrino, T.; Manna, L.; Kudera, S.; Licdl, T.; Koktysh, D.; Rogach, A. L.; Keller, S.; Raedler, J.; Natile, G.; Parak, W. J. *Nano Lett.* **2004**, *4*, 703–707.
199 Hanaki, K. -I.; Momo, A.; Oku, T.; Komoto, A.; Maenosono, S.; Yamaguchi, Y.; Yamamoto, K. *Biochem. Biophys. Res. Commun.* **2003**, *302*, 496–501.
200 Lingerfelt, B. M.; Mattoussi, H.; Goldman, E. R.; Mauro, J. M.; Anderson, G. P. *Anal. Chem.* **2003**, *75*, 4043–4049.
201 Goldman, E. R.; Clapp, A. R.; Anderson, G. P.; Uyeda, H. T.; Mauro, J. M.; Medinitz, I. L.; Mattoussi, H. *Anal. Chem.* **2004**, *76*, 684–688.
202 Lidke, D. S.; Nagy, P.; Heitzmann, R.; Arndt-Jovin, D. J.; Post, J. N.; Grecco, H. E.; Jares-Erijman, E. A.; Jovin, T. M. *Nat. Biotechnol.* **2004**, *22*, 198–203.
203 Dahan, M.; Levi, S.; Luccardini, C.; Rostanig, P.; Riveau, B.; Triller, A. *Science* **2003**, *302*, 442–445.

204 Zanchet, D.; Micheel, C. M.; Parak, W. J.; Gerion, D.; Williams, S. C.; Alivisatos, A. P. *J. Phys. Chem. B* **2002**, *106*, 11736–11758.
205 Sung, K. -M.; Mosley, D. W.; Peelle, B. R.; Zhang, S.; Jacobson, J. M. *J. Am. Chem. Soc.* **2004**, *126*, 5064–5065.
206 Liu, J.; Lu, Y. *J. Am. Chem. Soc.* **2003**, *125*, 6642–6643.
207 Sandstroem, P.; Boncheva, M.; Akerman, B. *Langmuir* **2003**, *19*, 7537–7543.
208 Gao, X.; Nie, S. *Trends Biotechnol.* **2003**, *21*, 371–373.
209 Alivisatos, A. P. *Nat. Biotechnol.* **2004**, *22*, 47–52.
210 Derfus, A. M.; Chan, W. C.W.; Bhatia, S. N. *Nano Lett.* **2004**, *4*, 11–18.
211 Mattheakis, L. C.; Dias, J. M.; Choi, Y. -J.; Gong, J.; Bruchez, M. P.; Liu, J.; Wang, E. *Anal. Biochem.* **2004**, *327*, 200–208.
212 Dubertret, B.; Skourides, P.; Norris, D. J.; Noireaux, V.; Brivanlou, A. H.; Libchaber, A. *Science* **2002**, *298*, 1759–1762.
213 Pankhurst, Q. A.; Connolly, J.; Jones, S. K.; Dobson, J. *J. Phys. D: Appl. Phys.* **2003**, *36*, R167–R181.
214 Alexiou, C.; Arnold, W.; Klein, R. J.; Parak, F. G.; Hulin, P.; Bergemann, C.; Erhardt, W.; Wagenpfeil, S.; Luebbe, A. S. *Cancer Res.* **2000**, *60*, 6641–6648.
215 Gu, H.; Zheng, R.; Zhang, X.; Xu, B. *J. Am. Chem. Soc.* **2004**, *126*, 5664–5665.
216 Wang, D.; He, J.; Rosenzweig, N.; Rosenzweig, Z. *Nano Lett.* **2004**, *4*, 409–413.
217 Storhoff, J. J.; Lazarides, A. A.; Mucic, R. C.; Mirkin, C. A.; Letsinger, R. L.; Schatz, G. C. *J. Am. Chem. Soc.* **2000**, *122*, 4640–4650.
218 Nam, J. -M.; Stoeva, S. I.; Mirkin, C. A. *J. Am. Chem. Soc.* **2004**, *126*, 5932–5933.
219 Fu, A.; Micheel, C. M.; Cha, J.; Chang, H.; Yang, H.; Alivisatos, A. P. *J. Am. Chem. Soc.* **2004**, *126*, 10832–10833.
220 Hoshino, A.; Hanaki, K. -I.; Suzuki, K.; Yamamoto, K. *Biochem. Biophys. Res. Commun.* **2004**, *314*, 46–53.
221 Khomutov, G. B. *Adv. Colloid Interface Sci.* **2004**, *111*, 79–116.
222 Verma, A.; Nakade, H.; Simard, J. M.; Rotello, V. M. *J. Am. Chem. Soc.* **2004**, *126*, 10806–10807.
223 Chen, D. -R.; Wendt, C. H.; Pui, D. Y.H. *J. Nanopart. Res.* **2000**, *2*, 133–139.
224 Reynolds, C. H.; Annan, N.; Beshah, K.; Huber, J. H.; Shaber, S. H.; Lenkinski, R. E.; Wortman, J. A. *J. Am. Chem. Soc.* **2000**, *122*, 8940–8945.
225 Santra, S.; Yang, H.; Holloway, P. H.; Stanley, J. T.; Mericle, R. A. *J. Am. Chem. Soc.* **2005**, *127*, 1656–1657.
226 Yang, H.; Holloway, P. H. *Appl. Phys. Lett.* **2003**, *82*, 1965–1967.
227 Pengo, P.; Polizzi, S.; Pasquato, L.; Scrimin, P. *J. Am. Chem. Soc.* **2005**, *127*, 1616–1617.
228 Shenhar, R.; Rotello, V. M. *Acc. Chem. Res.* **2003**, *36*, 549–561.
229 Drechsler, U.; Erdogan, B.; Rotello, V. M. *Chem. A Eur. J.* **2004**, *10*, 5570–5579.
230 Lucarini, M.; Franchi, P.; Pedulli, G. F.; Pengo, P.; Scrimin, P.; Pasquato, L. *J. Am. Chem. Soc.* **2004**, *126*, 9326–9329.
231 Pengo, P.; Broxterman, Q. B.; Kaptein, B.; Pasquato, L.; Scrimin, P. *Langmuir* **2003**, *19*, 2521–2524.
232 Ringler, P.; Schulz, G. E. *Science* **2003**, *302*, 106–109.
233 Zhang, S. *Nat. Biotechnol.* **2003**, *21*, 1171–1178.
234 Matsuura, K.; Yamashita, T.; Igami, Y.; Kimizuka, N. *Chem. Commun.* **2003**, *3*, 376–377.
235 Morikawa, M.; Yoshihara, M.; Endo, T.; Kimizuka, N. *J. Am. Chem. Soc.* **2005**, *127*, 1358–1359.
236 Kim, A. J.; Manoharan, V. N.; Crocker, J. C. *J. Am. Chem. Soc.* **2005**, *127*, 1592–1593.

237 Bruns, N.; Tiller, J. C. *Nano Lett.* **2005**, *5*, 45–48.
238 Sergeev, G. B., Komarov, V. S., Shabatin, V. P. The Method of Drug Substances Production. Patent RF 2195264, May 7, 2001.
239 Zheng, M.; Huang, X. *J. Am. Chem. Soc.* **2004**, *126*, 12047–12054.
240 Melikhov, I. V.; Komarov, V. S.; Severin, A. V.; Bozhevol'nov, V. E.; Rudin, V. N. *Doklady Phys. Chem.* **2000**, *373*, 125.
241 Suvorova, E. I.; Buffat, P. A. *Cryst. Reports* **2001**, *46*, 722–729.
242 Suvorova, E. I.; Polyak, P. E.; Komarov, V. F.; Melikhov, I. V. *Cryst. Reports* **2000**, *45*, 857–861.
243 Whaley, S. R.; English, D. S.; Hu, E. L.; Barbara, P. F.; Belcher, A. M. *Nature (London)* **2000**, *405*, 665–668.
244 Mirkin, C. A.; Taton, T. A. *Nature (London)* **2000**, *405*, 626–627.
245 Castner, D. G.; Ratner, B. D. *Surf. Sci.* **2002**, *500*, 28–60.
246 Tirrell, M.; Kokkoli, E.; Biesalski, M. *Surf. Sci.* **2002**, *500*, 61–83.
247 Xiao, S.; Liu, F.; Rosen, A. E.; Hainfeld, J. F.; Seeman, N. C.; Musier-Forsyth, K.; Kiehl, R. A. *J. Nanopart. Res.* **2002**, *4*, 317.
248 Gerion, D.; Chen, F.; Kannan, B.; Fu, A.; Parak, W. J.; Chen, D. J.; Majumdar, A.; Alivisatos, A. P. *Anal. Chem.* **2003**, *75*, 4766–4772.
249 Mertig, M.; Ciacchi, L. C.; Seidel, R.; Pompe, W.; De Vita, A. *Nano Lett.* **2002**, *2*, 841–844.
250 Pathak, S.; Choi, S. -K.; Arnheim, N.; Thompson, M. E. *J. Am. Chem. Soc.* **2001**, *123*, 4103–4104.
251 Coronini, R.; de Looze, M. -A.; Puget, P.; Bley, G.; Ramani, S. V. *Nat. Biotechnol.* **2003**, *21*, 21–29.
252 Mirkin, C. A. *J. Nanopart. Res.* **2000**, *2*, 121–122.
253 Rosi, N. L.; Merkin, C. A. Nanostructures in Biodiagnostics. *Chem. Rev.* **2005**, *105*, 1547–1562.
254 Lytton, A. K. R.; Lee, J. S. DNA-modified nanoparticles: gold and silver. In *Nanoscale Materials in Chemistry*, 2nd Ed.; Klabunde, K. J., Richards, Eds.; Wiley Publishers: New York, 2009; chapter 12, pp 405–440.
255 Mirken, C. A.; Letsinger, R. L.; Mucic, R. C.; Storhoff, J. J. *Nature* **1996**, *382* (6592), 607–609.
256 Alivisatos, A. P.; Johnson, K. P.; Peng, X. G.; Wilson, T. E.; Loweth, C. J.; Bruchez, M. P.; Schultz, P. G. *Nature* **1996**, *382*, 609–611.
257 Burda, C.; Chen, X.; Narayanan, R.; El-Sayed, M. A. *Chem. Rev.* **2005**, *105*, 1025–1102.
258 Lee, J. S.; Seferos, D. S.; Giljohann, D. A.; Mirken, C. A. *J. Am. Chem. Soc.* **2008**, *130* (16), 5430.
259 Park, S. Y.; Lytton-Jean, A. K.R.; Lee, B.; Weigand, S.; Schatz, G. C.; Mirken, C. A. *Nature* **2008**, *451*, 553.
260 Nykypanchuk, D.; Maye, M. M.; van der Lelie, D.; Gang, O. *Nature* **2008**, *451* (7178), 549–552.

Conclusion

At present, nanoscience and nanotechnology are being developed very rapidly.

Some researches of fine particles, particularly, of metal atoms and clusters in low-temperature matrices were already carried out back in the past century. The last decade of the twentieth century marks the transition to the active state of the development of studies on the synthesis of small-scale metal particles and investigation of their physicochemical properties. However, the impetus to the rapid development of nanotechnology was given in the beginning of the twenty-first century due to the admission of "the national nanotechnology initiative" by the USA.

Before the beginning of the twenty-first century, different terms were used. The word "nano" and the terms such as nanophysics, nanochemistry, nanomaterials, and nanobiology, which now make up the main content of nanoscience, were known and used by only a small number of scientists, while the majority of scientists used the term "ultrafine particles".

Today, nanoscience and nanotechnology are among the most important directions of modern natural science.

New programs, laboratories, centers, and faculties have appeared and are being developed further. The concepts of nanotechnology and nanomaterials are beginning to be reflected in school programs.

The transformations of substances studied by chemistry and nanochemistry serve as the basis for new processes used in nanotechnology and in the synthesis of nanomaterials. It can be said with confidence that the state of the art and prospects of nanochemistry largely determine the further development in other directions of nanotechnology associated, for instance, with medicine and material sciences. Some directions of nanochemistry are being developed very actively. Certain of these new directions are reflected in the second edition.

The experimental results allow formulating certain general conclusions and potential directions in studying chemical properties and reactivity of different elements in the periodic table as a function of the particle size.

Nanochemistry as an important part of nanoscience is divided into several separate directions. A traditional approach allows separate consideration of fundamental and applied directions. However, nowadays, the progress in fundamental scientific research virtually removes the boundaries and shortens the time between discovery of a new phenomenon and its practical application. Fundamental studies should be aimed at solving the definite problems of practice. At the same time, it is quite clear that concrete applied studies are impossible without serious, in-depth, and purely scientific investigations. The origination and development of nanoscience and nanochemistry as its most important component reflects the contemporary development of natural science.[942]

Yet another peculiarity associated with the development of nanoscience and nanochemistry, namely, their interdisciplinary character, deserves mention. The scope, approaches, and methods used in physics, chemistry, biology, and material sciences are closely interlinked. For such a situation, the successful development of various branches of nanoscience as a whole and nanochemistry in particular implies the organization of the cooperation of scientists in different fields within the frames of a common problem or program. The interdisciplinary character of nanoscience requires modernizing education and training of scientists in a new direction that will determine the development of natural science in the twenty-first century.

Turning back to nanochemistry, it should be mentioned that solution of many relevant problems is associated with using and elaborating various physical research methods. Thus, to find how the chemical properties and reactivity depend on the sizes of particles taking part in a reaction, it is necessary to employ methods that would allow one not only to determine the size of particles but also to follow the dynamics of their properties in the course of chemical reactions.

The experimental results obtained to date provide numerous concrete examples of unusual chemical reactions that occur with the participation of atoms, clusters, and nanoparticles for a wide range of elements in the periodic table.

These studies should be developed further. It seems quite probable that by carrying out the already known reactions of metals with the use of their nanoparticles, one can discover a lot of new unexpected transformations.

A certain exception is the reactions of particles obtained by multiple separation in the gas phase, which make it possible to follow the activity changes depending on the number of atoms.

Chemical reactions of nanoparticles with different sizes in liquid and gas phases were scarcely studied, to date, over a wide temperature interval. Such experiments would allow one to gain an insight into the effect the size of a particle exerts on the periodic changes in its chemical activity. This information in combination with high-level quantum-chemical calculations would allow us to make a first step from fragmentary explanations on how the number of atoms in a particle affects its activity toward the development of the general theory that would relate the size with chemical properties. However, at present, the experimentally feasible strategy is to consider reactions in different phases as merely separate problems of nanochemistry.

In the present stage, which reflects the traditional approach of chemistry based on the transition from reactions in the gas phase to those in the liquid and solid phases, the gas-phase processes that allow one to study reactions of ligand-free particles have gained importance. Dealing with ligandless particles brings up the problem of their assembling into blocks of certain size and the subsequent analysis of interactions of such blocks. To understand and simulate these processes, one has to determine the size and the energy state of all particles of different sizes. The introduction of stabilizing ligands brings up additional complications into this analysis due to the competition between different types of interactions. Such clusters play a special role in nanochemistry for synthesizing monodispersed particles and for studying the periodicity of their properties and peculiarities of the formation of more highly organized ensembles. Ligand-free particles are also a promising model for studying the quantum-size properties. New methods should be developed for synthesizing ligand-free particles and determining their structure. Such particles form the basis for building more complex assemblies.

As a rule, ligand-free clusters are obtained in nonequilibrium, metastable states. Elucidation of the mechanism of their synthesis would allow conducting a direct search for stabilizing ligands, materials-precursors, and ligands-spacers that regulate the distances between individual clusters. This would allow one to exercise control over self-organization processes and regulate the synthesis of materials with new properties that cannot be obtained by conventional methods.

Nonequilibrium metastable structures tend toward stabilization. Such states are realized owing to the formation of, as great as possible, the number of stable bonds. Probably, it is this process and the multielectron structure that determine the great number of free-energy minimums obtained in calculations. The development of quantum-chemical methods for analyzing and simulating ligand-free clusters is an intricate problem. Such experimental studies require very sophisticated and expensive equipment and may involve difficulties associated with temperature measurements.

In the liquid phase, the development of studies of chemical properties as a function of the size of particles taking part in the reaction has just started. As compared with the gas phase, the synthesis of metal nanoparticles in the liquid phase is a simpler and less expensive task. At the same time, it is difficult to exercise desired control over the size of synthesized particles, which often depends on the synthetic method.

In modern nanochemistry, attention is being shifted from inorganic nanoparticles (metals and semiconductors) to nanoparticles synthesized from various organic compounds. The development and refinement of methods for synthesizing organic nanoparticles is an individual problem of great importance for modification of drugs. For nanochemistry, the most important direction is the study of the interactions between organic and inorganic particles because this can lead to quite unexpected results and systems with new physicochemical properties.

Actually, the methods of synthesizing metal nanoparticles can be divided into "wet" and "dry". Each of them has its own advantages and drawbacks. Both the methods can be used for the synthesis of hybrid nanomaterials, which is of special importance for combining organic and inorganic compounds, preparation of symmetrical and asymmetrical core–shell nanosystems and multicomponent one- and two-dimensional structures and particles, which exhibit several valuable properties.

Particles with sizes from 1 to 10 nm and of sufficiently narrow size distributions were obtained. However, as already mentioned, only few studies on the effect of the particle size on their properties are available. From our viewpoint, this is determined by two factors. First, the works on regulating the number of atoms in particles are in a preliminary stage. The second and most important factor is that in the liquid phase, the particles as a rule consist of metal cores and stabilizing-ligand shells. Studying the reactivity of such particles with different chemicals entails a problem of separating the effects of cores and shells. The situation is complicated by the fact that the resulting size of a metal core depends on the chemical properties of molecules that constitute its stabilizing shell. Moreover, the core size defines the conditions of self-organization of stabilizing molecules. The complete separation of effects of cores and shells and the elucidation of the peculiarities mentioned above was therefore achieved only in few cases.

The approaches to solving this problem are associated with synthesizing new particles of a definite stoichiometric composition, developing methods for preparation of particles with different sizes under commensurable conditions, and on conducting a search for new stabilizers and ligands-spacers. The more active development of quantum-chemical methods for assessing the effect of ligand shell on the properties of metal cores is also necessary.

For liquid-phase reactions involving 1 nm particles, i.e. those that contain about 10 atoms, it is still difficult to write stoichiometric equations. This problem can be solved, if we solve the material balance between the consumption of starting reagents and the formation of final products. For the most part, the results of gas-phase reactions of metal particles containing several atoms were analyzed by merely determining their relative reactivities.

In nanochemistry, the reactions of the gas–solid kind have been increasingly gaining in popularity. They represent catalytic or sensoric processes that occur with participation of the solid phase on the surfaces of crystalline particles in a porous sample. Reactions in low-temperature condensates also proceed in the solid phase, and their realization depends on either the particle size or the cocondensate film thickness. Along with the particle size, the processes of accommodation, migration, and stabilization of metal particles also render substantial effects on the properties of solids. Moreover, the chemical nature of particles interacting with the surface can change the properties of the latter, while the changes in the conductivity form the basis for employing nanosize metal oxides as the sensor materials.

To create new materials for sensors and catalysis, nanosystems with well-developed surfaces should be actively elaborated. These are porous structures of various shapes such as wires and tubes, porous and doped oxides, and combinations of nanocrystals with different biological molecules.

The development of a new direction of nanochemistry is carried out in line with the cryochemical studies of metal particles on surfaces. By using low temperatures, the methods for synthesizing highly ordered nanosize structures on various organic and inorganic surfaces as well as those for incorporating metal nanoparticles into organic and polymeric matrices have been possible.

On the surface of nanoparticles, the atoms that interact with them are localized in sites with different coordination numbers, usually smaller as compared with compact materials. The shapes of deposited particles can be transformed under the effect of the surface, while the defects that exist on the nanoparticle surface and differ in both their nature and number can affect the migration kinetics, the aggregation, and the structure of assemblies formed. To study such effects, it is necessary to extend the possibilities of cryonanochemical synthesis on different surfaces. Low temperatures allow obtaining particles measuring less than 1 nm, the use of which requires extending the search for new highly effective stabilizers. By employing such particles, the new catalysts and sensors, anticorrosion coatings and protective films for optical devices, new drugs and dyes, and reagents for deactivation of noxious chemical and biological compounds can be manufactured. The successful solution of these problems requires developing new high-precision methods for analyzing the composition and structure, based on modern experimental techniques.

The use of low temperatures opens up new possibilities in synthesizing and studying the reactivity of condensed films with incorporated particles of metals and their oxides of various sizes. This provides a way to new chemoresistive nanosystems. The determination of the relationship between the number of atoms in a particle on a surface and their reactivity is among the most challenging problems of nanochemistry.

It is extremely important to develop new thermodynamic and kinetic models for describing the reactivity of particles with sizes less than 1 nm. The size of such particles can be considered as a thermodynamic quantity that performs a function of the temperature. The high chemical activity of nanosize metal particles allows considering such systems as a sort of chemical nanoreactors with stored energy, which can be liberated in an explosive process. Such systems give rise to new reactions, which cannot be realized under ordinary conditions.

A more thorough research on the stabilization and self-assembling of atoms and small clusters is required. To understand the stabilization processes associated with the interactions in the core–ligand system, it is necessary to develop the production of nanoparticles in a wide temperature range and extend the studies on the kinetics of ligand reactions with the nanoparticle core and the process of ligand self-organization. Of great importance for synthesizing assemblies of particles is the chemical reactivity of individual facets of nanocrystals. Experimental and theoretical studies should give definite answers to the questions of how the self-assembling of atoms proceeds, whether this process occurs by progressive addition of atoms to a particle or, e.g. a tetramer is formed at the interaction of two dimers. Moreover, it is vital to know whether the resulting assembly of particles retains the physicochemical properties of its individual components or whether they can change, and if so in what way. An insight into the peculiarities of self-assembling and self-organization of small particles into coarser assemblies will open up new possibilities for synthesizing materials with unusual properties. The least predictable chemical phenomena can be expected in the reactions of nanoparticles constituted by several metals.

When studying nanoparticles of various elements either in the volume or on the surface, the scientific problems are closely entangled with technological and engineering applications of such systems. The fundamental research is directed at elucidation of relationships between chemical and physical properties and between sizes and shapes of metallic particles. The technological goals are associated with using nanoparticles for creating new materials with unique optical, electrical, magnetic, mechanical, sensoric, and catalytic properties.

For technological applications of nanoparticles, the so-called problems of scaling acquire great importance. At present, various nanosize particles with unusual properties can be obtained only in milligram and even in nanogram amounts. Synthesizing greater amounts of these compounds, even several grams, leads to different, often poorly reproducible results. As a consequence, two trends are formed in nanochemistry. One of these trends is determined by the quest for new materials and their synthesis, even if that be in small amounts. These materials represent sensor materials and nanoelectronic devices. This approach can be named as "self-sufficient nanochemistry". The second trend is to use nanochemistry in a large-scale production of materials such as new commercial reagents, e.g. metal oxides and catalysts based on metal nanoparticles; powders, composites, and ceramics; and hybrid, consolidated, and other new nanomaterials.

In materials comprising small-scale particles, which are obtained under nonequilibrium conditions, the processes of relaxation, recrystallization, and homogenization that induce changes in their physicochemical and service properties occur. The stability of crystalline materials depends on the processes that control the increase or decrease in the sizes of particles formed in the course of synthesizing the material.

The size variations of particles involved in different processes are largely determined by their chemical activity. To elucidate the peculiarities of this relationship is among the most important problems of nanochemistry and, in our opinion, is directly related to the problem of stability of materials used in nanotechnology. The development of fundamental knowledge in the field of nanochemistry would allow us to gain a deeper insight into the processes that occur in various nanomaterials during their long-term service under different temperature conditions.

It is impossible to solve the problems of nanochemistry and study the physicochemical properties of nanosize particles without elaborating new experimental methods for their synthesis, as well as without new approaches to analyze their results. Procedures of extrapolating and analyzing the chemical activity from the top, i.e. from a compact system to a nanoparticle, have shown little promise for nanochemistry. An approach from the bottom, i.e. from individual

Conclusion

atoms and molecules that form the lower limit of synthesized particles, has proved to be more promising.

In analyzing the activity of metal clusters of different sizes, it should be realized that ligand-free particles in vacuum and ligand-stabilized particles are in fact of different formations.

For particles obtained in the liquid phase and used under the conditions of high vacuum, the active search for new stabilizers and spacers and the wider application of natural and synthetic, organic and inorganic mesoporous carriers are necessary. The potentialities of using micelles, microemulsions, and, particularly, dendrimers and polymers as nanoreactors are far from being exhausted.

The issue of how the shape of metal nanoparticles affects their activity is still open to a certain degree. This implies not only a transition from spherical particles to those of different shapes, which is of certain importance, but also to the activity of particles with one and the same number of atoms but with different shapes. The simplest test is to compare the chemical activity of particles consisting of three atoms but shaped either as triangles or linear chains.

Analyzing the activity of particles with different shapes is among the challenging problems for both experimental and theoretical studies of ligand-free clusters. Of yet greater importance is the development of methods for controlling the size and shape of nanoparticles synthesized in liquid and solid phases, when the particles are incorporated into a matrix or are surrounded by ligands. To understand the potentials and specifics of nanochemistry, it is necessary to extend the studies to subnanometers (less than 1 nm) and single molecules.

A study of the activity of metal nanoparticles in a wide temperature interval provides important information on the mutual effect exerted by temperature and size on the activity. The research of ligand-free clusters has already run into a problem of the enhanced activity of particles observed with a decrease in the temperature. For ligand-stabilized particles, there arises a problem of how to separate the effects exerted by the core size and by the nature of the ligand shells.

A special place in nanochemistry belongs to particles involved in the realization of various biological processes. The major direction of nanochemistry has been transferred to the field of nanobiology and nanomedicine. Such studies are associated with the problems of nutrition, environmental control, health, and life interval. The use of semiconductor quantum dots for labeling cells, tissues, bacteria, and viruses has actively developed. The studies on the transferring of nanosystems from organic media to aqueous solutions, the transforming of hydrophobic nanoparticles into hydrophilic ones, and using enzymes and other biomolecules as natural nanoreactors have gained wide acceptance. It is most important to elaborate new methods for building assemblies involving both biomolecules and inorganic materials. When using nanoparticles in biology and medicine, special attention should be paid to the problems of safety because of the possible toxicity of applied materials and to the yet incompletely clear trends in the fundamental properties of particles, associated with the decrease in their size. Nanochemistry plays the key role in realizing such processes, solving the problems of coassembling and self-assembling, and understanding the linkage between life sciences and material sciences.

The recent interest in organic nanoparticles is associated, on the one hand, with the fact that they are yet insufficiently understood and, on the other hand, with the high prospects of their application. The wider use of organic nanoparticles is explained by the development and solution of several problems. At present, the majority of synthesized organic nanoparticles involve aromatic and polycyclic compounds. Widening the range and chemical classes of organic substances convertible to nanoparticles can also extend their application field. The existing methods for synthesis and examination of organic nanoparticles should be improved and new methods should be invented. A general drawback of all organic nanoparticles, which limits their application, is the poor stability. Extension of the time and temperature intervals of stability for organic nanoparticles and also for nanoparticle-based diagnostic and film composites requires further development of research and fundamental studies. One of the approaches to solving the problem of regulating the activity and stability of small-scale organic particles may be the synthesis of multicomponent, particularly, multilayer, organic–inorganic (e.g. metal) nanoparticles. In such hybrid systems, new phenomena may appear, for instance, synergistic effects.

The aforementioned problems associated with organic nanoparticles concern all nanoparticles irrespective of goals and possible applications. Perhaps, nanodrugs should be considered separately. Besides the widely elaborated problem of the delivery of a drug to an affected organ, the problems of synthesis of new crystalline modifications of already known drugs and also of preparation of their earlier unknown polymer modifications probably deserve special mention. This is all the more important if nanoparticles of the original drug take part in the synthesis of new crystals and polymer modifications. In this case, one can expect some changes in physicochemical properties such as melting point, dissolution rate, solubility, hydrophobicity, hydrophilicity, and donor-acceptor properties.

The different physicochemical properties can, in turn, affect the drug activity. One can expect certain changes in the selectivity of drug interaction with cells and receptors of individual organs as well as in drug bioavailability and the time it stays in organism. In turn, this can entail certain modifications of therapeutic properties. A known drug used in curing certain diseases may find new applications. The modified therapeutic activity can result in a decrease in the drug dose, which in turn can reduce the costs for synthesizing this drug and, more so, eliminate the negative secondary and side effects.

Modern trends in the development of different directions of nanoscience allow us to argue that the role of nanochemistry will enhance in the immediate future, and its contribution to science and technology in the twenty-first century will constantly increase.

Index

Note: Page numbers with "*f*" denote figures; "*t*" tables.

A

Ab initio methods, 137–139, 143–145
Activation of small molecules, 93–96
Aerosol techniques, 33
Aggregates, reactivity of, 61
Aggregation
  of metal atoms or reactive molecules in low-temperature matrices/solvents, 56–61
  of silver particles, 24
Aluminum particles and reactions, 200–201, 200t
1-aminopyrene particles, 262
Analysis of spectral techniques, 85t
Androstenediol, 253
5-(2-anthryl)-1,3-diphenyl-2-pyrazoline particles, 258
Antisolvents for precipitation, 244–245
Application of organic nanoparticles, 257–269
Atomic-force microscopy, 80–81, 104, 157f, 210, 213, 266
Atovaquone (malorone), 239
Attraction forces, 215
Au-acetone particles, 64–65, 69f
Au–Ag alloy, 20–21
Au–Hg particles, 25

B

β-carotene, 242
β-polymorph, 255
Bimetallic Pd–Rh particles
  Ag–Pb particles, 107
  Au–Ni particles, 26
  Au–Pd particles, 25–26, 26t
  Au–Sn particles, 57–61
  Fe–Pt particles, 190
  Pd–Au particles, 21, 25
  Pd–Rh particles, 20
Binuclear compounds, 63
Biochemistry and nanotechnologies, 269
Biodegradation, 269
Biomolecules, 19, 183, 190, 319, 330–333, 335, 335f
9,10-bis(phenylethynyl)anthracene nanotubes, 262
Bismuth particles, 18

7-bromo-5-(2-chlorophenyl)-1,3-dihydro-2H-benzodiazepin-2-one (phenazepam), 255
Budesonide particles, 249, 250f

C

C545T particles, 249
Calcium particles, reactions, 127, 164–165, 164f, 165t
Carbamazepin, 256, 256f
Carbon aerosol gels, 231
Carbon nanotubes
  applications, 230
  filling, 134, 226–227, 328–329
  grafting, 227–229, 284–285
  intercalation, 143–145, 226, 229–230
Carvedilol, 253
Catalysis
  bimetallic Pt–Au, 302
  gold nanoparticle, 331–332
  influence of particle size, 307
  nanopens, 305
  palladium nanoparticles, 300, 309–310
  TiO_2 and Ag–TiO_2, 303–304
CdSe particles, 210, 315
CdTe particles, 63, 68f
Celecoxib, 263
Chemical methods, 242–257
Chemical reduction, 13–18, 22–27, 245–246
Chemical vapor deposition methods, 230
Chromium
  atoms, 55–56
  particles, 179
Cobalt particles, 15, 187–188
  reactions, 15
Co-condensates silver cyanobiphenyl, 111
Coercivity, 294, 294f
Complex of DNA and CdSe, 332
Conducting properties, 294–295
Copper particles, 16, 18–19, 33, 43, 63, 67f, 193, 193t, 195, 229
Core–ligand complex, 159–160
Core–shell
  Ag–Au particles, 21
  CdS–ZnS, 314
  Co–CdSe, 189–190

355

Core–shell *(Continued)*
  FePt–Fe$_3$O$_4$, 190
  Pd–Pt, Pd–Au, Pd–Rh, 20–21
Core–shell particles, structures, 17–18, 20–21, 45–46, 189, 212–213
Coulomb blockades, 295
Critical film thickness, 98
Cryochemical synthesis, 27–38, 251–257, 252f
Cryodrying, 251, 267
Cryospraying, 267
CuO particles, 69
Cyclosporine particles, 249

D

D5-androstene-3b,17b-diol, 253, 254f
Dendrimers, 18–22
Destruction of carbon tetrachloride, 96
Diffraction techniques, 76, 81–82, 213
Digestive ripening, 64–69
5-(4-dimethylaminophenyl)-3-(4-dimethylaminostyryl)-1-phenyl-2-pyrazoline particles, 242
1,3-diphenyl-5-pyrenyl-2-pyrazoline particles, 257–258
DNA-modified nanoparticles, 336–337, 337f
Drug modification, 334
Drug nanoparticles, 263–269
Dry grinding, 237–238
Dynamic light scattering (DLS), 84–85, 109

E

Electric explosion techniques, 196
Electrical properties, 294–295
Electrochemical dissolution, 16
Electron microscopy, 25–26, 44–46, 48, 77–78, 213
Electronic
  circuit, 317
  noses, 323
EPR method, 111
Ethanol, 59
EXAFS, 76, 82, 157f
Exchange reactions, kinetics, 289
Explosive reactions, 96–100

F

Fast crystallization, 7
Femtosecond lasers, 241
Films, conductivity, 107–108
Fluorescent organic nanoparticles, 258–259, 261f–262f

Forces between nanoparticles, 215–216
Fullerenes, 223–225, 229

G

Gabapentin, 253
GaP nanowires, 315
Germanium particles, 179
Gold
  DNA-Au particles, 336–337, 337f
  films, 212–213
  nanoshells, 197
  optical properties, 325
  particles, 5–6, 14–15, 20–21, 23–25, 61–62, 63f–65f, 64–67, 70f–71f, 84, 179, 193–197, 209–214, 269, 278, 289, 293, 295, 301–302, 301f, 313, 315, 319, 325–326, 330–332, 334–335. *See also* Silver
Gold–Tin (Au–Sn) bimetallic system, 57–61
  clusters in cold solvents, 58–61, 60f
Graphene, 230–231
Grignard reactions, 90–92
Griseofulvin, 264
Guanine, 258, 259f–260f

H

Helium nanodroplets, 34–36
Heterodimers, 17–18, 183
Heterogeneous films, 46
Hexogen nanoparticles, 36
Hollow structures, 19
Hybrid nanomaterials, 179, 323
Hydrocarbons chlorination, 167
Hydrosol, 242
Hydroxypropylmethylcellulose (HPMC), 264

I

Ibuprofen, 264
  grinding of, 238, 238f
  particles, 249, 266
Indium
  films, 279
  particles, 281
Indium nitride nanowires, 202
Indometacin, 240, 267–268
Interaction potentials, 215–216
Intraconazole, 264
Ion association, 246–247
Ionization potential, 163
Iron
  complexes, 187
  particles, 181–183, 182f, 186–187, 293
  reactions, 181–183, 182f
Itraconazole, 264, 266–267

Index

J
Jet milling, 238

K
Kinetic nanoparticles, 287–289, 323–324
Knudsen cells, 28–29
Knudson cell-mass spectroscopy, 58

L
Laser ablation, 239–241
Layering experiments, 59
Lead clusters, 101
Light beating spectroscopy. *See* Dynamic light scattering (DLS)
Light-emitting diode (LED) displays, 260
Lithium
　oxide, 312
　particles, 143–144
Lyophilization, 242

M
Magnesium
　atoms, 61, 123–131
　particles, 90–100
　reactions, 90–100
Magnetic properties, 293–294
Mass spectrometry, 83
Matrix-isolation method, 130–131
Mechanical grinding, 237–239
Medicine and nanotechnologies, 269
Mefenamic acid particles, 248
Metal atom chemistry, 63f
Metal nanoparticles, 55–74
Metal oxides nanoporous, 22
Metal-containing polymers, 19
Metal–nonmetal transition, 5–6
Mg–CO_2 complex, *ab initio* calculations, 125
Modification of particles, 251–257
Molybdenum
　particles, 174
　sulfide particles, 304–305

N
Nanobelts, 46, 321
Nanochemistry, definition, 6
Nanocomposites, definition, 2–4
Nanocrystals, 61
Nanomachining, 64–69
Nanoparticles classifications, 2
Nanopowders, 267
Nanorattles, 197–198
Nanoribbons, 46, 196
Nanorods, 69, 313, 316
Nanoscience, definition, 235
Nanostars, 69
Nanostructure definition, 2–4
Nanotubes
　AlN, GaN, 46
　FePb, Fe_3O_4, 45–46
　$NbSe_2$, 316
Nanowires, 69, 193
Neutron diffraction, 82
Nickel
　atoms, 56
　particles, 44, 57f, 183, 307
　reactions, 187
Niobium particles and reactions, 172–177, 172f
Nuclear magnetic resonance (nmr) spectroscopy, 83–84

O
Optical conductors, 262
Organic light-emitting diodes (OLED), 258–259
Organic nanoparticles, 235–274
Ostwald ripening, 67
Oxides, 17
　magnesium, absorption, 161
　reactions, 311–314

P
Palladium
　nanowires, 191
　particles, 25, 43–44, 179–180, 295, 309–310
1,2,3,4,5-pentaphenylcyclopenta-1,3-diene, 261
Perylene particles, 242, 243f, 245–246
Phenazepam, 255–256, 255f
Phenitoin, 240
Photochemical method, 248
Photochemical reduction, 11, 133
Photoelectron spectroscopy, 83
Photon correlation spectroscopy (PCS). *See* Dynamic light scattering (DLS)
Physical methods, 237–242
Piston-gap high-pressure homogenization, 239
Platinum particles, 15–16, 48
Polycarboxylic acids, 24, 132–133
Polydispersity index, 266
Poly-*p*-xylylene films, 101
Polypyrrole nanowires, 19, 319
Porous structure, 19, 22
Probe microscopy, 78–81
Properties of organic nanoparticles, 257–269
Propylparaben particles, 247

Q

Quantum dots, biolabels, 332–333
Quasi-elastic light scattering (QELS). *See* Dynamic light scattering (DLS)
Quasi-one-dimensional (1D) systems, 260–262
Quinacridone, 239
    particles, 240, 241f

R

Radiation reduction, 11, 24–25
Raman spectroscopy, 230–231
Rapid expansion of supercritical solutions, 27, 248–249
Rare-earth elements, reactions, 115–121
Reaction kinetics, 18–22
Reactions
    of magnesium particles, 90–100
    in micelles and emulsions, 18–22
Red–green–blue (RBG) emission, in displays, 261
Regenerative medicine, 269
Reprecipitation, 242–243
Rhodium particles, 18, 300
Ruthenium particles, 16

S

Samarium
    complexes, 120, 126, 131
    reactions, 116–120, 125, 130–131
Scotch tape method, 230
Self-organization, 275, 323, 325–326, 333, 335–336
Semiconductors, 314–323
Sensors
    NO_2, 321–322
    SO_2, 317–319
    temperature, 319
Shape-control, 17
Silica, mesoporous, 22
Silicon particles, 221–222
Silver
    clusters, 15, 111–112, 112f, 133–136, 141–142, 282–283
    particles, 15, 63, 66f, 84, 285, 303, 313. *See also* Gold
Simulation
    Ag_nPb_m clusters, 141
    heteroclusters, 141, 142f, 143–145
    systems, 143–149
Size, of particles, 266
Size effects, 235
    definition, 6, 275
    external, 4
    internal, 4–5
    optical spectra, 281–287
    organic vs. inorganic particles, 58
Sodium particles, 6–7
Sol–gel synthesis, 26
Solvated metal atom dispersion (SMAD), 55–74
    reactor, 56f
Solvated metal particles, 184
Solvent replacement, 242–244
Sonication, 264
Sonochemical method, 25–26
Spectra, 123
Spectral properties, 257–260
Spray drying, 242
Stability of drug particles, 264–267
Stabilization
    by mesogenes, 110–115
    by polymers, 101–110
    SMAD, 61
    by solvent, 184
Steric stabilization, 265
Sulfamethoxazole, 264
Supercritical fluid (SCF), 248–251
Supercritical solutions, 27
Superlattices, 216, 217f
Supermagnetism, 294
Supersaturation, 244–245

T

Tetrapod particles, 196
Theoretical simulation, 137, 149, 193
Thermodynamic consequences
    influence of pH, 300, 303, 313, 319
    melting point, 331
    shift chemical equilibrium, 290
Titanium, reactions, 169–170
Trans-1-cyano-1,2-bis(4'-methylbiphenyl-4-yl) ethylene particles, 243
Trapping, SMAD, 61
1,3,5-triphenyl-2-pyrazoline particles, 262
Tungsten particles, 177–178
Turbstratic graphite, 231

U

Ultra violet (UV)–visible spectrometry, 84

V

Vanadium particles, 178–179

Index

W
Water–oil emulsions, 247–248, 247f
Wet grinding, 238

X
X-ray diffraction (XRD), 18, 26, 45, 59, 115, 162, 184, 213, 231, 250–251, 253, 255
X-ray fluorescence spectroscopy, 82–83
X-ray photoelectron spectroscopy (XPS), 58, 157f

Z
Zeolite, 22
Zinc-nanooxide, reactions, 311
Zinc-selenium particles, 315

CPSIA information can be obtained at www.ICGtesting.com
Printed in the USA
BVOW01*2124300713

327039BV00008B/40/P